# Springer Complexity

Springer Complexity is a publication program, cutting across all traditional disciplines of sciences as well as engineering, economics, medicine, psychology and computer sciences, which is aimed at researchers, students and practitioners working in the field of complex systems. Complex Systems are systems that comprise many interacting parts with the ability to generate a new quality of macroscopic collective behavior through self-organization, e.g., the spontaneous formation of temporal, spatial or functional structures. This recognition, that the collective behavior of the whole system cannot be simply inferred from the understanding of the behavior of the individual components, has led to various new concepts and sophisticated tools of complexity. The main concepts and tools – with sometimes overlapping contents and methodologies – are the theories of self-organization, complex systems, synergetics, dynamical systems, turbulence, catastrophes, instabilities, nonlinearity, stochastic processes, chaos, neural networks, cellular automata, adaptive systems, and genetic algorithms.

The topics treated within Springer Complexity are as diverse as lasers or fluids in physics, machine cutting phenomena of workpieces or electric circuits with feedback in engineering, growth of crystals or pattern formation in chemistry, morphogenesis in biology, brain function in neurology, behavior of stock exchange rates in economics, or the formation of public opinion in sociology. All these seemingly quite different kinds of structure formation have a number of important features and underlying structures in common. These deep structural similarities can be exploited to transfer analytical methods and understanding from one field to another. The Springer Complexity program therefore seeks to foster cross-fertilization between the disciplines and a dialogue between theoreticians and experimentalists for a deeper understanding of the general structure and behavior of complex systems.

The program consists of individual books, books series such as "Springer Series in Synergetics", "Institute of Nonlinear Science", "Physics of Neural Networks", and "Understanding Complex Systems", as well as various journals.

# Springer Series in Synergetics



**SSSyn – An Interdisciplinary Series on Complex Systems**

The success of the Springer Series in Synergetics has been made possible by the contributions of outstanding authors who presented their quite often pioneering results to the science community well beyond the borders of a special discipline. Indeed, interdisciplinarity is one of the main features of this series. But interdisciplinarity is not enough: The main goal is the search for common features of self-organizing systems in a great variety of seemingly quite different systems, or, still more precisely speaking, the search for general principles underlying the spontaneous formation of spatial, temporal or functional structures. The topics treated may be as diverse as lasers and fluids in physics, pattern formation in chemistry, morphogenesis in biology, brain functions in neurology or self-organization in a city. As is witnessed by several volumes, great attention is being paid to the pivotal interplay between deterministic and stochastic processes, as well as to the dialogue between theoreticians and experimentalists. All this has contributed to a remarkable cross-fertilization between disciplines and to a deeper understanding of complex systems. The timeliness and potential of such an approach are also mirrored – among other indicators – by numerous interdisciplinary workshops and conferences all over the world.

Hermann Haken

# Information and Self-Organization

## A Macroscopic Approach to Complex Systems

Third Enlarged Edition

With 66 Figures

Springer

Professor Dr. Dr. h.c. mult. Hermann Haken
Institut für Theoretische Physik, Zentrum für Synergetik
Universität Stuttgart
Pfaffenwaldring 57/IV
70550 Stuttgart, Germany

The first edition appeared as Volume 40.

ISSN 0172-7389

ISBN 978-642-06957-4 e-ISBN 978-3-540-33023-3



Springer is a part of Springer Science+Business Media
springer.com


Printed in Germany



Cover-Design: Erich Kirchner, Heidelberg

*To the memory of Edith*

# Preface to the Third Edition

The widespread interest this book has found among professors, scientists and students working in a variety of fields has made a new edition necessary. I have used this opportunity to add three new chapters on recent developments. One of the most fascinating fields of modern science is cognitive science which has become a meeting place of many disciplines ranging from mathematics over physics and computer science to psychology. Here, one of the important links between these fields is the concept of information which, however, appears in various disguises, be it as Shannon information or as semantic information (or as something still different). So far, meaning seemed to be exorcised from Shannon information, whereas meaning plays a central role in semantic (or as it is sometimes called "pragmatic") information. In the new chapter 13 it will be shown, however, that there is an important interplay between Shannon and semantic information and that, in particular, the latter plays a decisive role in the fixation of Shannon information and, in cognitive processes, allows a drastic reduction of that information.

A second, equally fascinating and rapidly developing field for mathematicians, computer scientists and physicists is quantum information and quantum computation. The inclusion of these topics is a must for any modern treatise dealing with information. It becomes more and more evident that the abstract concept of information is inseparably tied up with its realizations in the physical world. I have taken care of these fundamental developments in two new chapters, 15 and 16, where I have tried not to get lost in too many mathematical and physical details. In this way I hope that the reader can get an easy access to these fields that carry great potentialities for future research and applications.

I thank Prof. Juval Portugali for his stimulating collaboration on the concepts of information. My thanks go to my secretary, Ms. Petra Mayer, for her very efficient typing (or rather typesetting) of the additional chapters. I thank Dr. Christian Caron of Springer company for the excellent cooperation.

Stuttgart, October 2005 Hermann Haken

# Preface to the Second Edition

Since the first edition of this book appeared, the study of complex systems has moved further still into the focus of modern science. And here a shift of emphasis can be observed, namely from the study of *states* of systems to that of *processes*. The systems may be of technical, physical, economic, biological, medical or other nature. In many cases, because of their complexity, it is impossible to theoretically derive the properties of complex systems from those of their individual parts. Instead we have to make do with experimentally observed data. But quite often these data are known only over a limited space-time domain. Even within such a domain, they may be scarce and noisy. In certain cases, the experimental observation cannot be repeated, for instance in astrophysics, or in some measurements of electro- and magneto-encephalograms, i. e. of electric and magnetic brain waves. In other cases, we are overwhelmed by the flood of data. Can we nevertheless, in all these cases, gain insight into the mechanisms underlying the observed processes? Can we make predictions about the system's behavior outside the observed time and/or space domain? Can we even learn how to control the system's behavior? These are questions that are relevant in many fields including robotics.

While in the first edition of this book the study of *steady states* of systems was in the foreground, in the present, second edition, the newly added material is concerned with the modelling and prediction of *processes* based on incomplete and noisy data. The vehicle I shall use is again Jaynes' maximum information (entropy) principle that I have developed further so as to deal with discrete and continuous Markov processes. This will allow us to make unbiased guesses for these processes. Finally I shall outline the links to chaos theory. Although I shall not state it explicitly in the additional chapters, there again the basic insight of synergetics may be important, namely that close to instability points many systems are governed by a low-dimensional, though noisy, dynamics.

I wish to thank Dr. Lisa Borland and Dr. Rudolf Friedrich for valuable discussions. I am grateful to my secretary, Ms. I. Möller, who typed (or rather typeset) the additions including their complicated formulas quickly and perfectly. Last but not least, I wish to express my gratitude to Springer-Verlag, in particular to Prof. W. Beiglböck, Dr. A. Lahee, Ms. B. Reichel-Mayer, and Ms. E. Pfendbach, for the excellent cooperation that has become a tradition.

Stuttgart, May 1999 Hermann Haken

# Preface to the First Edition

Complex systems are ubiquitous, and practically all branches of science ranging from physics through chemistry and biology to economics and sociology have to deal with them. In this book we wish to present concepts and methods for dealing with complex systems from a unifying point of view. Therefore it may be of interest to graduate students, professors and research workers who are concerned with theoretical work in the above-mentioned fields. The basic idea for our unified approach stems from that of synergetics. In order to find unifying principles we shall focus our attention on those situations where a complex system changes its macroscopic behavior qualitatively, or in other words, where it changes its macroscopic spatial, temporal or functional structure. Until now, the theory of synergetics has usually begun with a microscopic or mesoscopic description of a complex system.

In this book we present an approach which starts out from macroscopic data. In particular we shall treat systems that acquire their new structure without specific interference from the outside; i.e. systems which are self-organizing. The vehicle we shall use is information. Since this word has several quite different meanings, all of which are important for our purpose, we shall discuss its various aspects. These range from Shannon information, from which all semantics has been exorcised, to the effects of information on receivers and the self-creation of meaning.

Shannon information is closely related to statistical entropy as introduced by Boltzmann. A quite general formulation was given by Jaynes in the form of the maximum entropy principle which, for reasons to be explained in this book, will be called the "maximum information entropy principle". As was shown by Jaynes, this principle allows one to derive the basic relations of thermodynamics in a very elegant fashion and it provides one with an access to irreversible thermodynamics. Ingarden formulated what he called "information thermodynamics" introducing higher order temperatures. In spite of its success, the maximum information entropy principle has been criticized as being subjective because the choice of constraints under which entropy is maximized seems to be arbitrary. But with the aid of the results of synergetics we can solve this problem for a large class of phenomena, namely, for self-organizing systems which form a new structure via a "nonequilibrium phase transition". Thus our approach applies to many of the most interesting situations. We shall illustrate our general approach by means of examples from physics (lasers, fluid dynamics), computer science (pattern recognition by machines), and biology (morphogenesis of behavior). The last example in particular emphasizes the applicability of our approach to truly complex systems and shows how their behavior can be modeled by a well-defined procedure.

In this way our book will provide the reader with both a general theory and practical applications. I hope that it will turn out to become a useful tool both for studies and research work on complex systems. I am grateful to Prof. H. Shimizu for stimulating discussions on Sect. 1.6 and to my co-workers Dr. W. Banzhaf, M. Bestehorn, W. Lorenz, M. Schindel, and V. Weberruss for their proof-reading. I wish to thank Ms. A. Konz and Ms. I. Moeller for their perfect typing of various versions of the manuscript, and Mr. A. Fuchs and Mr. W. Lorenz for the preparation of the figures. The financial support of the Volkswagenwerk Foundation Hannover for the project *Synergetics* is gratefully acknowledged. Last but not least I express my gratitude to Springer-Verlag, in particular to Dr. Angela Lahee and Dr. Helmut Lotsch for the excellent cooperation.

Stuttgart, January 1988 *H. Haken*

# Contents

**1. The Challenge of Complex Systems** .......... 1
1.1 What Are Complex Systems? .......... 1
1.2 How to Deal with Complex Systems .......... 5
1.3 Model Systems .......... 7
1.4 Self-Organization .......... 10
1.5 Aiming at Universality .......... 11
1.5.1 Thermodynamics .......... 11
1.5.2 Statistical Physics .......... 12
1.5.3 Synergetics .......... 13
1.6 Information .......... 14
1.6.1 Shannon Information: Meaning Exorcised .......... 15
1.6.2 Effects of Information .......... 16
1.6.3 Self-Creation of Meaning .......... 23
1.6.4 How Much Information Do We Need to Maintain an Ordered State? .......... 29
1.7 The Second Foundation of Synergetics .......... 33

**2. From the Microscopic to the Macroscopic World ...** .......... 36
2.1 Levels of Description .......... 36
2.2 Langevin Equations .......... 37
2.3 Fokker-Planck Equation .......... 40
2.4 Exact Stationary Solution of the Fokker-Planck Equation for Systems in Detailed Balance .......... 41
2.4.1 Detailed Balance .......... 41
2.4.2 The Required Structure of the Fokker-Planck Equation and Its Stationary Solution .......... 42
2.5 Path Integrals .......... 44
2.6 Reduction of Complexity, Order Parameters and the Slaving Principle .......... 45
2.6.1 Linear Stability Analysis .......... 46
2.6.2 Transformation of Evolution Equations .......... 47
2.6.3 The Slaving Principle .......... 48
2.7 Nonequilibrium Phase Transitions .......... 49
2.8 Pattern Formation .......... 51

**3. ... and Back Again: The Maximum Information Principle (MIP)** .......... 53
3.1 Some Basic Ideas .......... 53
3.2 Information Gain .......... 57

3.3 Information Entropy and Constraints ........................ 58
3.4 Continuous Variables ........................................ 63

**4. An Example from Physics: Thermodynamics** .................... 65

**5. Application of the Maximum Information Principle to Self-Organizing Systems** ........................................ 69
5.1 Introduction ................................................ 69
5.2 Application to Self-Organizing Systems: Single Mode Laser ..... 69
5.3 Multimode Laser Without Phase Relations ...................... 71
5.4 Processes Periodic in Order Parameters ....................... 72

**6. The Maximum Information Principle for Nonequilibrium Phase Transitions: Determination of Order Parameters, Enslaved Modes, and Emerging Patterns** ........................................ 74
6.1 Introduction ................................................ 74
6.2 General Approach ............................................ 74
6.3 Determination of Order Parameters, Enslaved Modes, and Emerging Patterns ........................................ 76
6.4 Approximations .............................................. 77
6.5 Spatial Patterns ............................................ 78
6.6 Relation to the Landau Theory of Phase Transitions. Guessing of Fokker-Planck Equations ...................................... 79

**7. Information, Information Gain, and Efficiency of Self-Organizing Systems Close to Their Instability Points** ...................... 81
7.1 Introduction ................................................ 81
7.2 The Slaving Principle and Its Application to Information ...... 82
7.3 Information Gain ............................................ 82
7.4 An Example: Nonequilibrium Phase Transitions ................. 83
7.5 Soft Single-Mode Instabilities ............................... 84
7.6 Can We Measure the Information and the Information Gain? ..... 85
7.6.1 Efficiency ................................................ 85
7.6.2 Information and Information Gain .......................... 86
7.7 Several Order Parameters .................................... 87
7.8 Explicit Calculation of the Information of a Single Order Parameter ........................................ 88
7.8.1 The Region Well Below Threshold ........................... 89
7.8.2 The Region Well Above Threshold ........................... 90
7.8.3 Numerical Results ......................................... 93
7.8.4 Discussion ................................................ 94
7.9 Exact Analytical Results on Information, Information Gain, and Efficiency of a Single Order Parameter ...................... 95
7.9.1 The Instability Point ...................................... 97
7.9.2 The Approach to Instability ............................... 98
7.9.3 The Stable Region ......................................... 99
7.9.4 The Injected Signal ....................................... 100

7.9.5 Conclusions ... 101
7.10 The $S$-Theorem of Klimontovich ... 102
7.10.1 Region 1: Below Laser Threshold ... 104
7.10.2 Region 2: At Threshold ... 104
7.10.3 Region 3: Well Above Threshold ... 105
7.11 The Contribution of the Enslaved Modes to the Information Close to Nonequilibrium Phase Transitions ... 107

**8. Direct Determination of Lagrange Multipliers** ... 115
8.1 Information Entropy of Systems Below and Above Their Critical Point ... 115
8.2 Direct Determination of Lagrange Multipliers Below, At and Above the Critical Point ... 117

**9. Unbiased Modeling of Stochastic Processes: How to Guess Path Integrals, Fokker-Planck Equations and Langevin-Ito Equations** ... 125
9.1 One-Dimensional State Vector ... 125
9.2 Generalization to a Multidimensional State Vector ... 127
9.3 Correlation Functions as Constraints ... 130
9.4 The Fokker-Planck Equation Belonging to the Short-Time Propagator ... 132
9.5 Can We Derive Newton's Law from Experimental Data? ... 133

**10. Application to Some Physical Systems** ... 135
10.1 Multimode Lasers with Phase Relations ... 135
10.2 The Single-Mode Laser Including Polarization and Inversion ... 136
10.3 Fluid Dynamics: The Convection Instability ... 138

**11. Transitions Between Behavioral Patterns in Biology, An Example: Hand Movements** ... 140
11.1 Some Experimental Facts ... 140
11.2 How to Model the Transition ... 141
11.3 Critical Fluctuations ... 147
11.4 Some Conclusions ... 151

**12. Pattern Recognition. Unbiased Guesses of Processes: Explicit Determination of Lagrange Multipliers** ... 153
12.1 Feature Selection ... 153
12.2 An Algorithm for Pattern Recognition ... 159
12.3 The Basic Construction Principle of a Synergetic Computer ... 161
12.4 Learning by Means of the Information Gain ... 163
12.5 Processes and Associative Action ... 165
12.6 Explicit Determination of the Lagrange Multipliers of the Conditional Probability. General Approach for Discrete and Continuous Processes ... 169
12.7 Approximation and Smoothing Schemes. Additive Noise ... 174

12.8 An Explicit Example: Brownian Motion .................. 181
12.9 Approximation and Smoothing Schemes. Multiplicative (and Additive) Noise .................. 184
12.10 Explicit Calculation of Drift and Diffusion Coefficients. Examples 185
12.11 Process Modelling, Prediction and Control, Robotics .......... 187
12.12 Non-Markovian Processes. Connection with Chaos Theory ..... 189
12.12.1 Checking the Markov Property .................. 189
12.12.2 Time Series Analysis .................. 190

**13. Information Compression in Cognition: The Interplay between Shannon and Semantic Information** .................. 195
13.1 Information Compression: A General Formula .................. 195
13.2 Pattern Recognition as Information Compression: Use of Symmetries .................. 197
13.3 Deformations .................. 199
13.4 Reinterpretation of the Results of Sects. 13.1–13.3 .................. 201

**14. Quantum Systems** .................. 203
14.1 Why Quantum Theory of Information? .................. 203
14.2 The Maximum Information Principle .................. 205
14.3 Order Parameters, Enslaved Modes and Patterns .................. 211
14.4 Information of Order Parameters and Enslaved Modes ......... 214

**15. Quantum Information** .................. 216
15.1 Basic Concepts of Quantum Information. Q-bits .................. 216
15.2 Phase and Decoherence .................. 218
15.3 Representation of Numbers .................. 219
15.4 Register .................. 220
15.5 Entanglement .................. 221

**16. Quantum Computation** .................. 222
16.1 Classical Gates .................. 222
16.2 Quantum Gates .................. 223
16.3 Calculation of the Period of a Sequence by a Quantum Computer . 227
16.4 Coding, Decoding and Breaking Codes .................. 229
16.4.1 A Little Mathematics .................. 230
16.4.2 RSA Coding and Decoding .................. 230
16.4.3 Shor's Approach, Continued .................. 231
16.5 The Physics of Spin 1/2 .................. 233
16.6 Quantum Theory of a Spin in Mutually Perpendicular Magnetic Fields, One Constant and One Time Dependent .................. 235
16.7 Quantum Computation and Self-Organization .................. 241

**17. Concluding Remarks and Outlook** .................. 242

**References** .................. 244

**Subject Index** .................. 251

# 1. The Challenge of Complex Systems

The aim of this book is to develop concepts and methods which allow us to deal with complex systems from a unifying point of view. The book is composed of two parts: The introductory chapter deals with complex systems in a qualitative fashion, while the rest of the book is devoted to quantitative methods. In Chap. 1 we shall present examples of complex systems and some typical approaches to dealing with them, for instance thermodynamics and synergetics. We shall discuss the concept of self-organization and, in particular, various aspects of information. The last section of this chapter gives an outline of our new theory, which may be viewed as a *macroscopic* approach to synergetics. In Chap. 2 a brief outline of the *microscopic* approach to synergetics is presented, while Chap. 3 provides the reader with an introduction to the maximum information entropy principle, which will play an important role in our book. Chapter 4 illustrates this principle by applying it to thermodynamics.

The remainder of the book will then be devoted to our quantitative method and its applications; detailed examples from physics and biology will be presented. Finally, it will be shown that an important approach in the field of pattern recognition is contained as a special case in our general theory so that indeed a remarkable unification in science is achieved. Readers who are not so much interested in a qualitative discussion may skip this introductory chapter and proceed directly to Chaps. 2, or 3 and 4, or 5, depending on their knowledge.

But now let us start with some basics.

## 1.1 What Are Complex Systems?

First of all we have to discuss what we understand by complex systems. In a naive way, we may describe them as systems which are composed of many parts, or elements, or components which may be of the same or of different kinds. The components or parts may be connected in a more or less complicated fashion. The various branches of science offer us numerous examples, some of which turn out to be rather simple whereas others may be called truly complex.

Let us start with some examples in physics. A gas is composed of very many molecules, say of $10^{22}$ in a cubic centimeter. The gas molecules fly around in quite an irregular fashion, whereby they suffer numerous collisions with each other (Fig.1.1). By contrast, in a crystal the atoms or molecules are well-arranged and undergo only slight vibrations (Fig.1.2). We may be interested in specific properties,

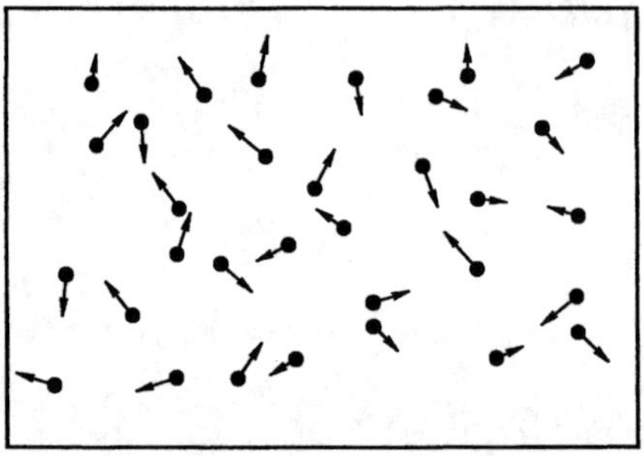

Fig. 1.1. Gas atoms moving in a box

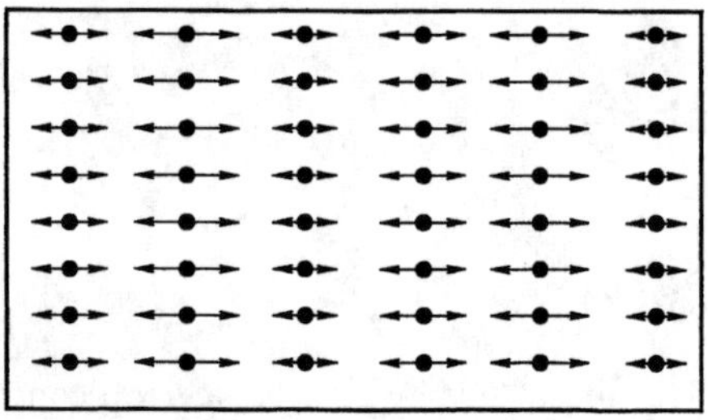

Fig. 1.2. Atoms in a crystal

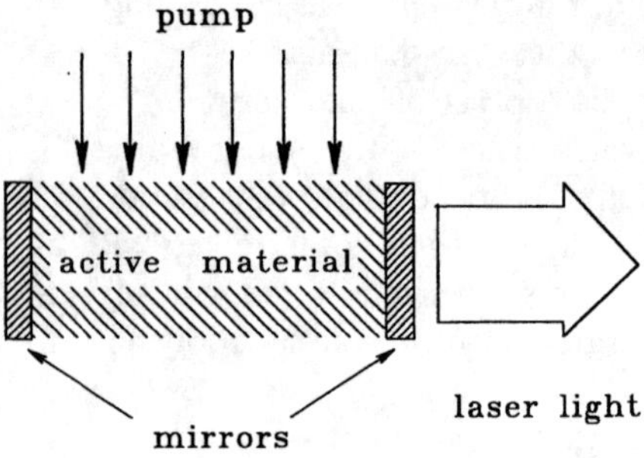


Fig. 1.3. Schematic drawing of a laser

such as the pressure or temperature of a gas or the compressibility of a crystal. Or we may consider these systems with a view to their serving a purpose, e.g. a gas like water vapor may be used in a steam engine, a crystal may be used as a conductor of electricity etc. Some physical systems are primarily devised to serve a purpose, e.g. a laser (Fig.1.3). This new light source is constructed to produce a specific type of light.

In chemistry we are again dealing with complex systems. In chemical reactions, very many molecules participate, and lead to the formation of new molecules. Biology abounds with complex systems. A cell is composed of a complicated cell membrane, a nucleus and cytoplasm, each of which contain many further components (Fig.1.4). In a cell between a dozen and some thousand metabolic processes may go on at the same time in a well-regulated fashion. Organs are composed of many cells which likewise cooperate in a well-regulated fashion. In turn organs serve specific purposes and cooperate within an animal. Animals themselves form animal societies (Fig.1.5). Probably the most complex system in the world is the human brain composed of $10^{10}$ or more nerve cells (Fig.1.6). Their cooperation allows us to recognize patterns, to speak, or to perform other mental functions.

In the engineering sciences we again have to deal with complex systems. Such systems may be machines, say an engine of an automobile, or whole factories, or power plants forming an interconnected network. Economy with its numerous

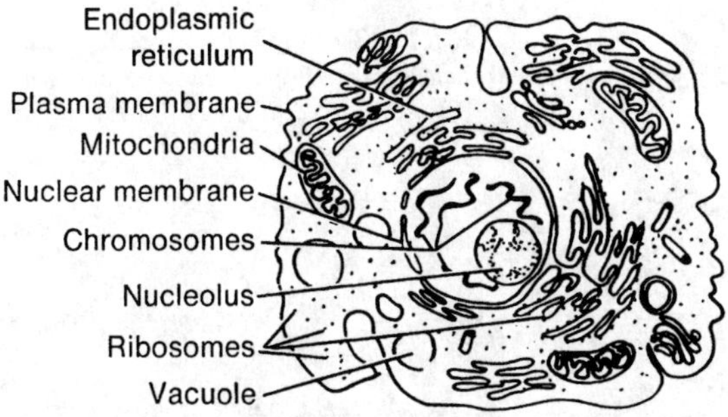


**Fig. 1.4.** A typical biological cell [from D.A. Anderson, R.J. Sobieski: *Introduction to Microbiology* (C.V. Mosby Company 1980)]

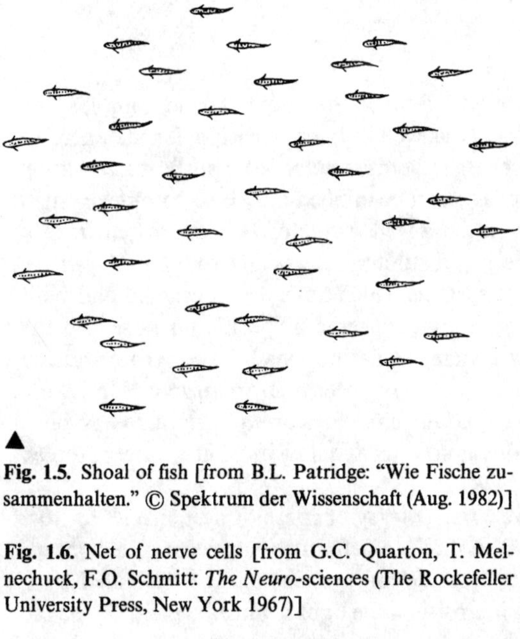

**Fig. 1.5.** Shoal of fish [from B.L. Patridge: "Wie Fische zusammenhalten." © Spektrum der Wissenschaft (Aug. 1982)]

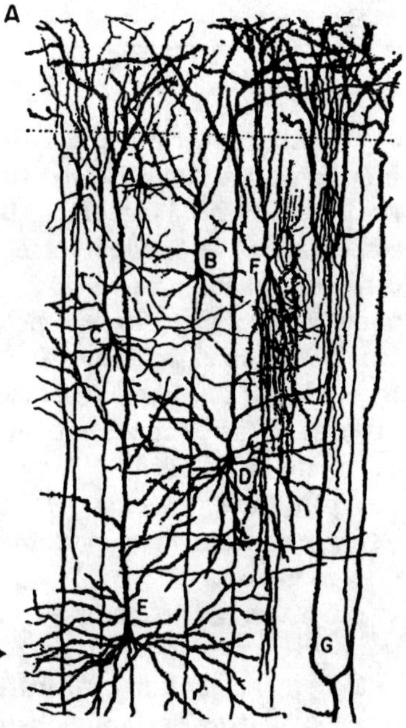


**Fig. 1.6.** Net of nerve cells [from G.C. Quarton, T. Melnechuck, F.O. Schmitt: *The Neuro*-sciences (The Rockefeller University Press, New York 1967)]

participants, its flows of goods and money, its traffic, production, consumption and storage of goods provides us with another example of a complex system. Similarly, society with its various human activities and their political, religious, professional, or cultural habits is a further example of such a system. Computers are more and more conceived as complex systems. This is especially so with respect to computers of the so-called 5th generation, where knowledge processing will be replacing the number crunching of today's computers.

Systems may not only be complex as a result of being composed of so many parts but we may also speak of complex behavior. The various manifestations of human behavior may be very complex as is studied e.g. in psychology. But on the other hand, we also admire the high coordination of muscles in locomotion, breathing etc. (Fig. 1.7). Finally, modern science itself is a complex system as is quite evident from its enormous number of individual branches.

**Fig. 1.7.** Trotting horse [from E. Kolb: *Lehrbuch der Physiologie der Haustiere* (Fischer-Verlag, Stuttgart 1967)]

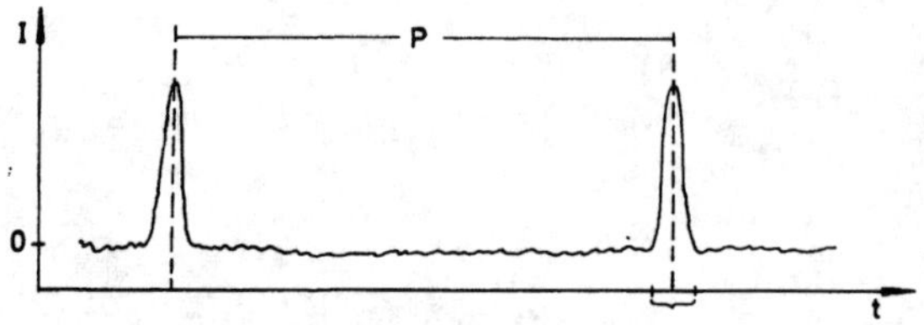


**Fig. 1.8.** Light pulses from a pulsar [from Weigert, Wendke: *Astronomie und Astrophysik* (Physik-Verlag, Weinheim 1982)]

We may now ask the question of why numerous systems are so complex and how they came into existence. In biology as well as in engineering, for instance, we readily see the need for complexity. These systems serve specific purposes and upon scrutinization we find that these purposes can be fulfilled only by a complex system composed of many parts which interact in a well-regulated fashion. When we talk about their coming into existence, we may distinguish between two types of systems: On the one hand we have man-made systems which have been designed and built by people so that these machines or constructs serve a specific purpose. On the other hand there are the very many systems in nature which have been produced by nature herself, or in other words, which have been self-organized. Here, quite evidently, the evolutionary vision, i.e. Darwinism, plays an important role in biology, where an attempt is made to understand, why and how more and more complex systems evolve.

After this rather superficial and sketchy survey of complex systems, let us now try to give a more rigorous definition. A modern definition is based on the concept of algebraic complexity. At least to some extent, systems can be described by a sequence of data, e.g. the fluctuating intensity of the light from stars (Fig.1.8), or the fever curve of a sick person where the data are represented by numbers. So, let us consider a string of numbers and let us try to define the complexity of such a string. When we think of specific examples, say of numbers like 1, 4, 9, 16, 25, 36, ..., we realize that such a string of data can be produced by a simple law, namely in this case by the law $n^2$ where $n$ is an integer. Therefore, whenever a string of data is presented, we may ask whether there is a computer program and a set of initial data which then allow us to compute the whole set of data by means of this program. Of course, depending on the construction of the computer, one computer program may be longer than that of another.

Therefore, in order to be able to compare the length of programs, we must introduce a universal computer. Without going into details we may state that such a universal computer can be constructed, at least in a thought experiment as was shown by Turing (Fig.1.9). Therefore, we shall call such universal computer a Turing machine. The idea then is to try to compress the program and the initial set of data to a minimum. The minimum length of a program and of the initial data is a measure

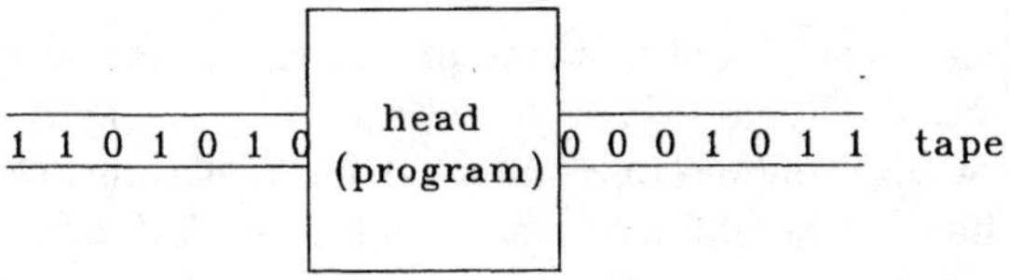


**Fig. 1.9.** Schematic of a Turing machine

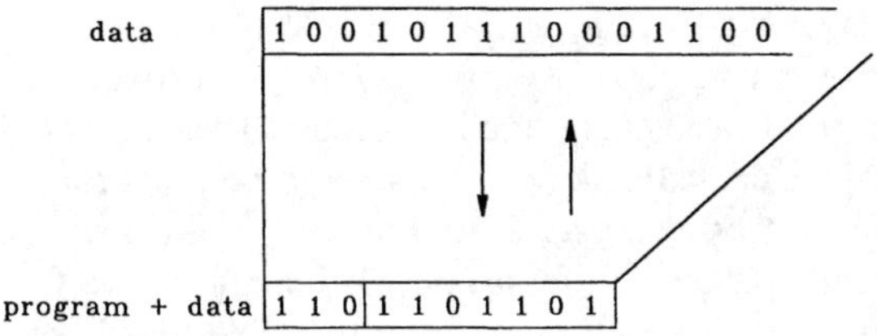


**Fig. 1.10.** Compression of a string of data into a minimal set of a program and data

of the algebraic degree of complexity (Fig.1.10). However, such a definition has a drawback. As can be shown by means of a famous theorem by Goedel, this problem of finding a minimum program and a minimum number of initial data cannot be solved in a universal fashion. In other words, there is no general algorithm available which could solve this problem. Rather we can develop such algorithms only in special cases. Indeed, occasionally one can construct shortcuts. Let us consider a gas. There, one might attempt to follow up the paths of the individual particles and their collisions and then derive the distribution function of the velocity of the individual particles. This problem has not been solved yet, when one starts from a microscopic description. Nevertheless, it has been possible in statistical mechanics to derive this distribution function, known as the Boltzman distribution, in a rather simple and elegant fashion without invoking the microscopic approach, but using the concept of entropy (see below). A number of similar examples can be formulated which show us that there exist shortcuts by which an originally very complicated problem can be solved in a rather direct fashion. Thus, we realize that the concept of complexity is a very subtle one. Indeed the main purpose of our book will be to provide such shortcuts from a unifying point of view which will then allow us to deal with complex systems.

A complex system may be considered from various points of view. For instance, we may treat a biological system at the macroscopic level by studying its behavior, or at an intermediate level by studying the functioning of its organs, or finally we could study the chemistry of DNA. The data to be collected often seem to be quite inexhaustible. In addition it is often impossible to decide which aspect to choose a priori, and we must instead undergo a learning process in order to know how to cope with a complex system.

## 1.2 How to Deal with Complex Systems

The more science becomes divided into specialized disciplines, the more important it becomes to find unifying principles. Since complex systems are ubiquitous, we are confronted with the challenge of finding unifying principles for dealing with such systems. In order to describe a complex system at a microscopic level, we need an enormous amount of data which eventually nobody, even not a society, is able to handle. Therefore, we have to introduce some sort of economy of data collecting or of thinking. In addition, we can hope to obtain deep insights when we find laws which can be applied to a variety of quite different complex systems.

When we look for universal laws it is wise to ask at which level we wish to formulate them; be it microscopic or macroscopic. Accordingly we may arrive at a quite different description of a system. For instance at a microscopic level, a gas is entirely disordered, whereas at the macroscopic level it appears to be practically homogeneous, i.e. structureless. In contrast, a crystal is well ordered at the microscopic level, whereas again at the macroscopic level it appears homogeneous. In biology we deal with a hierarchy of levels which range from the molecular level through that of cells and organs to that of the whole plant or animal. This choice of levels may be far too rough, and an appropriate choice of the level is by no means a trivial problem. In addition, "microscopic" and "macroscopic" become relative concepts. For instance a biomolecule may be considered as "macroscopic" as compared to its atomic constituents, whereas it is "microscopic" as compared to a cell. Incidentally, at each level we are confronted with a specific kind of organization or structure.

The method of modern western science can certainly be described as being analytical. By decomposing a system into its parts we try to understand the properties of the whole system. In a number of fields we may start from first principles which are laid down in fundamental laws. The field in which this trend is most pronounced is, of course, physics and in particular elementary particle physics. Usually it is understood that the parts and their properties are "objectively" given and that then one needs "merely" to deduce the properties of the total system from the properties of its parts. Two remarks are in place. First, strictly speaking, we in fact infer microscopic events from macroscopic data, and it is an interesting problem to check whether different microscopic models can lead to the same macroscopic set of data. Second, the analytic approach is based on the concept of reducibility, or in the extreme case on reductionism. But the more we are dealing with complex systems, the more we realize that reductionism has its own limitations. For example, knowing chemistry does not mean that we understand life. In fact, when we proceed from the microscopic to the macroscopic level, many new qualities of a system emerge which are not present at the microscopic level.

For instance, while a wave can be described by a wavelength and an amplitude, these concepts are alien to an individual particle such as an atom. What we need to understand is not the behavior of individual parts but rather their orchestration. In order to understand this orchestration, we may in many cases appeal to model systems in which specific traits of a complex system can be studied in detail. We shall discuss a number of model systems in Sect. 1.4. Another approach to dealing with complex systems is provided by a macroscopic description. For example, we do not describe a gas by listing all the individual coordinates of its atoms at each instant, but rather in terms of macroscopic quantities such as pressure and temperature. It is a remarkable fact that nature herself has provided us with means of measuring or sensing these quantities.

In order to deal with complex systems, we quite often still have to find adequate variables or relevant quantities to describe the properties of these systems. In all cases, a macroscopic description allows an enormous compression of information so that we are no more concerned with the individual microscopic data but rather with global properties. An important step in treating complex systems consists in establishing relations between various macroscopic quantities. These relations are a consequence of microscopic events which, however, are often unknown or only partially known. Examples of such relations are provided by thermodynamics where, for instance, the law relating pressure and temperature in a gas is formulated, and derived by statistical mechanics from microscopic laws. In general we have to guess the nature of the microscopic events which eventually lead to macroscopic data.

In this book we want to show how such guesses can be made for systems belonging to quite different disciplines. At the same time we shall see that at a sufficiently abstract level there exist profound analogies between the behavior of complex systems; or, in other words, complicated behavior can be realized on quite different substrates. Very often we recognize that the more complex a system is, the more it acquires traits of human behavior. Therefore, we are led, or possibly misled, into describing the behavior of complex systems in anthropomorphic terms. In the natural sciences it has become a tradition to try to exorcise anthropomorphisms as far as possible and to base all explanations and concepts on a more or less mechanistic point of view. We shall discuss this dilemma: mechanistic versus anthropomorphic in later sections of this chapter, in particular when we come to discuss information and the role of meaning and purpose.

Let me conclude this section with a general remark. Not so long ago it was more or less generally felt that a great discrepancy exists between, say physics or natural science on the one hand and humanistics on the other, the latter dealing with truly complex behavior and complex systems. Physics was for a long time revered because of its ability to predict events within an unlimited future. As we shall see, the more physics has to deal with complex systems, the more we realize that new concepts are needed. Some of the characteristics which were attributed to physics such as the capability of making precise predictions are losing their hold.

## 1.3 Model Systems

The great success of physics rests on its methodology. In this, complex systems are decomposed into specific parts whose behavior can be studied in a reproducible fashion, whereby only one or very few parameters are changed. Famous examples of this method are the experiments by Galileo on falling bodies, or Newton's analysis of the motion of the planets by means of considering only a system composed of the sun and one planet. Or in other words, he treated a one-, or at maximum, a two-body problem. This approach gave rise to Newtonian mechanics. From its formulation it was deduced, e.g. by Laplace, that Newtonian mechanics implies total predictability of the future, once the velocity and positions of the individual particles

Fig. 1.11

Fig. 1.12

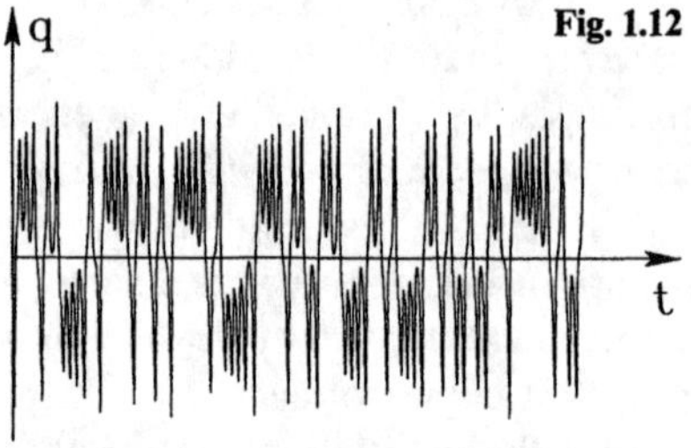


**Fig. 1.11.** Steel ball falling on a razor blade. Depending on its initial position, the ball is deflected along a wide trajectory to the left or to the right

**Fig. 1.12.** Time variation of a quantity in a chaotic system

of a system are known at an initial time. The concept of predictability has been shaken twice in modern physics. Quantum mechanics tells us that we are not able to measure the velocity and the position of a particle at the same time both with infinite precision and, therefore, that we are not able to make accurate predictions of the future path of a particle.

More recently, the theory of so-called deterministic chaos has shown that even in classical mechanics predictability cannot be guaranteed with absolute precision. Consider the following very simple example of a steel ball falling on a vertical razor blade (Fig.1.11). Depending on its precise position with respect to the razor blade, its trajectory may be deflected to the left or to the right. That means the future path of the particle, i.e. the steel ball, depends in a very sensitive fashion on the initial condition. A very small change of that condition may lead to quite a different path. Over the past years numerous examples in physics, chemistry, and biology have been found where such a sensitivity to initial conditions is present (Fig.1.12). But in spite of these remarks the general idea of finding suitable model systems for a complex system is still valid.

Here we wish to list just a few well-known examples of model systems. The light source laser has become a paradigm for the self-organization of coherent processes because in the laser the atoms interact in a well-regulated fashion so to produce the coherent laser wave (Fig.1.13). Another example for the self-organized formation of macroscopic structures is provided by fluids. For instance, when a fluid is heated from below, it may show specific spatial patterns such as vortices or honeycombs (Fig.1.14). Or when a fluid is heated more, it may show spatio-temporal patterns, e.g. oscillations of vortices. Chemical reactions may give rise to macroscopic patterns, e.g. chemical oscillations where a change of colour occurs periodically, for instance from red to blue to red etc. Other phenomena are spiral patterns or concentric waves (Fig.1.15). In biology the clear water animal hydra has become a model system for morphogenesis. When a hydra is cut into two parts, a new head is readily formed where there was only a foot left and vice versa, a foot is formed where only a head was left (Fig.1.16).

Detailed experiments may allow us to draw conclusions on the mechanism of this restoration on the basis of the concept of chemical fields which are formed by

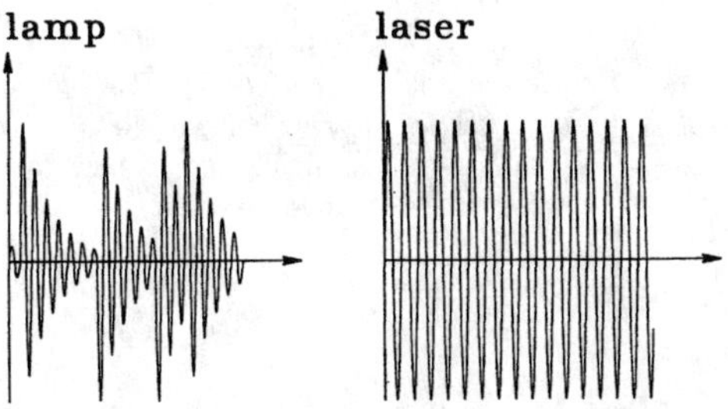


**Fig. 1.13.** The basic difference between the light from a lamp and from a laser. In both cases the electric field strength of the field amplitude is plotted versus time. On the left hand side the light from a lamp consists of individual uncorrelated wave tracks. On the right hand side in the laser the light wave consists of a single practically infinitely long sinusoidal wave

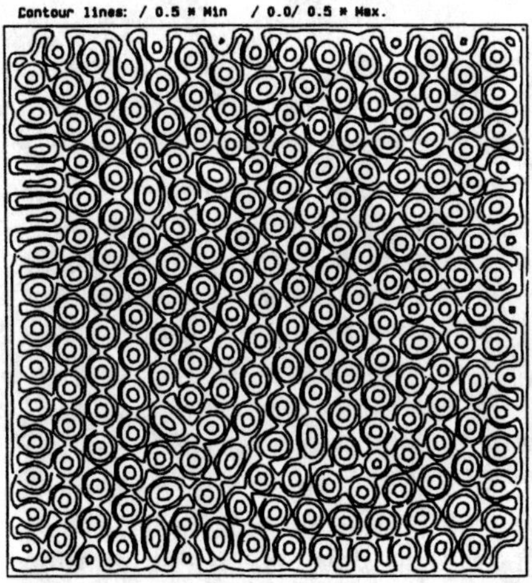


**Fig. 1.14.** A pattern in fluid dynamics

**Fig. 1.15.** Spiral waves in the Belousov-Zhabotinsky reaction [from S.C. Müller, T. Plesser, B. Hess (unpublished)]

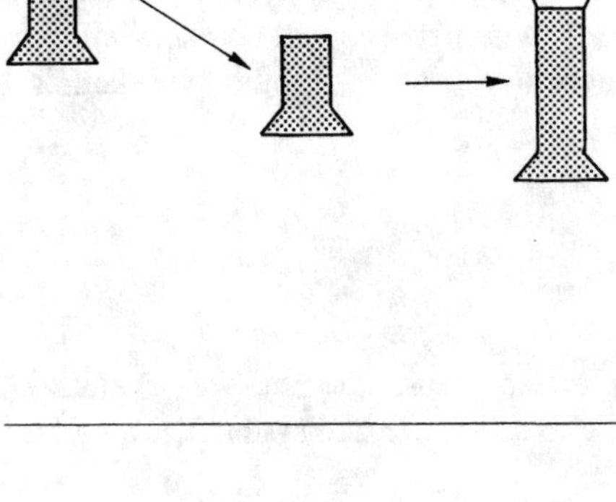

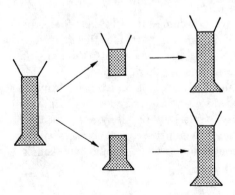

**Fig. 1.16.** An experiment on hydra reveals that in this species the information on the differentiation of cells cannot be laid down in the genes. From left to right: Intact hydra with head and tail is cut in the middle into two pieces. After a while the upper part regenerates by forming a tail, the lower part regenerates by forming a head

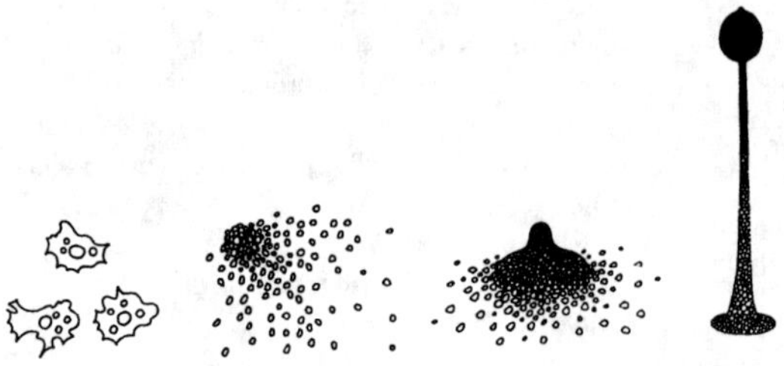

**Fig. 1.17.** Developmental stages of slime mold. From left to right: The individual cells assemble, aggregate more and more, and finally differentiate to form the mushroom

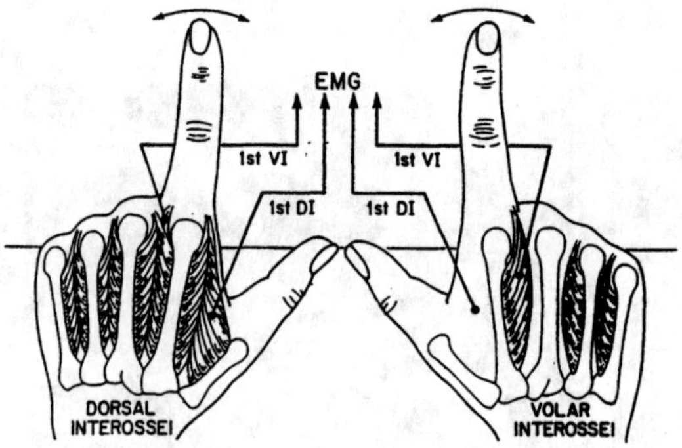


**Fig. 1.18.** Experimental setup to study involuntary changes of hand movements [from J.A.S. Kelso: "Dynamic Patterns in Complex, Biological Systems: Experiment and Synergetic Theory" (preprint)]

production and diffusion of chemicals. Another example of self-organization in morphogenesis is provided by slime mold (Fig.1.17). This little fungus usually exists in form of individual cells which live on a substrate. But then within the lifecycle of slime mold, these individual cells assemble at a point, differentiate and form the mushroom which then eventually spreads its spores, whereupon the life cycle starts again. Another model system is the squid axon used to study nerve conduction, or the well-known example of Drosophila in genetics where the giant chromosomes, the rapid multiplication rate and the possibility of causing mutations make this little animal an ideal object of study in this field.

More recently human hand movements have become a model system for studying the coordination of muscles and nerve cells and in particular the transitions between various kinds of movement (Fig.1.18). We shall come back to this example and to other examples later in the book. The involuntary change of hand movements is strongly reminiscent of the change of gaits of horses, cats or other quadrupeds. Quite generally speaking, these model systems allow us to develop new concepts which can first be checked against a variety of relatively simple systems and then later applied to truly complex systems. In this way our subsequent chapters will be devoted to the development of such new concepts whose applicability will then be illustrated by a number of explicit examples.

## 1.4 Self-Organization

As mentioned before, we may distinguish between man-made and self-organized systems. In our book we shall be concerned with self-organized systems. It may be mentioned, however, that the distinction between these two kinds of systems is not completely sharp. For instance humans may construct systems in such a way that by building in adequate constraints the system will be enabled to find its specific function in a self-organized fashion. A typical example mentioned before is the laser where the specific set-up of the laser by means of its mirrors allows the atoms to produce a specific kind of light. Quite evidently in the long run it will be desirable to construct computers which do programming in a self-organized fashion.

For what follows it will be useful to have a suitable definition of self-organization at hand. We shall say that a system is self-organizing if it acquires a spatial, temporal or functional structure without specific interference from the outside. By "specific" we mean that the structure or functioning is not impressed on the system, but that the system is acted upon from the outside in an nonspecific fashion. For instance the fluid which forms hexagons is heated from below in an entirely uniform fashion, and it acquires its specific structure by self-organization. In our book we shall mainly be concerned with a particular kind of self-organization, namely so-called nonequilibrium phase transitions.

As we know, systems in thermal equilibrium can show certain transitions between their states when we change a parameter, e.g. the temperature. For instance, when we heat ice it will melt and form a new state of a liquid, namely water. When we heat water up more and more, it will boil at a certain temperature and form vapor. Thus, the same microscopic elements, namely the individual molecules, may give rise to quite different macroscopic states which change abruptly from one state to another. At the same time new qualities emerge, for example ice has quite different mechanical properties to those of a gas.

In the following we shall be concerned with similar changes in the state of systems far from thermal equilibrium. Examples have been provided in Sect. 1.4, for instance by the liquid which forms a particular spatial pattern, by the laser which emits a coherent light wave, or by biological tissues which undergo a transition towards a differentiation leading to the formation of specialized organs.

## 1.5 Aiming at Universality

### 1.5.1 Thermodynamics

Thermodynamics is a field which allows us to deal with arbitrarily complex systems from a universal point of view. For instance we may ascribe temperature to a stone, to a car, to a painting, or to an animal. We further know that important properties of systems change when we change their temperature. Just think of melting of ice at the melting temperature, or of the importance of measuring the temperature of an ill person. However, this example illustrates at the same time that temperature alone is certainly not sufficient to characterize a car or a painting in many other respects. A stone, a car, a dress and a painting have the properties of being in thermal equilibrium. Such a state is reached when we leave a system entirely on its own, or when we couple it to another system which is in thermal equilibrium at a specific temperature.

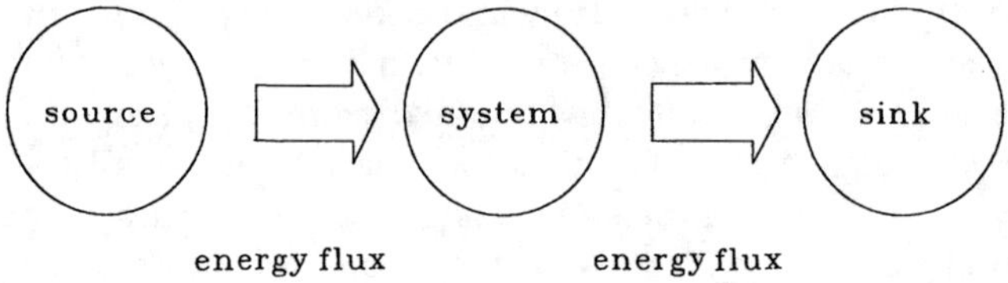


**Fig. 1.19.** Scheme of an open system which receives its energy from a source and dissipates the rest of the energy into a sink

Another important and even central concept of thermodynamics is that of entropy. Entropy is a concept which refers to systems in thermal equilibrium which can be characterized by a temperature $T$. The change of entropy is then given by the well-known formula $dS = dQ_{rev}/T$. Here, $T$ is the absolute temperature and $dQ_{rev}$ is the amount of heat which is reversibly added to or removed from the system. The general laws of thermodynamics are:

1) The first law which states that in a closed system energy is conserved, whereby energy may acquire various forms, such as the internal energy, work being done, or heat. So a typical form relating the changes $dU$, $dA$, $dQ$ to one another reads

$$dU = dQ - dA \ . \tag{1.1}$$

2) The second law tells us that in a closed system entropy can never decrease, but can only increase until it reaches its maximum. As we shall see later, the conservation laws, e.g. for energy, together with the so-called maximum entropy principle, allow us to derive certain microscopic properties of a system from macroscopic data. For instance we may derive the velocity distribution function of a gas in a straight forward manner. In the present book we shall be practically exclusively concerned with *open systems* (Fig.1.19). These are systems which are maintained in their specific states by a continuous influx of energy and/or matter. As we shall see, traditional thermodynamics is not adequate for coping with these systems; instead we have to develop some new kind of thermodynamics which will be explained in detail in the following chapters.

Thermodynamics can be considered as a macroscopic phenomenological theory. Its foundations lie in statistical physics upon which we shall make a few comments in the next section.

### 1.5.2 Statistical Physics

In this field an attempt is made, in particular, to derive the phenomenological, macroscopic laws of thermodynamics by means of a microscopic theory. Such a microscopic theory may be provided by the Newtonian mechanics of the individual gas particles, or by quantum mechanics. By use of appropriate statistical averages the macroscopic quantities are then derived from the microscopic laws. A central concept is again entropy, $S$. According to Boltzmann, it is related to the number $W$

of the different microscopic states which give rise to the *same macroscopic* state of the system, by means of the law

$$S = k \ln W \quad (1.2)$$

where $k$ is Boltzmann's constant. A crucial and not yet entirely solved problem is that of explaining why macroscopic phenomena may be irreversible while all fundamental laws are reversible. For instance the laws of Newtonian mechanics are invariant under the reversal of time, i.e. when we let a movie run backwards, all the processes shown there in reverse sequence are allowed in Newtonian mechanics. On the other hand it is quite evident that in macroscopic physics processes are irreversible. For instance when we have a gas container filled with gas molecules and we open a valve the gas will go to a second vessel and fill both vessels more or less homogeneously. The reverse process, i.e. that one vessel is emptied spontaneously and all the molecules return to the original vessel is never observed in nature.

Despite the difficulty in rigorously deriving irreversibility, by means of statistical physics we can explain a number of the phenomena of irreversible thermodynamics, such as relaxation processes, heat conduction, diffusion of molecules, etc.

### 1.5.3 Synergetics

The third approach to formulating universal laws valid for complex systems is that of synergetics. In this field we study systems that can form spatial, temporal or functional structures by means of self-organization. In physics, synergetics deals with systems far from thermal equilibrium. Typical examples are fluids heated from below, or lasers. Systems from chemistry and biology can also be conceived as physical systems and can be treated again by synergetics. But synergetics deals also with other systems, such as those in economy or sociology. In synergetics we focus our attention on qualitative, macroscopic changes, whereby new structures or new functions occur. This restriction to qualitative, macroscopic changes is the price to be paid in order to find general principles.

We shall remind the reader of the main principles of synergetics in Chap. 2. There we shall see that in physics, synergetics starts from a microscopic formulation, for example from the microscopic equations of motion. In other cases such as biology or chemistry a mesoscopic approach may be appropriate where we start from suitable subsystems, for instance adequate properties of a total cell in biology. It is assumed that the system under consideration is subject to external constraints, such as a specific amount of energy being fed into the system. Then when this control parameter is changed, an instability may occur in which the system tends to a new state.

As is shown in synergetics, at such an instability point, in general just a few collective modes become unstable and serve as "order parameters" which describe the macroscopic pattern. At the same time these macroscopic variables, i.e. the order parameters, govern the behavior of the microscopic parts by the "slaving principle". In this way the occurrence of order parameters and their ability to enslave allows the system to find its own structure. When control parameters are changed over a

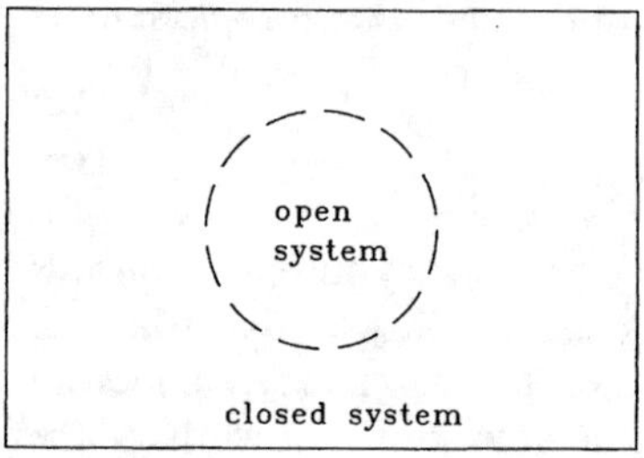


◀ **Fig. 1.20.** An open system embedded in a closed system

**Fig. 1.21**

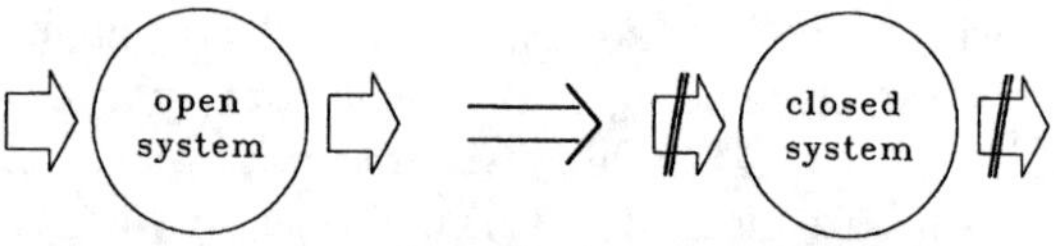


**Fig. 1.21.** A closed system may be considered as the limiting case of an open system into and out of which energy fluxes are cut

wide range, systems may run through a hierarchy of instabilities and accompanying structures.

Synergetics is very much an open-ended field in which we have made only the very first steps. In the past one or two decades it has been shown that the behavior of numerous systems is governed by the general laws of synergetics, and I am convinced that many more examples will be found in the future. On the other hand we must be aware of the possibility that still more laws and possibly still more general laws can be found.

As mentioned earlier, thermodynamics deals with systems in thermal equilibrium, whereas synergetics deals with systems far away from thermal equilibrium. But here quite a peculiar situation arises. On the one hand we can always embed an open system into a larger closed one (Fig.1.20). Earth for example, is an open system because it is fed with energy from the sun and it emits its energy during night into the universe. But taking the sun and, say, part of the universe as a whole system, we may consider the whole system as a closed one to which the laws of thermodynamics apply. In so far we see that the laws of synergetics must not be in contradiction to those of thermodynamics. But on the other hand any open system may be considered in the limiting case, where the energy or matter fluxes tend to zero so that eventually we deal with a closed system (Fig.1.21). Therefore, the general laws of thermodynamics must be obtainable as limiting cases of those of synergetics.

As the reader will notice this program is not yet finished but leaves space for future research. Until now, synergetics has started from the microscopic or mesoscopic level. In this book however, we shall attempt to present a second foundation of synergetics which we shall discuss in some detail in Chaps. 5–7 and then in greater detail in the following chapters. The starting point for this macroscopic approach is the concept of information and we shall deal with some of its most important aspects in the next section.

## 1.6 Information

The use of the word information is connected with considerable confusion. This is caused by the fact that the word information is used with many quite different meanings. In every day language, information is used in the sense of message or instruction. A letter, a television transmission or a telephone call all carry information. In the following we shall be concerned with the scientific use of the word information. We shall start with the concept of Shannon information where information is used without any meaning. Then we shall briefly study information with respect to messages and finally we shall be concerned with the problem of the self-creation of meaning.

Quite evidently, when dealing with physical systems, we wish to eliminate all kinds of anthropomorphisms because we wish to describe a physical system in an as objective manner as possible. But in biology too, this trend is quite obvious so that eventually we have a more or less physical or even mechanistic picture of biological systems. But strangely enough it appears with respect to the development of modern computers, e.g. those of the fifth generation, that we wish to reintroduce meaning, relevance, etc. Therefore, in this section we wish to discuss ways in which we can return from a concept of information from which meaning was exorcised to the act of self-creation of meaning.

### 1.6.1 Shannon Information: Meaning Exorcised

We shall discuss the concept of Shannon information in detail in Chap. 3, but in order to have a sound basis for our present discussion we shall elucidate the concept of Shannon information by means of some examples. When we toss a coin we have two possible outcomes. Or when we throw a die we have six possible outcomes. In the case of the coin we have two kinds of information, head or tail; in the case of the die we have the information that one of the numbers from one to six has appeared. Similarly, we may have answers "yes" or "no", etc. The concept of Shannon information refers simply to the number of possibilities, $Z$, which in the case of a coin are two, in the case of a die are six. As we shall see later, a proper measure for information is not the number $Z$ itself but rather its logarithm where usually the logarithm to base 2 is taken, i.e. information is defined by

$$I = \log_2 Z \ . \tag{1.3}$$

This definition can be cast into another form which we shall come across time and again in this book. Consider for example a language such as English. We may label its letters $a, b, c, \ldots$ by the numbers $j = 1, 2, \ldots$ i.e. $a \to 1, b \to 2$ etc. Then we may count the frequencies $N_j$ of occurrence of these letters in a particular book or in a library perhaps. We define the relative frequency of a letter labeled by $j$ as

$$p_j = \frac{N_j}{N} \tag{1.4}$$

where $N$ is the total number of letters counted, $N = \sum N_j$. Then the average information per letter contained in that book (or library) is given by

$$i = -\sum_j p_j \log_2 p_j \ . \tag{1.5}$$

For a derivation of this formula see Chap. 3. Shannon used his concept to study the capacity of a communication channel to transfer information even under the impact of noise. Two features of the Shannon information are of importance in what follows. 1) Shannon information is not related to any meaning. So concepts such as meaningful or meaningless, purposeful etc. are not present. 2) Shannon information refers to closed systems. There is only a fixed reservoir of messages, whose number is $Z$.

### 1.6.2 Effects of Information

In this section we wish to introduce a new approach which is a step towards a concept of information which includes semantics. We are led to the basic idea by the observation that we can only attribute a meaning to a message if the response of the receiver is taken into account. In this way we are led to the concept of "relative importance" of messages which we want to demonstrate in the following.

Let us consider a set of messages each of which is specified by a string of numbers. The central problem consists in modeling the receiver. We do this by invoking modern concepts of dynamic systems theory or, still more generally, by concepts of synergetics. We model the receiver as a dynamic system. Though we shall describe such systems mathematically in the next chapter, for our present purpose a few general remarks will suffice. We consider a system, e.g. a gas, a biological cell, or an economy, whose states can be characterized at the microscopic, mesoscopic or macroscopic level by a set of quantities, $q$, which we shall label by an index $j$, i.e. $q_j$. In the course of time, the $q_j$'s may change. We may lump the $q_j$'s together into a state vector $\boldsymbol{q}(t) = [q_1(t), q_2(t), \dots, q_N(t)]$. The time evolution of $\boldsymbol{q}$, i.e. the dynamics of the system, is then determined by differential equations of the form

$$\frac{d\boldsymbol{q}}{dt} = \boldsymbol{N}(\boldsymbol{q}, \alpha) + \boldsymbol{F}(t) \tag{1.6}$$

where $\boldsymbol{N}$ is the deterministic part and $\boldsymbol{F}$ represents fluctuating forces. All we need to know, for the moment, is the following: If there are no fluctuating forces, once the value of $\boldsymbol{q}$ at an initial time is set, and the so-called control parameters $\alpha$ are fixed, then the future course of $\boldsymbol{q}$ is determined uniquely. In the course of time, $\boldsymbol{q}$ will approach an attractor. To visualize a simple example of such an attractor consider a miniature landscape with hills and valleys modeled by paper (Fig. 1.22,23). Fixing $\alpha$ means a specific choice of the landscape, in which a ball may slide under the action of gravity (and under a frictional force). Fixing $\boldsymbol{q}$ at an initial time means placing the ball initially at a specific position, for instance on the slope of a hill (Fig. 1.22). From there it will slide down until it arrives at the bottom of the valley: this is then an attractor. As the experts know, dynamic systems may possess also other kinds of attractors, e.g. limit cycles, where the system performs an indefinite oscillation, or still more complicated are attractors such as "chaotic attractors". In the following, it will be sufficient to visualize our concepts by considering the attractor as the bottom of a valley (a so-called fixed point). When fluctuations $\boldsymbol{F}$ are present, the ball may jump from one attractor to another (Fig. 1.24).

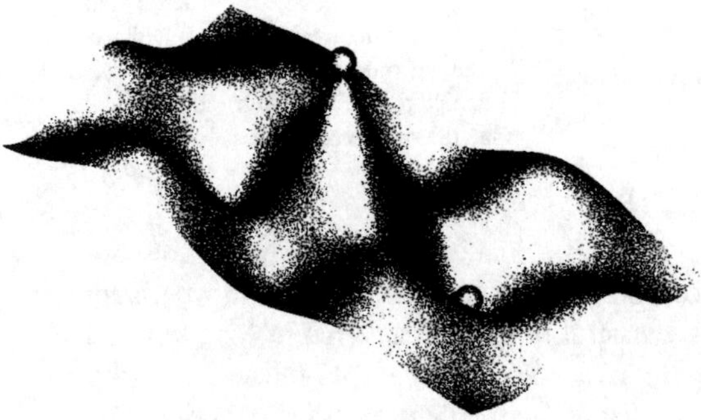

**Fig. 1.22.** Visualization of a dynamical system with fixed point attractors by means of a miniature landscape formed of deformed paper

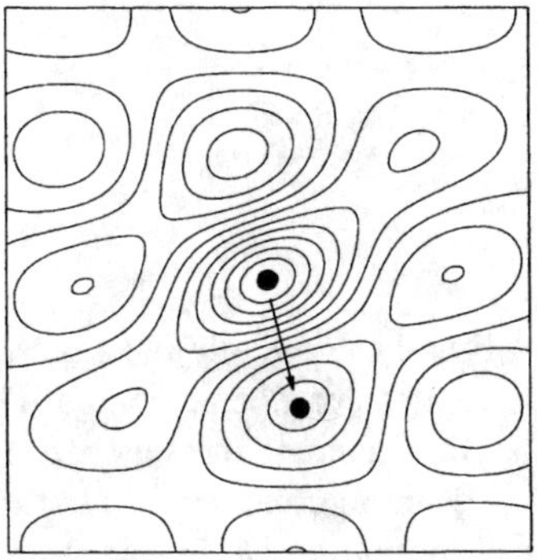

**Fig. 1.23.** Isobases belonging to the landscape of Fig.1.22

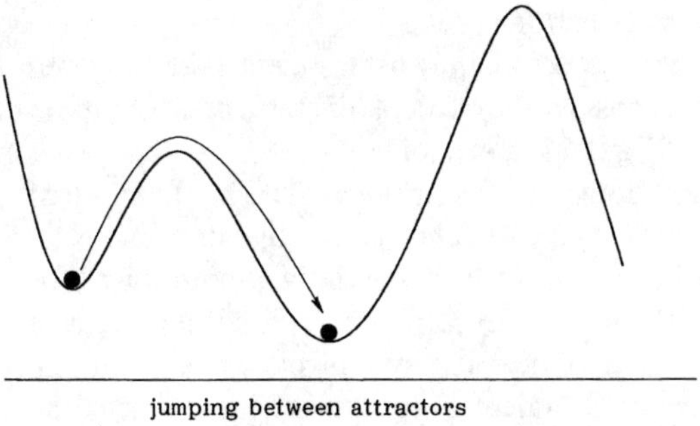


**Fig. 1.24.** Illustration of the jumping of a system between two fixed point attractors

After these preparatory remarks let us return to our original problem, namely to attribute a meaning to a message. We assume that the receipt of a message by the system means that the parameters $\alpha$ and the initial value of $\boldsymbol{q}$ are set by the message. For the time being we shall assume that these parameters are then uniquely fixed. An extension of the theory to an incomplete message is straightforward (see below). We first ignore the role of fluctuations. We assume that before the message arrives the system has been in an attractor which we shall call the neutral state. The attractor may be a resting state i.e. a fixed point, but it could equally well be a limit cycle, a torus or a strange attractor, or a type of attractor still to be discovered by dynamic systems theory. We shall call this attractor $\boldsymbol{q}_0$. After the message has been received and the parameters $\alpha$ and the initial value $\boldsymbol{q}$ are newly set, in principle two things may happen. Let us assume that we are allowed to wait for a certain measuring time so that the dynamic system can be assumed to be in one of its possible attractors. Then either the message has left the system in the $\boldsymbol{q}_0$ state. In such a case the message is evidently useless or meaningless.

The other case is that the system goes into a new attractor. We first assume that this attractor is uniquely determined by the incident message. Clearly, different messages can give rise to the same attractor. In this case we will speak of redundancy of the messages.

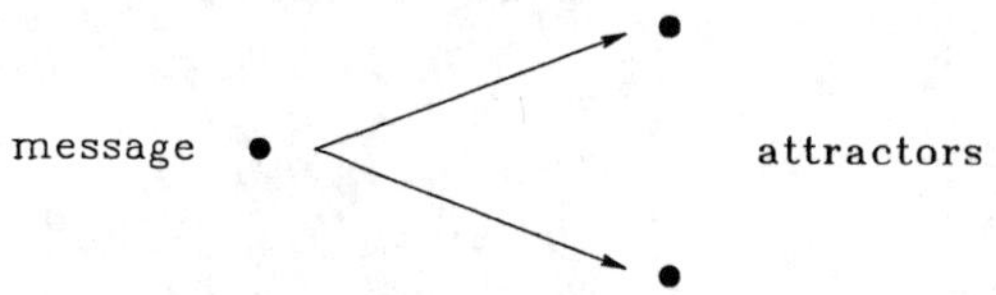


**Fig. 1.25.** A message can reach two different attractors by means of internal fluctuations of the system by a process depicted in Fig.1.24. In this way two attractors become accessible

Finally, especially in the realm of biology it has been a puzzle until now how information can be generated. This can be easily visualized, however, if we assume that the incident message produces the situation depicted in Fig.1.25, which is clearly ambiguous. Two new stable points (or attractors) can be realized depending on a fluctuation within the system itself. Here the incident message contains information in the ordinary sense of the word, which is ambiguous and the ambiguity is resolved by a fluctuation of the system. Loosely speaking, the original information is doubled because now two attractors become available. In the case of biology these fluctuations are realized in the form of mutations. In the realm of physics however, we would speak of symmetry breaking effects.

Taking all these different processes together we may list the elementary schemes shown in Fig.1.26. Of course, when we consider the effect of different messages, more complicated schemes such as those of Fig.1.27 may evolve.

We shall now treat the question of how we can attribute values to the incident messages or, more precisely speaking, we want to define a "relative importance of the messages". To this end we first have to introduce a "relative importance" for the individual attractors. In reality, the individual attractors will be the origin of new messages which are then put into a new dynamical system and we can continue this process ad infinitum. However, for practical purposes, we have to cut the hierarchical sequence at a certain level and at this level we have to attribute values of the relative importance to the individual attractors. Since our procedure can already be clearly demonstrated if we have a one-step process, let us consider this process in detail.

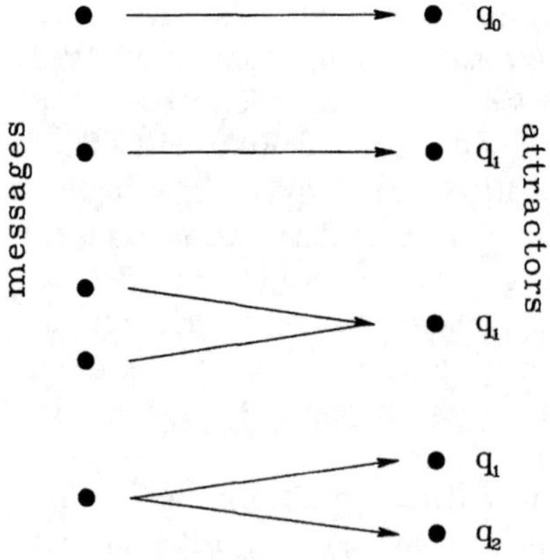


**Fig. 1.26.** Various possibilities for a message to reach attractors

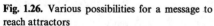

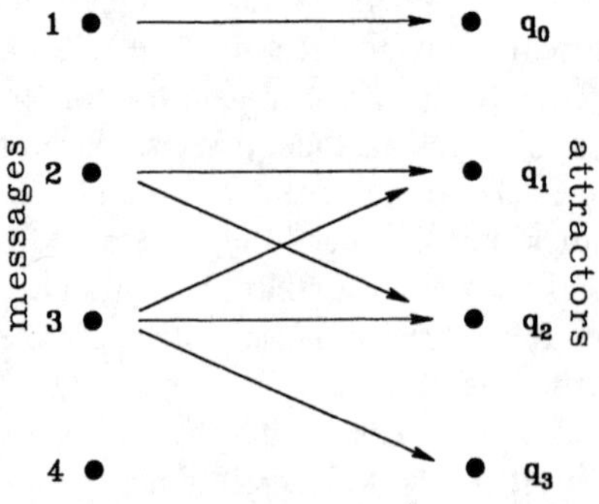


**Fig. 1.27.** Another example of how messages can reach attractors

Let us attribute a "relative importance" to the individual attractors where the attractor 0 with $q_0$ has the value 0, while the other attractors may have values $0 \le p'_j \le 1$, which we normalize to

$$\sum_j p'_j = 1 \ . \tag{1.7}$$

The assignment of $p'_j$ depends on the task that the dynamic system has to perform. We may think of a specific task which can be performed just by a single attractor or we may think of an ensemble of tasks whose execution is of a given relative importance. Clearly the relative importance of the messages $p_j$ does not only depend on the dynamic system but also on the tasks it must perform. The question is now: What are the values $p_j$ of the incident messages? To this end we consider the links between a message and the attractor into which the dynamical system is driven after receipt of this message. If an attractor $k$ (including the 0 attractor) is reached after receipt of the message $j$ we attribute to this process the matrix element $M_{jk} = 1$ (or $= 0$). If we allow for internal fluctuations of a system, a single message can drive the system via fluctuations into several different attractors which may occur with branching rates $M_{jk}$ with $\sum_k M_{jk} = 1$. We define the "relative importance" $p_j$ by

$$p_j = \sum_k L_{jk} p'_k = \sum_k \frac{M_{jk}}{\sum_{j'} M_{j'k} + \varepsilon} p'_k \ , \tag{1.8}$$

where we let $\varepsilon \to 0$. (This is to ensure that the ratio remains determined even if the denominator and nominator vanish simultaneously.) We first assume that for any $p'_k \ne 0$ at least one $M_{jk} \ne 0$. One readily convinces oneself that $p_j$ is normalized which can be shown by the steps

$$\sum_j p_j = \sum_{kj} \frac{M_{jk}}{\sum_{j'} M_{j'k} + \varepsilon} p'_k \tag{1.9}$$

$$= \sum_k \left( \sum_j \frac{M_{jk}}{\sum_{j'} M_{j'k} + \varepsilon} \right) p'_k \tag{1.10}$$

$$= \sum_k p'_k = 1 \tag{1.11}$$

where the bracket in (1.10) is equal to 1.

Now consider the case where for some $k$-values, for which $p'_k \ne 0$, all $M_{jk} = 0$. In this case in the sums over $k$ in (1.9) and (1.10) some coefficients of $p'_k \ne 0$ vanish and, since $\sum_k p'_k = 1$, we obtain $\sum_j p_j < 1$. If this inequality holds, we shall speak of an *information deficiency*.

In a more abstract way we may adopt the left hand side of (1.8) as a basic definition where we assume

$$\sum_j L_{jk} \le 1 \ , \tag{1.12}$$

where the equality sign holds only when there is no information deficiency.

We note that instead of the requirement $M_{jk} = 1$ as in case of a single final attractor for an incident message, $M_{jk}$ can be generalized to

$$0 < M_{jk} \leq 1 \ . \tag{1.13}$$

The form (1.8), left hand side, immediately allows us to write down the formulas for several systems that are coupled one after the other. For instance in the two step process we immediately obtain

$$p_j = \sum_k L_{jk}^{(1)} p'_k = \sum_{kk'} L_{jk}^{(1)} L_{kk'}^{(2)} p''_{k'} \tag{1.14}$$

where one can convince oneself very easily that $\sum_j p_j = 1$ provided $\sum_k p'_k = 1$ and $\sum_j L_{jk} = 1$. The individual steps read

$$\sum_j p_j = \sum_j \sum_{kk'} L_{jk}^{(1)} L_{kk'}^{(2)} p''_{k'} = \sum_{kk'} \underbrace{\left( \sum_j L_{jk}^{(1)} \right)}_{=1} L_{kk'}^{(2)} p''_{k'} \tag{1.15}$$

$$= \sum_{k'} \underbrace{\sum_k L_{kk'}^{(2)}}_{=1} p''_{k'} = 1 \ . \tag{1.16}$$

We may define

$$L'_{jk'} = \sum_k L_{jk}^{(1)} L_{kk'}^{(2)} \ . \tag{1.17}$$

Because the $L$'s are positive we find

$$L'_{jk} \geq 0 \tag{1.18}$$

and because of the normalization properties (in case of no information deficiency)

$$\sum_j L'_{jk'} = \sum_k \sum_j L_{jk}^{(1)} L_{kk'}^{(2)}$$

$$= \sum_k L_{kk'}^{(2)} = 1 \tag{1.19}$$

we readily obtain

$$L'_{jk} \leq 1 \tag{1.20}$$

so that $L'_{jk}$ obeys the inequality

$$0 \leq L'_{jk} \leq 1 \ . \tag{1.21}$$

We mention that the recursion from $p''$ or still higher order $p^{(n)}$ to $p$ may depend on the paths.

Our above approach not only introduces the new concept of relative importance of a message but it also provides us with an algorithm to determine $p_j$ which has some conceptual and practical consequences. With a given task or ensemble of tasks, this algorithm allows us to select the message to be sent, namely the one with the biggest $p_j$. If there are several $p_j$ of the same size it does not matter which message is sent. From the conceptual point of view we may then decide whether a dynamical system annihilates, conserves or generates information. To this end we make use of the concept of information in the sense of conventional information theory. But instead of the information content due to the relative frequency of symbols we use the relative importance within a set of messages, i.e. we introduce the quantities

$$S^{(0)} = -\sum_j p_j \ln p_j \tag{1.22}$$

$$S^{(1)} = -\sum_k p'_k \ln p'_k \;, \tag{1.23}$$

where $p_j$ and $p'_k$ have been defined above in the text. If $\sum_k p'_k = 1$, as is always assumed here, and $\sum_j p_j < 1$, an information deficiency is present. In the case $\sum_j p_j = 1$ we shall speak of annihilation of information if

$$S^{(1)} < S^{(0)} \tag{1.24}$$

of conservation of information if

$$S^{(1)} = S^{(0)} \tag{1.25}$$

and of generation of information if

$$S^{(1)} > S^{(0)} \;. \tag{1.26}$$

The meaning of this definition quickly becomes clear when we treat special cases. If, for instance, two messages lead to the same attractor there is a redundancy in the system and the information content (in the traditional technical sense of the word) becomes smaller. It is reduced from

$$S^{(0)} = -K[\tfrac{1}{2}\ln(\tfrac{1}{2}) + \tfrac{1}{2}\ln(\tfrac{1}{2})] = K \ln 2 \tag{1.27}$$

to

$$S^{(1)} = -K \cdot 1 \cdot \ln 1 = 0 \;. \tag{1.28}$$

In the case of a one-to-one mapping of $p_j$ onto $p'_k$ we find the transfer of $\{p_j\}$ into the same set $\{p'_k\}$, except maybe for the permutation of indices, i.e. for different numbering of the states. In such a case (1.25) clearly holds. Finally, in the case (1.26), the $p_j$, where one $p_j = 1$ and all others $= 0$, are transferred into e.g. $p' = p'' = \frac{1}{2}$ and all others are equal to 0. Then $S^{(0)} = -K \cdot 1 \cdot \ln 1 = 0$ is enlarged to

$$S^{(1)} = -K[\tfrac{1}{2}\ln(\tfrac{1}{2}) + \tfrac{1}{2}\ln(\tfrac{1}{2})] = K \ln 2 \;. \tag{1.29}$$

Of course these examples are not meant to prove the definitions (1.24–26) but rather to illustrate their meaning.

Our approach based on synergetics has some further nice features. Semantics has become the problem of studying the response (attractors) of the dynamic system. The system may be error-correcting (or may supplement partial information). If the incident message does not set the initial state $\boldsymbol{q}$ *on the attractor* (i.e. *not correctly*), it may set the initial state $\boldsymbol{q}$ within the *basin of the attractor* i.e. on the slope of the hill surrounding the bottom of the specific valley which represents the attractor (fixed point). In this way the system pulls the state vector into the attractor corresponding to that basin, i.e. into the *correct* state. It will be an interesting problem to determine the minimum number of bits required to realize a given attractor (or to realize a given value of "relative importance").

Within our present scheme, the learning process of a system can also be modeled. A system can be "sensitized" or "desensitized" with respect to messages $j$ e.g. by letting more or fewer parameters react to specific messages.

In the above treatment we have assumed that the value of the messages is measured with respect to the *same* initial state of the receiver. In the next step of our considerations we may assume that messages apply to a receiver in *another* initial state which has been set for instance by a previous message.

In such a way we obtain an interference of messages and the relative importance of a message depends on the messages previously delivered to the receiver. In the general case, the relative importance of a message will depend in a non-commuting way on the sequence of the messages. In this way the receiver is transformed by messages again and again and clearly the relative importance of messages will become a function of time.

Another remark might be useful, particularly in relation to synergetic processes. A synergetic system not only needs to be a dynamical system showing e.g. limit cycle or chaotic behavior, but it might also be one in which irreversible processes leading for instance from a disorganized liquid state into a structured solid state occur.

Let us conclude this part with a comment on pattern recognition which will be eludicated from various points of view in this book. Pattern recognition can be considered as a processing of incoming messages by a receiver, e.g. the brain or a machine. It is therefore an interesting task to discuss pattern recognition using the ideas just outlined. I suggest that pattern recognition, at least in general, is a multistep process in which the receiver takes an active part. In the first step, the pattern is received at a global level where, in general, several attractors can be reached. Then, the sensory system is requested to focus its attention on the exploration of additional features so that a finer set of attractors can be selected. To be more explicit: In the first step for instance the global shape of the contour lines of an object are determined e.g. close to a circle, rectangular, etc. Then, in the case of a circle, there are several attractors: apple, face, wheel, tree. Then the receiver asks back for further details, e.g. color, vertical lines (nose?), etc. In this way the process can be continued.

Note that our interpretation of pattern recognition differs from the "traditional" approach to which we shall come in Chap. 12. There the pattern is first decomposed into its "primitives" or "features". Here we start from the *global pattern* (contour line) and then proceed to more and more details.

This approach offers us an explanation (or at least a hint) as to why, in human pattern recognition, even interrupted contour lines are supplemented such that a continuous line is "seen".

### 1.6.3 Self-Creation of Meaning

As was mentioned previously, synergetics may be considered as a theory of the emergence of new qualities at a macroscopic level. By means of a suitable interpretation of the results of synergetics, we may thus study the *emergence of meaning* as the emergence of a new quality of a system, or in other words the *self-creation of meaning*. In order to study how this happens we want to compare a physical system, namely the laser, with several model systems of biology. Let us start with some general remarks on the role of information in biological systems.

One of the most striking features of any biological system is the enormous degree of coordination among its individual parts. In a cell, thousands of metabolic processes may go on at the same time in a well-regulated fashion. In animals, millions to billions of neurons and muscle cells cooperate to bring about well-coordinated locomotion, heartbeat, breathing or blood flow. Recognition is a highly cooperative process, and so are speech and thought in humans. Quite clearly, all these well-coordinated, coherent processes become possible only through the exchange of information, which must be produced, transmitted, received, processed, transformed into new forms of information, communicated between different parts of the system and at the same time, as we shall see, between different hierarchical levels. We are thus led to the conclusion that information is a crucial element of the very existence of life.

The concept of information is a rather subtle one and it will be the goal of this section to further elucidate some of its aspects. As we shall see, information is linked not only with channel capacity or with orders given from a central controller to individual parts of a system – it can acquire also the role of a "medium" to whose existence the individual parts of a system contribute and from which they obtain specific information on how to behave in a coherent, cooperative fashion. At this level, semantics may come in.

Let us first have a look at physics. In closed systems the second law of thermodynamics tells us that structures decay and systems become more and more homogeneous, at least on a macroscopic level. At the microscopic level complete chaos may occur. For these reasons information cannot be generated by systems in thermal equilibrium; in closed systems thermal equilibrium is eventually reached. But a system in thermal equilibrium cannot even *store* information. Let us consider a typical example, namely a book. At first sight, it seems to be in thermal equilibrium, and indeed we can measure its temperature. But in spite of that, it has not reached its final state of complete thermal equilibrium. In the course of time, the printer's ink in the individual letters will diffuse away until a homogeneous state is reached.

This simple example teaches us that any memory consisting of a closed system is out of thermal equilibrium and it is always necessary to ask *how long* the

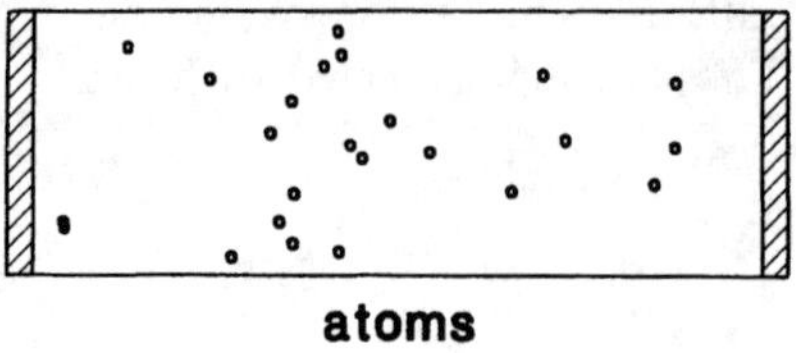


Fig. 1.28. Laser active atoms embedded in a crystal of a laser setup

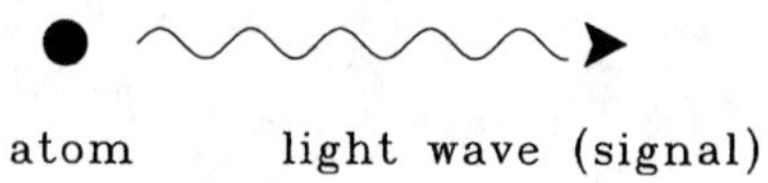


Fig. 1.29. An excited atom emits a light wave (signal)

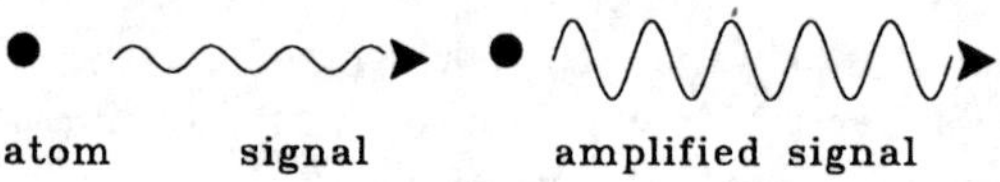


Fig. 1.30. When the light wave hits an excited atom it may cause the atom to amplify the original light wave

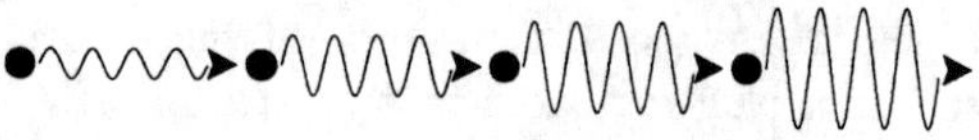

Fig. 1.31. A cascade of amplifying processes

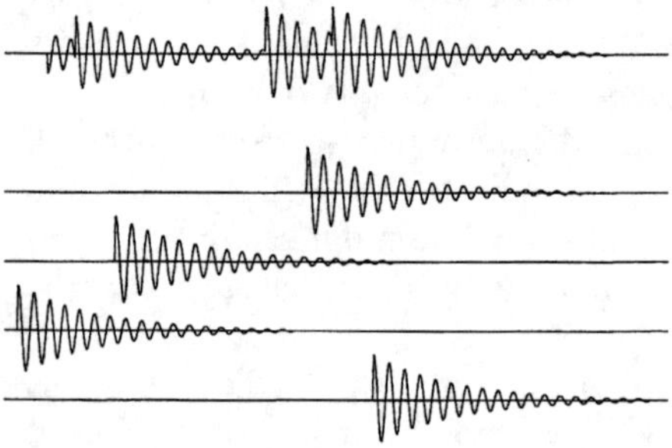

Fig. 1.32. The incoherent superposition of amplified light waves produces a still rather irregular light emission

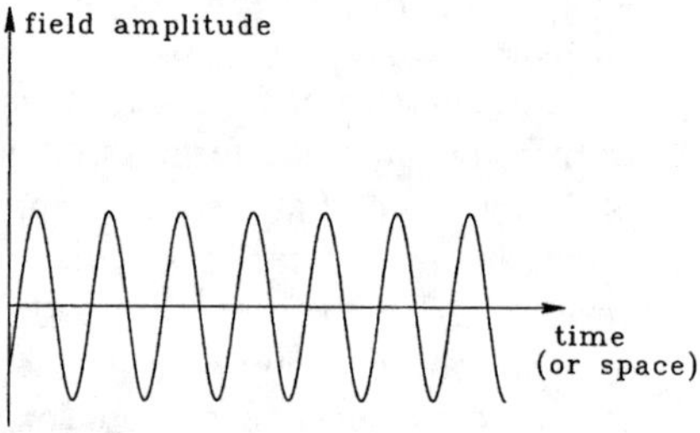


Fig. 1.33. In the laser the field amplitude is represented by a sinusoidal wave with a practically stable amplitude and only small phase fluctuations

information can be stored in each specific case. Let us therefore consider open systems which are kept far from thermal equilibrium by an influx of energy and/or matter into the system. As was mentioned before, in open systems, even in the inanimate world, specific spatial or temporal structures can be generated in a self-organized fashion. Examples are provided by the laser which produces coherent light, by fluids which can form specific spatial or temporal patterns, or by chemical reactions which can show continuous oscillations, or spatial spirals, or concentric waves. Even at this level we can speak to some extent of creation or storage of information. On the other hand, we can hardly attribute words like relevance, purpose or meaning to these processes.

Let us discuss the laser in some more detail because it allows us to introduce a terminology which is also useful for biological and other systems. In the laser a number of atoms are embedded, for instance, in a crystal such as ruby (Fig.1.28). After excitation from the outside, these atoms may emit individual light wave trains

(Fig.1.29). Thus, each atom emits a signal, i.e. it creates information which is carried by the light field. In the laser cavity the emitted wave trains may hit another excited atom and cause it to amplify the original wave (Fig.1.30). In this way, the information serves the purpose of enhancing the signal (Fig.1.31). Because the individual excited atoms may emit light waves indepently of each other and these may then be amplified by other excited atoms, a superposition of uncorrelated, though amplified wave trains results and a quite irregular pattern is observed (Fig.1.32).

But when the signal reaches a sufficiently high amplitude, an entirely new process starts. The atoms begin to oscillate coherently and the field itself becomes coherent, i.e. it is no longer composed of individual uncorrelated wave tracks but has become a practically infinitely long sinusoidal wave (Fig.1.33).

We have here a typical example of self-organization where the temporal structure of the coherent wave emerges without interference from the outside. Order is established. The detailed mathematical theory shows that the emerging coherent light wave serves as order parameter which forces the atoms to oscillate coherently, or in other words it enslaves the atoms (Fig.1.34). Note that we are dealing here with circular causality: On the one hand the order parameter enslaves the atoms, but on the other hand it is itself generated by the joint action of the atoms (Fig.1.35).

From the viewpoint of information, the order parameter serves a double role: it informs the atoms how to behave, and in addition, it informs the observer about the macroscopic ordered state of the system. While an enormous amount of information is needed to describe the states of the individual atoms, once the ordered state is established, only a single quantity, namely the phase of the total light field is necessary, i.e. we have an enormous compression of information. We may call the

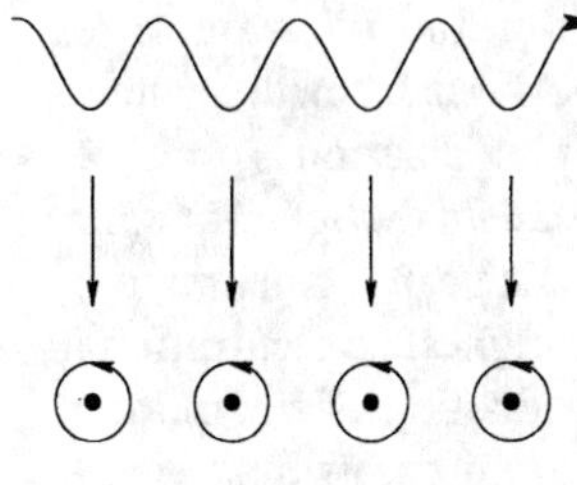


**Fig. 1.34.** Illustration of the slaving principle. The field acts as order parameter and prescribes the motion of the electrons in the atoms. In other words the motion of the electrons becomes slave to the field

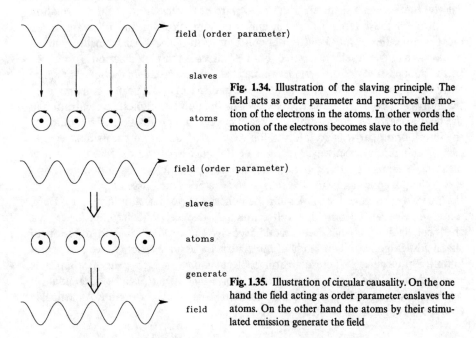

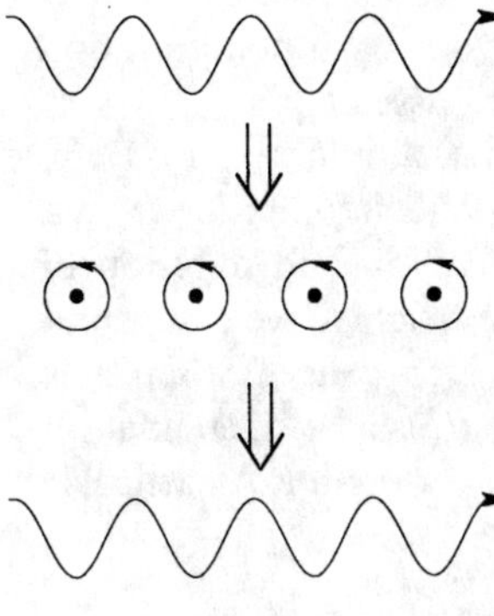


**Fig. 1.35.** Illustration of circular causality. On the one hand the field acting as order parameter enslaves the atoms. On the other hand the atoms by their stimulated emission generate the field

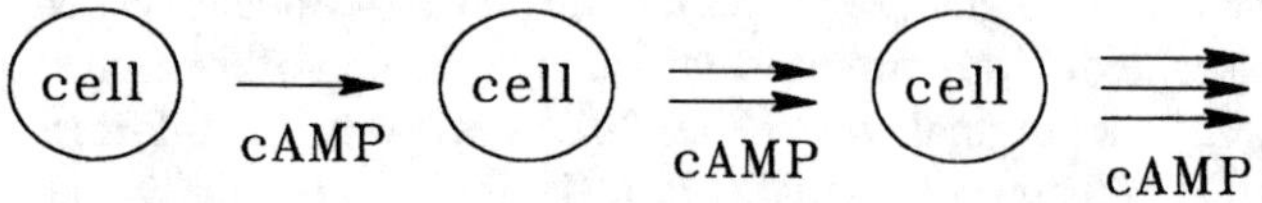


**Fig. 1.36.** Illustration of the amplification of the number of cAMP molecules in the cells of slime mold

**Fig. 1.37.** Schematic illustration of the concentration wave of cAMP in slime mold formation

order parameter an "informator". Over the past years, it has been shown that these concepts apply to a large number of quite different physical, chemical and biological systems.

To elucidate the role of information exchange at the level we are presently considering, let us take the example of slime mold (*dictiostelium discoideum*). Usually its cells live individually on a substrate but when food becomes scarce they assemble at a particular point. The mechanism of this kind of self-assembly is as follows:

The individual cells start to emit the substance, cyclic Adenosinemonophosphate (cAMP); thus they send out a signal or a message, i.e. information. Once cAMP molecules hit other cells, these are induced to increase their production in much the same way as the laser atoms amplify the incoming signal (Fig.1.36). Quite clearly, the elements themselves are not aware of the meaning of the information but through the interplay between emission, amplification and diffusion of the cAMP molecules, a spiral pattern of concentration of cAMP is formed, i.e. information at a higher level is generated (Fig.1.37). Because this information is produced by the cooperativity of the system, we may call it *synergetic information.* The spiral waves form some kind of gradient field (the informator) which can be measured by the individual cells which then move towards the point of highest concentration in the field. Clearly we can distinguish here between the production of information, the information carrier and information receiver which in our case would be cell, cAMP and cell, respectively. However at the next level, we observe that a new meaning has arisen, namely the established pattern of a molecular concentration serves the purpose of guiding the cells to the center of their assembly.

Basically the same idea holds for the concept of positional information. Here, it is assumed that the individual cell within a tissue receives its information from a chemical field which has been established by the production and diffusion of chemicals. In general, two kinds of molecules are assumed, namely activator and inhibitor molecules. Where activator molecules have a high concentration it is assumed that specific genes can be switched on which then cause the differentiation of a cell. In this way the chemical field plays the role of the informator. A particular model system has been hydra.

It is useful to recall what we have established so far. Quite evidently, there is a hierarchy of informational levels. At the lowest level, the individual parts can emit information which hits other parts of the system. Such an information transfer can take place between specific pairs of elements or the information can be transferred by a general carrier. An example for the first case are nerve fibers each connecting two neurons; examples of the second case are provided by hormones released to the blood, or by pheromones released into air.

Although, in all these cases the exchange of information may initially occur at random, a competition or cooperation between different kinds of signals sets in, and eventually a new collective state is reached which differs qualitatively from the disordered or uncorrelated state present before. Thus, a new state is described by an order parameter or a set of order parameters or equivalently by one or several informators. The states of the individual parts are determined by means of the slaving principle. But one may describe this process in another way, namely that a specific consensus was reached among the individual parts of the system or that self-organization has happened. At the same time information compression takes place. The information appears manifest at a macroscopic level and, in many cases, increases the reliability and/or efficiency of the system, or serves other purposes as mentioned above.

This new collective level becomes observable to the outer world and by establishing this relationship or context a new semantic level is reached. By the way, the context may be established with the outer world but equally well within the same system. Here then words like useful, useless, or relevance can be applied. This is quite evident from the example of the laser where the cooperative state reaches a high efficiency. In the analogous case of a biological system, such behavior is then useful for the whole system. Beyond instability points the system can acquire different possible states and it needs additional information on which state to choose. One possibility is that this information is provided genetically, or by constraints established by other parts of the system. But often in such a case of degeneracy, the surroundings play an important role, or in other words, it is the context which judges the value of the kind of state to be established. In the opinion of the author it is here that information in the biological sense starts. Through instability a collective state is formed but it acquires its meaning only with respect to the surroundings and, in a way, with respect to its value for the survival of the whole system.

These remarks also apply to the genetic code, though its very origin is not yet too well clarified. One may speculate that at first fluctuations occur which create some biological macromolecule with specific properties. The most important of these is that it can multiply in an autocatalytic fashion. The value of the information conveyed by this molecule to its phenotype is then judged by the environment to which other molecules with their phenotypes may also belong. By the interplay of mutation and selection new types of molecules and their corresponding phenotypes are then generated and in this way we observe the creation of new information. But whether this information is useful or not can be checked upon only by the interaction of the particular species with its environment.

In the considerations above we described the first steps of the formation of ordered or structured collective states. But in contrast to the physical systems mentioned above, such as lasers, fluid dynamics, or chemical reactions, a new feature appears in biology, namely a solidification. For instance, when the genes of a cell are switched on by activator molecules, the cell differentiates into a specific cell which now is no longer modifiable or can no longer be transformed back into the original cell. In a way dynamical processes may lead to solid structures like bones or organs. In a similar way, information is laid down in a rigid manner in DNA, i.e. in the genetic code. It appears that lower animals are constructed more or less by the rules given by the genetic code with a rather rigid "wiring" of their nervous systems.

On the other hand in higher animals, in addition to rigid wiring of the nervous system a good deal of self-organization appears. The interaction of the system with its environment, together with the genetic information laid down in the system leads to the formation of new information. Through the continuous testing of the new information stored and created in the brain by the environment, new contexts are established and thus a new kind of semantics occur. But we may also expect that "solidification" occurs at various hierarchical levels of semantic information and serves for making the system more reliable, and to store information (memory). While the concept of Hebb's synapse, one which is strengthened by its use, may be a correct concept, the building up of semantics requires a high degree of cooperativity within the system and a repeated interaction with the outside world. In this respect, semantic information is not a static property, but rather a process in which contexts and relevance are checked, reinforced or dismissed again and again. By the way, I believe that consciousness is not a static state, but a process in which information is continuously transferred between various parts of the brain and repeatedly processed there.

At this point a word on pattern recognition may be in order. Lower animals immediately react to stimuli such as light flashes and only few criteria are needed, such as theshold of intensity, in order to respond to a signal. In higher animals, however, the incoming information will certainly be compared with stored information. However, our picture of how this comparison is done is changing slightly.

Quite often it is assumed that the incoming pattern is compared with templates. However, the storage of a template would require quite a large amount of information. Therefore, one might imagine in the sense of synergetics, that only specific characteristic features are stored in the form of order parameters which may then be called upon to generate a detailed picture. In this sense then, pattern recognition becomes an active process in which new patterns are formed in a self-organized fashion by the brain which, using certain hypotheses, checks them repeatedly against the incoming patterns. For instance, it is well known that when people look at faces, they focus their attention on specific parts like eyes, or nose, or mouth and look at them again and again.

Let us finally discuss a point which applies specifically to humans. In contrast to animals, human beings can transfer information not only by the genetic code, but also by teaching which in the world of animals takes place only in a very limited way. So a good deal of our culture is based on this new way of transferring information from one generation to the next. But here an enourmous difficulty arises because of the tremendous amount of knowledge which has been accumulated by humanity. Therefore, quite in the spirit of synergetics, it will be important to find unifying ideas and principles to cope with this large amount of information.

In addition our approach provides us with a picture rather different from those conventionally drawn from biological systems. There, it is assumed that there exists one single command center, say in the brain which then organizes all the behavior. The model that we are strongly supporting calls rather for processes of self-organization, and more recently we were able to prove this hypothesis by our quantitative theory of specific experiments on the correlation of hand movements and their changes. In these experiments, performed by S. Kelso, test persons were asked to oscillate their fingers in parallel. At an increased oscillation frequency an involuntary change to an antiparallel oscillation occurred. The way in which this transition occurs can be represented in all its details by the assumption of self-organization of the behavior of neurons and muscles.

This is certainly an extreme case and in general the information production and transfer in biological systems must be considered in two ways: the one is the conventional one in which specific motor programs serve for specific actions, whereas other phenomena occur in an entirely self-organized fashion. We may hypothesize that self-organization in information processing in biological systems plays a widespread and major role. This is borne out by the great flexibility of biological systems and their adaptability and plasticity.

In my opinion, the study of information in biological systems is also of interest to modern society whose proper functioning relies on the adequate production, transfer, and processing of information. Perhaps the most important aspect which has emerged is that of circular causality which results in a collective state which in sociology may represent a social climate, a general public opinion, a democracy or a dictatorship.

### 1.6.4 How Much Information Do We Need to Maintain an Ordered State?

Let us consider our standard example, namely the laser. Let us assume that there are atoms in the laser each having two levels. The total number of atoms in the lower state will be denoted by $N_1$, the number of atoms in the upper state by $N_2$. We have the relation

$$N_1 + N_2 = N \ . \tag{1.30}$$

In the sense of quantum mechanics we may relate the occupation numbers $N_1$ and $N_2$ to the occupation probability

$$p_j = \frac{N_j}{N} \, , \qquad j = 1, 2 \tag{1.31}$$

for a single atom. Thus, the information per atom is given by

$$i = -p_1 \ln p_1 - p_2 \ln p_2 \tag{1.32}$$

and then for all atoms by

$$I = -N(p_1 \ln p_1 + p_2 \ln p_2) . \tag{1.33}$$

As we know, an exited atom may emit a photon either by spontaneous emission, or if other photons are already present, by so-called stimulated emission. We may identify a single photon with a symbol that carries an element of information. As we know, photons can escape through the mirrors. Therefore, we may ask the question of what production rate of photons is necessary in order to maintain a coherent state?

According to laser theory we must not only introduce the number of photons $n$ as a variable, but in addition the inversion which is defined as the difference between the occupation numbers of the upper and lower state:

$$D = N_2 - N_1 . \tag{1.34}$$

According to the theory the production rate of photons is given by the equation

$$\frac{dn}{dt} = WDn - 2\kappa n . \tag{1.35}$$

The first term on the right-hand side describes the production rate of photons, where $W$ is a rate constant for this production, whereas the second term describes the escape of photons through the mirrors so that the actual production is diminished. As is shown in laser theory, (1.35) describes the production of coherent photons; the production of incoherent photons is neglected. Because of the laser process the inversion also changes in time. Its rate of change is given by the equation

$$\frac{dD}{dt} = \frac{D_0 - D}{T} - 2WDn . \tag{1.36}$$

Here $D_0$ is the inversion produced by the pump process and relaxation processes which do not give rise to laser light emission. $T$ is the time in which any deviation of the inversion relaxes towards the inversion $D_0$. The last term in (1.36) stems from the laser process in which photons are produced. In general, the decay constant $\kappa$ is much smaller than the rate constant $1/T$. This allows us to apply the so-called adiabatic approximation in which we may write

$$\frac{dD}{dt} \approx 0 . \tag{1.37}$$

Using (1.37) in (1.36) we can immediately solve (1.36) for $D$ thus obtaining

$$D = \frac{D_0}{1 + 2TWn} . \tag{1.38}$$

When the laser is not too far above the onset of laser action, we may expand the denominator as a power series in the photon number $n$ so that in the leading approximation we obtain

$$D \approx d_0 - 2D_0 TWn \ . \tag{1.39}$$

Inserting this result into the equation for the production rate of photons (1.35), we readily obtain

$$\frac{dn}{dt} = (WD_0 - 2\kappa)n - 2TW^2 D_0 n^2 \ . \tag{1.40}$$

While the second term in (1.40) will always lead to a decrease in the production rate of the photons, the first term will give rise to a positive production rate provided the inequality

$$WD_0 - 2\kappa > 0 \tag{1.41}$$

holds. Equation (1.41) is identical with the laser condition. Thus (1.41) guarantees a positive net production rate of photons, or in other words, a *positive net production rate of signals.* This is necessary for the maintainance of a nonzero flux of photons and thus for the ordered state of the laser. According to (1.41) this can be established only if the inversion $D_0$ which is achieved by pumping is sufficiently high. From

$$n = \frac{WD_0 - 2\kappa}{2TW^2 D_0} \tag{1.42}$$

we may deduce that the condition (1.41) guarantees a non-vanishing number of photons which must be present in the laser at all times.

Let us now study the behavior of the information (1.32) or (1.33) when we increase the pump rate or in other words the inversion $D_0$. To this end we insert (1.42) in (1.39) and obtain

$$D \approx \frac{2\kappa}{W} = \text{const. !} \qquad \text{for } n \geq 0 \text{ or} \tag{1.43}$$

$$D = D_0 \tag{1.44}$$

for $n = 0$. In other words when we start from a low pump rate, $D_0$ is small and no photons are present. Then $D$ increases at the same rate as $D_0$. But once laser action sets in, the inversion $D$ becomes a constant according to (1.43) and shows no further increase. All the additional energy fed into the laser is transformed into coherent photons. Using (1.34) and (1.30) we have the relations

$$N_1 = \tfrac{1}{2}(N - D) \tag{1.45}$$

$$N_2 = \tfrac{1}{2}(N + D) \tag{1.46}$$

which can then be transformed according to (1.31) into the occupation probabilities

$$p_1 = \frac{1}{2}\left(1 - \frac{D}{N}\right) \tag{1.47}$$

$$p_2 = \frac{1}{2}\left(1 + \frac{D}{N}\right) . \tag{1.48}$$

Inserting (1.47) and (1.48) into (1.32) we obtain

$$i = -\frac{1}{2}\left(1 - \frac{D}{N}\right)\ln\frac{1}{2}\left(1 - \frac{D}{N}\right) - \frac{1}{2}\left(1 + \frac{D}{N}\right)\ln\frac{1}{2}\left(1 + \frac{D}{N}\right) . \tag{1.49}$$

In the following we shall use the parameter $\gamma$ defined by

$$\gamma = \frac{D}{N} . \tag{1.50}$$

Because $D$ takes values in the range

$$D\colon -N, \ldots, +N , \tag{1.51}$$

$\gamma$ must lie in the range

$$\gamma\colon -1 \ldots +1 . \tag{1.52}$$

The behavior of $i$ as a function of $\gamma$ is plotted in Fig.1.38. The result (1.49) jointly with (1.43) and (1.44) allows us to study the change in the information of an individual atom when we increase the pump rate $D_0$. According to Fig. 1.39 the information first rises, goes through a maximum and then saturates. Our present approach is not capable of dealing with the information contained in the light field because so far we have not considered any fluctuations, i.e. any probability distribution over the photon numbers $n$. One of the main objectives of our book will it be to study the information as a function of pump strength not only for the atoms but also for the photons.

Indeed we shall see that a surprising result is obtained, namely that the interesting information close to the point where laser action starts is contained in the photons rather than in the atoms.

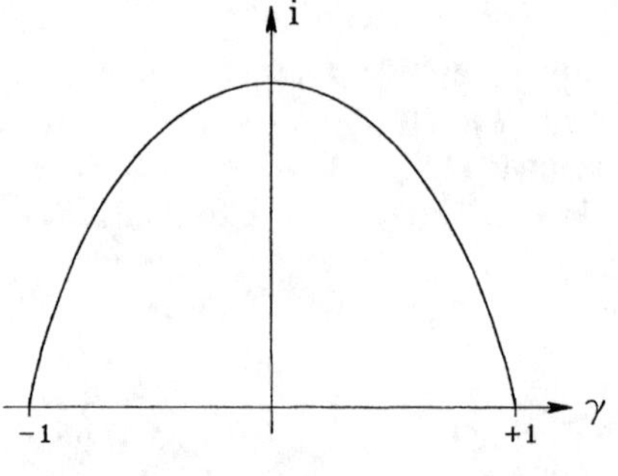


Fig. 1.38. The information of an atom versus the parameter $\gamma$

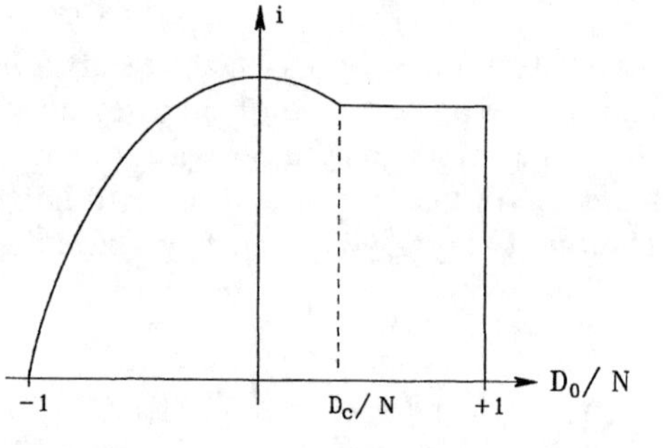


Fig. 1.39. The behavior of information of a single atom in the laser

## 1.7 The Second Foundation of Synergetics

After having discussed the qualitative aspects of information and self-organization in the previous sections, we now wish to come to the hard core of our approach which will then be followed up in the remainder of this book. Let us briefly recall what we have been doing in the field of synergetics so far. There, we started from the microscopic or mesoscopic level at which we formulated equations. Then, by using the concepts of instability, order parameters, and slaving, which can be cast into a rigorous mathematical form, we could show the emergence of structures and, concomitantly, of new qualities at a macroscopic level.

In a way parallels can be drawn between the latter approach and that of statistical mechanics. We wish now to develop an approach which can be put in analogy with that of thermodynamics. Namely, we wish to treat complex systems by means of macroscopically observed quantities. Then we shall try to guess the microscopic structure of the processes which give rise to the macroscopic structure or the macroscopic behavior. The vehicle we shall use for this purpose is the maximum entropy principle, or the maximum information entropy principle which was developed quite generally by Jaynes.

We shall give a detailed presentation of this principle in Chap. 3. Here it will suffice to summarize the basic idea. We start from macro-observables which may fluctuate and whose mean values are known. We distinguish the macro-variables by an index $k$ and denote their mean values by $f_k$. We wish then to make a guess at the probability distribution $p_j$ of the system over states labeled by the index $j$. This is achieved under the maximization of the information

$$i = -\sum_j p_j \ln p_j \tag{1.53}$$

under the constraint that

$$\sum_j p_j f_j^{(k)} = f_k \; . \tag{1.54}$$

Evidently, $f_j^{(k)}$ is the contribution of state $j$ to the macro-variable labeled by $k$. Furthermore we require

$$\sum_j p_j = 1 \; , \tag{1.55}$$

i.e. that the probability distribution $p_j$ is normalized to unity. As was shown by Jaynes and as will be demonstrated in Chap. 4, this principle allows us to derive the basic formulas of thermodynamics in a very short and elegant fashion. For this derivation the constraints refer to the conserved quantities of a closed system, i.e. energy, particle numbers etc. The crux of the problem of extending this maximum entropy principle to systems far from thermal equilibrium or even to non-physical systems lies in the adequate choice of constraints.

As we shall see, the constraints which have been used so far, of energy conservation or even of regulated energy fluxes into the system, are inadequate to treat open systems and especially to treat the transition from a structureless to a structured state as occurs in non-equilibrium phase transitions. The maximum entropy principle has been criticized occasionally because the choice of the constraints seems to introduce a certain subjectivity in that the constraints are said to be chosen arbitrarily at the will of the observer rather than by objective criteria.

This criticism has been debated by Jaynes in detail, but I should like to add here another point of view. Scientific progress relies on a general consensus being reached within the scientific community; results are made objective by general agreement. One might call this attitude "relative objectivism". This is actually the most which can be said about any physical theory because in the natural sciences a theory can never be verified but only falsified, a point quite correctly made by Popper. Thus, what we have to adopt is a learning process based on the correct choice of adequate constraints. This is in fact what has happened in thermodynamics where by now we all know that the adequate constraints are the conservation laws.

In the field of non-equilibrium phase transitions, or more generally speaking, of open systems, we wish to make the first steps by showing what these constraints are. Indeed, when we confine our analysis to non-equilibrium phase transitions, we find complete agreement between the macroscopic approach by the maximum (information) entropy principle and the results derived from a microscopic theory for all cases where the microscopic distribution functions are known. Therefore, I am sure that a consensus can be found here too.

There is another aspect important from the mathematical point of view. Namely, when we prescribe specific constraints which are given experimental mean values, the maximum (information) entropy principle will always provide us with distribution functions which reproduce these mean values. In this sense we are dealing here with a tautology. Then, however, in the next step we may infer new mean values by means of the probability distribution and then predictions are made which can be checked experimentally. If these predictions are not fulfilled, we may choose these new experimental data as additional constraints which then give rise to altered probability distribution functions. In this way an infinite process has been started. But in spite of this cautioning remark, we may find a consensus on the proper choice of a limited set of constraints, provided we confine our analysis to specific classes of phenomena.

One such class is, as mentioned, closed (thermodynamical) systems with their appropriate constraints. Another class consists of non-equilibrium phase transitions which will be treated here. As we shall see, this class comprises numerous phenomena in various fields, such as the emergence of spatial patterns, of new types of information and even of oscillatory phenomena. The appropriate choice of constraints for processes leading to deterministic chaos remains, at least partly, a task for the future.

As we shall see, the main new insight which we are gaining by our approach into the constraints is the following: In a first step one may guess that the adequate constraints must include the macroscopic variables, or in other words, the order parameters. But it is known that in non-equilibrium phase transitions critical fluctuations of the order parameters occur, i.e. that their fluctuations become macroscopic variables. Indeed, it will turn out that the inclusion of the fluctuations of the order parameters is the crucial step in finding adequate constraints for this class of phenomena.

Using the results of the microscopic theory as a guide, we are then able to do much more; namely, we can do without the order parameters from the outset. Instead our approach will start from correlation functions, i.e. moments of observed variables from which we may then reconstruct the order parameters and the enslaved modes. Incidentally, we can also construct the macroscopic pattern, or in other words we may automatize the recognition of the evolving patterns which are produced in a non-equilibrium phase transition.

In conclusion, let us return to the discussion of the relation between the analytical (or microscopic) approach and the holistic (or macroscopic) approach, and let us make a further point in favor of a macroscopic approach. Quite often, even the subsystems become very complicated so that it is difficult or even impossible to formulate the microscopic or mesoscopic equations explicitly. When we go to the extreme case, namely the human brain, the subsystems are, for instance, the nerve cells (neurons) which are themselves complicated systems. A nerve cell contains its soma, an axom and up to 80 thousand dendrites by which the cell is connected with other nerve cells. In the human brain there are about 10 billion nerve cells. It is proposed in synergetics that despite this enormous complexity, a number of behavioral patterns can be treated by means of the order parameter concept, where the order parameter equations are now established in a phenomenological manner.

Recently, we were able to find a paradigm, namely the coordination of hand movements and especially involuntary changes between hand movements. Though the relevant subsystems are quite numerous and consist of nerve cells, muscle cells and other tissue, the behavior can be represented by a single order parameter. We shall describe these experiments in this book and elucidate our general approach by this example. Further examples will be taken from laser physics and fluid dynamics. Looking at the numerous examples treated from the microscopic point of view in synergetics, it is not difficult to find many more applications of our new macroscopic approach. In addition, numerous examples, especially in biology, can be found where only the macroscopic approach is applicable.

We conclude with the remark that we shall use the expression “maximum information principle” exchangeably with “maximum entropy principle”. But, as will transpire, the term information is the more appropriate to the situation in non-equilibrium systems.

# 2. From the Microscopic to the Macroscopic World...

## 2.1 Levels of Description

In this chapter I present the basic concepts and methods which have been used in synergetics to study self-organization by means of a microscopic approach. Readers familiar with this approach may skip this chapter and proceed directly to Chap. 3. On the other hand those readers who are unfamiliar with these concepts and methods and who wish to penetrate more deeply into them are advised to read my books "Synergetics. An Introduction" and "Advanced Synergetics" where all these concepts are explained in great detail.

When we deal with a system, we first have to identify the variables or quantities by which we wish to describe the system. Such a description can be done at various levels which are, however, interconnected with each other. Let us discuss the example of a fluid (Fig.2.1). At the microscopic level, the fluid can be described as being composed of individual molecules. Thus, for a complete description of the fluid, we have to deal with the positions and velocities of the individual molecules. However, for many purposes it is sufficient to deal with the fluid at a mesoscopic level. Here, we start from volume elements which are still small compared to the total size of the fluid, but which are so large that we may safely speak of densities, velocity fields, or local temperatures. Finally, our concern will be the macroscopic level at which we wish to study the formation of structures, or of patterns, for instance a hexagonal pattern in a fluid which is heated from below. A similar subdivision of levels can also be made with respect to biological systems. But as the reader will recognize quickly, we have here a much greater arbitrariness in choosing our levels.

Let us consider the example of a biological tissue. At the microscopic level we may speak of biomolecules or equally well of organelles and so forth. At the mesoscopic level we may speak of cells and finally at the macroscopic level we may be concerned with whole tissues or organs formed by these cells. Quite clearly, for the transition from the mesoscopic to the macroscopic level we must neglect many detailed features of the cells and their constituents. Instead we must pick out those features which are relevant for the formation of organs. Therefore, at this mesoscopic level we are already dealing with an enormous compression of information.

In this chapter we shall mainly be concerned with the transition from the mesoscopic to the macroscopic level, though in a number of cases a direct transition from the microscopic to the macroscopic level can also be performed. Within

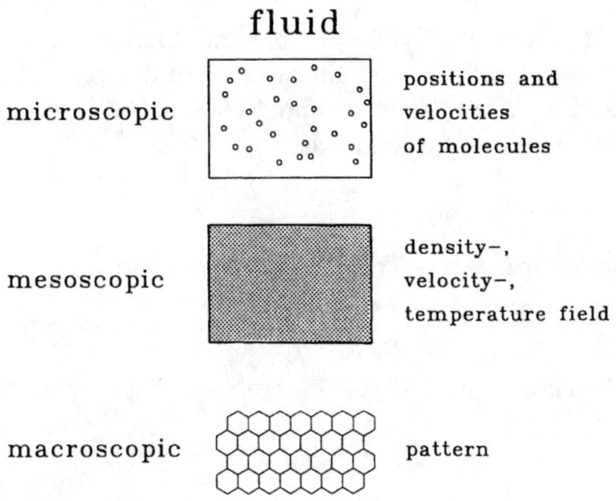


**Fig. 2.1.** Illustration of the microscopic, mesoscopic and macroscopic approach by means of the example of a fluid

**Table 2.1.** Examples of mesoscopic variables

| Field of Study | Variables |
|---|---|
| chemical reactions, solidification | densities of molecules in different phases |
| fluids | velocity fields |
| flames | temperature fields |
| plasmas | electric and magnetic fields |
| lasers, parametric oscillators | atomic polarization, inversion |
| solid state physics, Gunn oscillator, filamentation | densities of electrons and holes |
| morphogenesis | densities of cells in tissues |
| pre-biotic evolution | numbers of biomolecules |
| population dynamics | numbers of animals |
| neuronal nets | firing rates of neurons |
| locomotion | elongation and contraction of muscles |
| economy | monetary flows |
| sociology | number of people with specific attitudes |
| synergetic computers | activation of elements |

physics, the mesoscopic level can be reached by means of statistical mechanics where certain relevant variables are then introduced and treated. In most cases, however, such as in chemistry and biology we shall use phenomenological equations for the corresponding variables.

Let us give the reader an impression of the variety of problems to be treated by means of listing a number of examples of mesoscopic variables (Table 2.1).

## 2.2 Langevin Equations

A quite typical example for the description at the mesoscopic level is provided by the Brownian motion of a particle, say a dust particle, which is immersed in a fluid. Its motion is described by the Langevin equation where the variable $q$ is then to be identified with the velocity of the particle. The microscopic motion of all the molecules of the liquid has two effects. On the one hand it leads to a damping of the velocity and on the other hand it leads to random impulses delivered to the particle under consideration. The Langevin equation reads

$$\dot{q} = K(q) + F(t) \tag{2.1}$$

where in the case of the Brownian particle $K$ is given by

$$K(q) = -\gamma q \ . \tag{2.2}$$

A form which we are quite often concerned with in synergetics is given by the non-linear expression

$$K(q) = \alpha q - \beta q^3 \ . \tag{2.3}$$

The fluctuating forces $F$ are characterized by the properties

$$\langle F(t) \rangle = 0 \quad \text{and} \tag{2.4}$$

$$\langle F(t) F(t') \rangle = Q \delta(t - t') \tag{2.5}$$

where the average is taken over the stochastic process. When we deal with several variables $q_1, \dots, q_N$ which are lumped together into a state vector $\boldsymbol{q}$, the Langevin equation reads

$$\dot{\boldsymbol{q}} = \boldsymbol{K}(\boldsymbol{q}) + \boldsymbol{F}(t) \tag{2.6}$$

and the fluctuating forces are assumed to possess the properties

$$\langle F_j(t) \rangle = 0 \tag{2.7}$$

$$\langle F_j(t) F_{j'}(t') \rangle = Q_j \delta_{jj'} \delta(t - t') \ . \tag{2.8}$$

Note that $\boldsymbol{q}$ may be a vector in a high dimensional space thus representing a very complicated system. In a number of cases the fluctuating forces are themselves dependent on the state variable $\boldsymbol{q}$. In such a case a number of specific problems arise which were solved by the Îto or Stratonovich calculus. In the *Îto calculus* the Langevin equation (2.1) must be replaced by the following equation

$$dq(t) = K(q(t))\, dt + g(q(t))\, dw(t) \ . \tag{2.9}$$

Here $K$ and $g$ are in general non-linear functions of $q$ whereas $dw$ describes a stochastic process where we make the assumptions

$$\langle dw \rangle = 0 \tag{2.10}$$

$$\langle dw(t)\, dw(t') \rangle = \delta(t - t')\, dt \ . \tag{2.11}$$

In the Îto formulas it is assumed that $q(t)$ and $dw$ which occur in the last term of (2.9) are statistically uncorrelated. When we deal with a multidimensional state vector with components $q_1, \ldots, q_m$ the Îto equation reads

$$dq_l(t) = K_l(\boldsymbol{q}(t))\,dt + \sum_m g_{lm}(\boldsymbol{q}(t))\,dw_m(t) \tag{2.12}$$

where $dw_m$, which describes the stochastic process, has the properties

$$\langle dw_m \rangle = 0 \tag{2.13}$$

$$\langle dw_m(t)\,dw_l(t') \rangle = \delta_{lm}\delta(t-t')\,dt \ . \tag{2.14}$$

To exhibit the special features of the Îto procedure let us consider an arbitrary differentiable function

$$\boldsymbol{u} = \boldsymbol{u}(\boldsymbol{q}) \tag{2.15}$$

and its differential. Because of the property (2.14) we have to go up to the second derivative according to

$$du_j = \sum_k \frac{\partial u_j}{\partial q_k} dq_k + \frac{1}{2} \sum_{kl} \frac{\partial^2 u_j}{\partial q_k \partial q_l} dq_k\,dq_l \ . \tag{2.16}$$

Inserting (2.12) into (2.16) and keeping the terms including $dt$ but neglecting all higher order terms we arrive at

$$du_j = \sum_k \frac{\partial u_j}{\partial q_k} \left[K_k(q)\,dt + \sum_m g_{km}\,dw_m(t)\right] + \frac{1}{2} \sum_{kl} \frac{\partial^2 u_j}{\partial q_k \partial q_l} \left[\sum_{mn} g_{km} g_{ln}\,dw_m\,dw_n\right] . \tag{2.17}$$

Let us now briefly remind the reader of the *Stratonovich approach.* The stochastic equation for a single variable reads

$$dq = K(q)\,dt + g(q)\,dw(t) \tag{2.18}$$

and for the components of a state vector

$$dq_l = K_l(\boldsymbol{q})\,dt + \sum_m g_{lm}(\boldsymbol{q})\,dw_m(t) \ . \tag{2.19}$$

In contrast to the Îto calculus, the last terms in (2.18) or (2.19) are now interpreted differently, namely they have to be evaluated according to the midpoint rule in which $g(q(t_i))\,dw(t_i)$ is replaced by

$$g\left(q\left(\frac{t_i + t_{i-1}}{2}\right)\right) dw(t_i) \ , \tag{2.20}$$

i.e. $q$ and $dw$ are no more statistically independent. As we shall see later, we can recover the Îto equation by means of a macroscopic approach.

## 2.3 Fokker-Planck Equation

For many applications, especially when the problems are non-linear, i.e. when $K$ is a non-linear function of $q$, it is advantageous to proceed to the Fokker-Planck equation which is formulated for the distribution function $f(q,t)$. It describes the probability of finding the variable $q$ in the interval $q \to q + dq$ at time $t$.

The Fokker-Planck equation belonging to the Langevin equation (2.1) reads

$$\frac{\partial f}{\partial t} = -\frac{\partial}{\partial q}(K(q)f) + \frac{Q}{2}\frac{\partial^2}{\partial q^2}f \tag{2.21}$$

where the first term on the right-hand side is denoted as the drift term, and the second term is called the diffusion term.

The stationary solution obeying

$$\frac{\partial f}{\partial t} = 0 \tag{2.22}$$

can easily be found, provided the boundary condition

$$f(q) \to 0 \qquad \text{for} \qquad q \to \mp\infty \tag{2.23}$$

is fulfilled. Then the stationary solution reads

$$f = N \exp\left[-\int_{q_0}^{q} \frac{2K(q')}{Q} dq'\right] \tag{2.24}$$

where $N$ is the normalization factor so that the integral over $f$ is equal to unity. In the case of a multidimensional state vector, we have to find the Fokker-Planck equation for

$$f(\boldsymbol{q}, t) \ . \tag{2.25}$$

This equation reads

$$\frac{\partial f}{\partial t} = -\sum_j \frac{\partial}{\partial q_j}(K_j f) + \frac{1}{2}\sum_{jk} Q_{jk}\frac{\partial^2}{\partial q_j \partial q_k} f \ . \tag{2.26}$$

Explicit solutions of (2.26) are available only in special cases, e.g. if $K$ is linear in the variables $q$ and $Q_{jk}$ is independent of $\boldsymbol{q}$. In such a case the time dependent and stationary solutions can be constructed explicitly (cf. my book *Advanced Synergetics*). The stationary solution of (2.26) can also be constructed explicitly, provided the so-called rule of detailed balance is fulfilled (see below, Sect. 2.4).

When we start from the *Îto differential equation*, the corresponding Fokker-Planck equation has the following form

$$\frac{\partial f}{\partial t} = -\sum_k \frac{\partial}{\partial q_k}[K_k(\boldsymbol{q})f] + \frac{1}{2}\sum_{kl}\frac{\partial^2}{\partial q_k \partial q_l}\left[\sum_m g_{km} g_{lm} f\right] \ . \tag{2.27}$$

In the case of the *Stratonovich calculus*, however, the Fokker-Planck equation can be shown to read

$$\frac{\partial f}{\partial t} = -\sum_l \frac{\partial}{\partial q_l}\left\{\left[K_l(q) + \frac{1}{2}\sum_{kj}\frac{\partial g_{lj}}{\partial q_k}g_{kj}\right]f\right\}$$
$$+\frac{1}{2}\sum_{lm}\frac{\partial^2}{\partial q_l \partial q_m}\left(\sum_i g_{li}g_{mi}f\right) . \quad (2.28)$$

## 2.4 Exact Stationary Solution of the Fokker-Planck Equation for Systems in Detailed Balance

In this section we show that under the condition of detailed balance the stationary solution of the Fokker-Planck equation may be found explicitly by quadratures.

While the principle of detailed balance is expected to hold for practically all systems in thermal equilibrium, this need not be so in systems far from thermal equilibrium. Thus each individual case requires a detailed discussion (e.g., by symmetry considerations) as to whether this principle is applicable. Also, an inspection of the structure of the Fokker-Planck equation will enable us to decide whether detailed balance is present.

### 2.4.1 Detailed Balance

We denote the set of variables $q_1, \ldots, q_N$ by $\boldsymbol{q}$ and the set of the variables under time reversal by

$$\tilde{\boldsymbol{q}} = \{\varepsilon_1 q_1, \ldots, \varepsilon_N q_N\} , \quad (2.29)$$

where $\varepsilon_i = -1\,(+1)$ depending on whether the coordinate $q_i$ changes sign (does not change sign) under time reversal. Furthermore, $\boldsymbol{\lambda}$ stands for a set of externally determined parameters. The time reversed quantity is denoted by

$$\tilde{\boldsymbol{\lambda}} = \{\nu_1 \lambda_1, \ldots, \nu_M \lambda_M\} , \quad (2.30)$$

where $\nu_i = -1\,(+1)$ depends on the inversion symmetry of the external parameters under time reversal. We denote the joint probability of finding the system at $t_1$ with coordinates $\boldsymbol{q}$ and at $t_2$ with coordinates $\boldsymbol{q}'$ by

$$P(\boldsymbol{q}', \boldsymbol{q}; t_2, t_1) . \quad (2.31)$$

In the following, we consider a stationary system so that the joint probability depends only on the time difference $t_2 - t_1 = \tau$. Thus (2.31) may be written as

$$P(\boldsymbol{q}', \boldsymbol{q}; t_2, t_1) = W(\boldsymbol{q}', \boldsymbol{q}; \tau) . \quad (2.32)$$

We now formulate the principle of detailed balance. The following two definitions are available.

1) The principle of detailed balance (first version)

$$W(\boldsymbol{q}', \boldsymbol{q}; \tau, \lambda) = W(\tilde{\boldsymbol{q}}, \tilde{\boldsymbol{q}}'; \tau, \tilde{\lambda}) \; . \tag{2.33}$$

The joint probability may be expressed by the stationary distribution $f(\boldsymbol{q})$ multiplied by the conditional probability $P$, where stationarity is exhibited by writing

$$P = P(\boldsymbol{q}' | \boldsymbol{q}; \tau, \lambda) \; . \tag{2.34}$$

Therefore, we may reformulate (2.33) as follows:

$$P(\boldsymbol{q}' | \boldsymbol{q}; \tau, \lambda) f(\boldsymbol{q}, \lambda) = P(\tilde{\boldsymbol{q}} | \tilde{\boldsymbol{q}}'; \tau, \tilde{\lambda}) f(\tilde{\boldsymbol{q}}', \tilde{\lambda}) \; . \tag{2.35}$$

Here and in the following we assume that the Fokker-Planck equation possesses a unique stationary solution. One may then show directly that

$$f(\boldsymbol{q}, \lambda) = f(\tilde{\boldsymbol{q}}, \tilde{\lambda}) \tag{2.36}$$

holds. We define the transition probability per second by

$$w(\boldsymbol{q}', \boldsymbol{q}; \lambda) = [(d/d\tau) P(\boldsymbol{q}' | \boldsymbol{q}; \tau, \lambda)]_{\tau=0} \; . \tag{2.37}$$

Taking the derivative with respect to $\tau$ on both sides of (2.35) and putting $\tau = 0$ (but $\boldsymbol{q} \neq \boldsymbol{q}'$), we obtain

2) the principle of detailed balance (second version)

$$w(\boldsymbol{q}', \boldsymbol{q}; \lambda) f(\boldsymbol{q}, \lambda) = w(\tilde{\boldsymbol{q}}, \tilde{\boldsymbol{q}}'; \tilde{\lambda}) f(\tilde{\boldsymbol{q}}', \tilde{\lambda}) \; . \tag{2.38}$$

This obviously has a very simple meaning. The left-hand side describes the total transition rate out of the state $\boldsymbol{q}$ into a new state $\boldsymbol{q}'$. The principle of detailed balance then requires that this transition rate is equal to the rate in the reverse direction for $\boldsymbol{q}'$ and $\boldsymbol{q}$ with reverse motion, e.g., with reverse momenta. It can be shown that the first and second version are equivalent.

### 2.4.2 The Required Structure of the Fokker-Planck Equation and Its Stationary Solution

Using the conditional probability $P$ (which is nothing but the Green's function) we write the Fokker-Planck equation in the form

$$\frac{d}{d\tau} P(\boldsymbol{q}' | \boldsymbol{q}; \tau, \lambda) = L(\boldsymbol{q}', \lambda) P(\boldsymbol{q}' | \boldsymbol{q}; \tau, \lambda) \tag{2.39}$$

where we assume that the operator $L$ has the form

$$L(\boldsymbol{q}) = -\sum_i \frac{\partial}{\partial q_i} K_i(\boldsymbol{q}, \lambda) + \frac{1}{2} \sum_{ik} \frac{\partial^2}{\partial q_i \partial q_k} Q_{ik}(\boldsymbol{q}, \lambda) \; . \tag{2.40}$$

We may always assume that the diffusion coefficients are symmetric

$$Q_{ik} = Q_{ki} \ . \tag{2.41}$$

We define the following new coefficients:

a) the irreversible drift coefficients

$$D_i(\boldsymbol{q}, \lambda) = \tfrac{1}{2}[K_i(\boldsymbol{q}, \lambda) + \varepsilon_i K_i(\tilde{\boldsymbol{q}}, \tilde{\lambda})] \equiv D_i^{ir} \ ; \tag{2.42}$$

b) the reversible drift coefficients

$$J_i(\boldsymbol{q}, \lambda) = \tfrac{1}{2}[K_i(\boldsymbol{q}, \lambda) - \varepsilon_i K_i(\tilde{\boldsymbol{q}}, \tilde{\lambda})] \equiv D_i^{r} \ . \tag{2.43}$$

$J_i$ transforms as $q_i$ under time reversal.

We write the stationary solution of the Fokker-Planck equation in the form

$$f(\boldsymbol{q}, \lambda) = \mathcal{N} \mathrm{e}^{-\Phi(\boldsymbol{q}, \lambda)} \ , \tag{2.44}$$

where $\mathcal{N}$ is the normalization constant and $\Phi$ may be interpreted as a generalized thermodynamic potential. The necessary and sufficient conditions for the principle of detailed balance to hold read

$$Q_{ik}(\boldsymbol{q}, \lambda) = \varepsilon_i \varepsilon_k Q_{ik}(\tilde{\boldsymbol{q}}, \tilde{\lambda}) \ , \tag{2.45}$$

$$D_i - \frac{1}{2} \sum_k \frac{\partial Q_{ik}}{\partial q_k} = -\frac{1}{2} \sum_k K_{ik} \frac{\partial \Phi}{\partial q_k} \ , \tag{2.46}$$

$$\sum_i \left( \frac{\partial J_i}{\partial q_i} - J_i \frac{\partial \Phi}{\partial q_i} \right) = 0 \ . \tag{2.47}$$

If the diffusion matrix $Q_{ik}$ possesses an inverse, (2.46) may be solved for the gradient of $\Phi$

$$\frac{\partial \Phi}{\partial q_i} = \sum_k (Q^{-1})_{ik} \left( \sum_l \frac{\partial Q_{kl}}{\partial q_l} - 2D_k \right) \equiv A_i \ . \tag{2.48}$$

This shows that (2.48) implies the integrability condition

$$\frac{\partial}{\partial q_j} A_i = \frac{\partial}{\partial q_i} A_j \ , \tag{2.49}$$

which is a condition on the drift and diffusion coefficients as defined by the right-hand side of (2.48). Substituting $A_i$ and $A_j$ from (2.48), the condition (2.47) acquires the form

$$\sum_i \left[ \frac{\partial J_i}{\partial q_i} - J_i \sum_k (Q^{-1})_{ik} \left( \sum_l \frac{\partial Q_{kl}}{\partial q_l} - 2D_k \right) \right] = 0 \ . \tag{2.50}$$

Thus the conditions for detailed balance to hold are given finally by (2.45,49,50). Equation (2.46) or equivalently (2.48) then allows us to determine $\Phi$ by pure quadratures, i.e., by a line integral. Thus the stationary solution of the Fokker-Planck equation may be determined explicitly.

## 2.5 Path Integrals

The time-dependent solutions of the Fokker-Planck equation can be represented in the form of path integrals. For the sake of simplicity we shall be concerned in the following with the case where the diffusion constant $Q$ is independent of $\boldsymbol{q}$. Let us first treat the one dimensional case, i.e. that $\boldsymbol{q}$ is a single variable. Let us split the time interval $t$ into equidistant steps

$$t_0, t_1 = t_0 + \tau, \dots, t_N = t_0 + \tau N \,. \tag{2.51}$$

The distribution function $f$ at time $t$ can be constructed as a multiple integral over all the intermediate positions $q_0, q_1, q_2, \dots$ (cf. Fig.2.2). The explicit form of the path integral then reads

$$f(q,t) = \lim_{\substack{N\to\infty \\ N\tau=t}} \int \cdots \int Dq\, \mathrm{e}^{-G/2} f(q', t_0) \tag{2.52}$$

where we have used the abbreviations

$$Dq = (2Q\tau\pi)^{-N/2}\, dq_0, \dots, dq_{N-1} \quad \text{and} \tag{2.53}$$

$$G = \sum_\nu \frac{\tau\left[\dfrac{q_\nu - q_{\nu-1}}{\tau} - K(q_{\nu-1})\right]^2}{Q} \,. \tag{2.54}$$

The explicit derivation of formula (2.52) is presented in *Synergetics. An Introduction.* Let us now consider the generalization of (2.52) to the case of a multi-dimensional vector $\boldsymbol{q}$ which has $n$ components. We still assume that $Q_{mn}$ is independent of the state variable $\boldsymbol{q}$. We then obtain the following results

$$f(\boldsymbol{q},t) = \lim_{\substack{N\to\infty \\ N\tau=t}} \int \cdots \int D\boldsymbol{q}\, \mathrm{e}^{-G/2} f(\boldsymbol{q}', t_0) \tag{2.55}$$

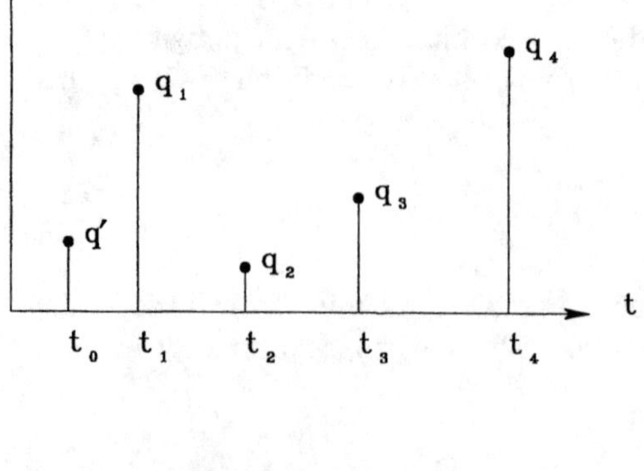


**Fig. 2.2.** The positions of a particle or of the state of a system in the course of time

where we have used the following abbreviations

$$Dq = \prod_{\mu=0}^{N-1} \{(2\pi\tau)^{-n/2}(\det Q)^{-1/2}\}(dq_1, \ldots, dq_n)_\mu \tag{2.56}$$

$$q_N = q \; ; \qquad q_0 = q' \tag{2.57}$$

$$G = \tau \sum (\dot{q}_\nu^T - K_{\nu-1}^T)Q^{-1}(\dot{q}_\nu - K_{\nu-1}) \tag{2.58}$$

$$\dot{q}_\nu = \tau^{-1}(q_\nu - q_{\nu-1}) \, , \qquad K_{\nu-1} = K(q_{\nu-1}) \, , \tag{2.59}$$

where $T$ denotes the transposed vector.

Let us finally remind the reader of the master equation. Let us consider a discrete state space vector $\boldsymbol{m}$. Then we are interested in the probability distribution $P(\boldsymbol{m}, t)$. Provided we are dealing with a Markovian process, $P$ obeys the master equation.

$$\frac{dP(\boldsymbol{m}, t)}{dt} = \sum_{\boldsymbol{n}} w(\boldsymbol{m}, \boldsymbol{n})P(\boldsymbol{n}) - \sum_{\boldsymbol{n}} w(\boldsymbol{n}, \boldsymbol{m})P(\boldsymbol{m}) \, . \tag{2.60}$$

Again, it is difficult to find explicit solutions of (2.60). In the special case of detailed balance, the stationary probability distribution can be constructed explicitly. When detailed balance is present, $P$ fulfills the following relation

$$w(\boldsymbol{n}, \boldsymbol{m})P(\boldsymbol{m}) = w(\boldsymbol{m}, \boldsymbol{n})P(\boldsymbol{n}) \, . \tag{2.61}$$

Then the steady state solution of (2.60) can be written down explicitly in the form

$$P(\boldsymbol{m}) = N \, \mathrm{e}^{\Phi(\boldsymbol{m})} \tag{2.62}$$

where $\Phi(\boldsymbol{m})$ is defined by

$$\Phi(\boldsymbol{m}) = \Phi(\boldsymbol{n}_0) + \sum_{j=0}^{N-1} \ln \left\{ \frac{w(\boldsymbol{n}_{j+1}, \boldsymbol{n}_j)}{w(\boldsymbol{n}_j, \boldsymbol{n}_{j+1})} \right\} \quad \text{and} \tag{2.63}$$

$$\boldsymbol{m} \equiv \boldsymbol{n}_N \, . \tag{2.64}$$

## 2.6 Reduction of Complexity, Order Parameters and the Slaving Principle

In this section we treat systems which are composed of many parts. We wish to study qualitative changes in the behavior of the system. To this end we make several transformations of the variables and their equations. Then in Sects. 2.7,8 we shall present some important applications. We start from a state vector $\boldsymbol{q}$ which describes the total system at the microscopic or mesoscopic level.

$$\boldsymbol{q} = \boldsymbol{q}(x, t) \, . \tag{2.65}$$

For what follows we shall assume that the state vector is a function of the space coordinate $\boldsymbol{x}$ so that quite generally we shall assume that $\boldsymbol{q}$ obeys an evolution equation of the type

$$\dot{\boldsymbol{q}} = \boldsymbol{N}(\boldsymbol{q}, \alpha) + \boldsymbol{F}(t) \ . \tag{2.66}$$

In this equation $\boldsymbol{N}$ is a nonlinear function of $\boldsymbol{q}$ that may also contain differential operators, for instance the Laplace operator differentiating $\boldsymbol{q}$ which respect to the spatial coordinates. $\alpha$ is a control parameter, e.g. the power input into a laser, or the amount of heating of a fluid, or certain signals impinging on a biological system, and $\boldsymbol{F}(t)$ is a fluctuating force. We now proceed in several steps.

### 2.6.1 Linear Stability Analysis

In the following we shall assume that for a fixed value of the control parameter, $\alpha_0$, the solution of the deterministic equation is known, i.e. that $\boldsymbol{q}_0$ solves the equation

$$\dot{\boldsymbol{q}} = \boldsymbol{N}(\boldsymbol{q}, \alpha_0) \ . \tag{2.67}$$

We then study the behavior of the solution when the control parameter $\alpha$ is changed. To this end we make the hypothesis

$$\alpha\colon \boldsymbol{q} = \boldsymbol{q}_0 + \boldsymbol{w} \ . \tag{2.68}$$

We assume that $\boldsymbol{q}_0$ changes smoothly with $\alpha$

$$q_0 = q_0(\alpha) \ . \tag{2.69}$$

We wish to study the stability of that solution $\boldsymbol{q}_0$. We thus insert the hypothesis (2.68) into the equation (2.66) where $\boldsymbol{F}$ however, is dropped. We then obtain

$$\dot{\boldsymbol{q}}_0 + \dot{\boldsymbol{w}} = \boldsymbol{N}(\boldsymbol{q}_0 + \boldsymbol{w}, \alpha) \ . \tag{2.70}$$

Under the assumption that $\boldsymbol{w}$ is a small quantity, we may expand the right-hand side of (2.70) into a power series in $\boldsymbol{w}$ and keep only the two leading terms

$$\dot{\boldsymbol{q}}_0 + \dot{\boldsymbol{w}} = \boldsymbol{N}(\boldsymbol{q}_0, \alpha) + \boldsymbol{L}(\boldsymbol{q}_0)\boldsymbol{w} + \cdots \ . \tag{2.71}$$

On account of (2.67), the first term on the l.h.s. of (2.70) cancels with the first term on the r.h.s. of (2.71) so that we are left with the equation

$$\dot{\boldsymbol{w}} = \boldsymbol{L}(\boldsymbol{q}_0)\boldsymbol{w} \ . \tag{2.72}$$

Note that $L$, which depends on $\boldsymbol{q}_0$, may still contain differential operators acting on the space coordinates in $\boldsymbol{w}$. Nevertheless the general solution of (2.72) can be written in the form

$$\boldsymbol{w}(t) = \mathrm{e}^{\lambda t}\boldsymbol{v} \ . \tag{2.73}$$

where $\boldsymbol{v}$ is a time-independent vector.

We note that more general cases have also been treated, namely those where $\boldsymbol{q}_0$ is a periodic or quasiperiodic function. For the detailed results I must refer the reader to my book *Advanced Synergetics*. For what follows it is important to distinguish between the so-called unstable and stable modes. The unstable modes are those for which

$$\lambda > 0 \tag{2.74}$$

holds. They shall be denoted by

$$\lambda_u, \boldsymbol{v}_u \ . \tag{2.75}$$

The stable modes are characterized by

$$\lambda < 0 \tag{2.76}$$

and shall be denoted by

$$\lambda_s, \boldsymbol{v}_s \ . \tag{2.77}$$

Note that the terms "unstable" and "stable" refer only to the linear analysis. In fact, it will turn out that in general the so-called unstable modes will become stabilized by means of their interaction with the stable modes. Note further, that our approach is a fully nonlinear one and the linear stability analysis serves only to find an adequate frame of reference in which to represent the desired solution $\boldsymbol{q}$ of (2.66).

### 2.6.2 Transformation of Evolution Equations

In order to solve (2.66) in the nonlinear and stochastic case, we make the hypothesis

$$\boldsymbol{q} = \boldsymbol{q}_0 + \sum_u \xi_u(t)\boldsymbol{v}_u + \sum_s \xi_s(t)\boldsymbol{v}_s \ . \tag{2.78}$$

In the case where $L$ contains differential operators acting on space variables, $\boldsymbol{v}$ is a space dependent function

$$\boldsymbol{v} = \boldsymbol{v}(\boldsymbol{x}) \ . \tag{2.79}$$

When we insert (2.78) into (2.66) and project both sides onto the expansion functions $\boldsymbol{v}_u$ and $\boldsymbol{v}_s$, we obtain equations for the mode amplitudes $\xi_u$ and $\xi_s$

$$\dot{\xi}_u = \lambda_u \xi_u + N_u(\xi_u, \xi_s) + \boldsymbol{F}_u(t) \tag{2.80}$$

$$\dot{\xi}_s = \lambda_s \xi_s + N_s(\xi_u, \xi_s) + \boldsymbol{F}_s(t) \ . \tag{2.81}$$

The indices $u$ and $s$ serve two purposes. Firstly, they indicate whether we are dealing with the amplitudes of the unstable or of the stable modes. Secondly, they serve to number the individual components of $\xi_u$ and $\xi_s$. For instance we may let $u$ and $s$ take the values $u = 1, \ldots, M$ and $s = M + 1, \ldots$. The context will show which meaning the indices $u$ or $s$ have to be given. The amplitudes $\xi_u$ will be called order parameters.

### 2.6.3 The Slaving Principle

The transformation (2.78) of the equation (2.66) does not reduce the complexity, i.e. the equations (2.80) and (2.81) are fully equivalent to the equations (2.66). The slaving principle of synergetics allows us, however, to eliminate from (2.80) and (2.81) the slaved mode amplitudes by means of an explicit formula

$$\xi_s(t) = f_s[\xi_u(t), t] \ . \tag{2.82}$$

The explicit construction of $f_s$ is described in my book *Advanced Synergetics* and in special cases also in my book *Synergetics. An Introduction.* Here, we just illustrate the contents of (2.82) by means of a simple explicit example where we present the slaving principle in its leading term. To this end let us consider the equations for the amplitudes $\xi_u$ and $\xi_s$, namely (2.80,81), in the following form

$$\dot{\xi}_u = \lambda_u \xi_u + h_u(\xi_u, \xi_s) + F_u(t) \tag{2.83a}$$

$$\dot{\xi}_s = \lambda_s \xi_s + g_s(\xi_u) + q_s k_s(\xi_u) + F_s(t) \ . \tag{2.83b}$$

Here it is assumed that $h_u$ is a nonlinear function which starts with powers of at least second order in $\xi_u$. Similarly, $g_s$ is a function starting at the same power. It may then be shown that $\xi_s$ starts with powers of at least second order in $\xi_u$. In its simplest form, the slaving principle amounts to putting $\dot{\xi}_s$ in (2.83b) equal to zero.

Keeping the leading orders we readily obtain the result

$$\xi_s \approx -\frac{1}{\lambda_s} g_s(\xi_u) - \frac{1}{\lambda_s} F_s(t) \ . \tag{2.84}$$

This result can be proven rigorously to lowest order in $F_s$ and $\xi_u$. We now wish to study what the slaving principle means for the solution of the Fokker-Planck equation. For this purpose we transform the Fokker-Planck equation from the old state vector $\boldsymbol{q}$ into the new variables $\xi_u$, $\xi_s$

$$\boldsymbol{q} \to \xi_u, \xi_s \ . \tag{2.85}$$

The Fokker-Planck equation then acquires the general form

$$\dot{P}(\xi_u, \xi_s; t) = L(\xi_u, \xi_s) P(\xi_u, \xi_s; t) \tag{2.86}$$

where $L$ is a linear operator. Let us now consider the steady state solution of (2.86). It can always be written in the form

$$P(\xi_u, \xi_s) = P(\xi_s | \xi_u) f(\xi_u) \tag{2.87}$$

where the l.h.s. is a joint probability whereas $P$ on the r.h.s. is a conditional probability. $f$ is a distribution function for the order parameters alone. The slaving principle, in its leading approximation, now means that the conditional probability $P$ on the r.h.s. of (2.87) can be written more specifically as

$$P(\xi_s | \xi_u) = \prod_s P_s(\xi_s | \xi_u) \ . \tag{2.88}$$

Our result (2.84) can now be used to give us an explicit example of what (2.88) may look like in this lowest order approximation of the slaving principle. The fluctuating forces are, as usual, Gaussian distributed, i.e. the probability of finding $F_s$ within the interval $F \to F + dF$ is given by

$$P(F \leq F_s \leq F + dF) = N' \exp(-F_s^2/Q')\,dF \ . \tag{2.89}$$

Now we may solve the relation (2.84) for $F_s$

$$F_s = -\lambda_s \left[ \xi_s + \frac{1}{\lambda_s} g_s(\xi_u) \right] \ . \tag{2.90}$$

This allows us to determine the conditional probability by using (2.89). In this way we obtain

$$P(\xi_s|\xi_u) = N \exp\left\{ -\left[ \xi_s + \frac{1}{\lambda_s} g_s(\xi_u) \right]^2 \Big/ Q \right\} d\xi_s \ . \tag{2.91}$$

where we have used the abbreviations

$$Q^{-1} = Q'^{-1} \lambda_s^2 \tag{2.92}$$

$$N = N' \lambda_s \ . \tag{2.93}$$

## 2.7 Nonequilibrium Phase Transitions

In many cases of practical interest, the number of order parameters may be very small or even one, whereas the number of slaved modes is still very large. Let us consider the case of a single order parameter and let us drop the index $u$ for simplicity

$$\xi_u \to \xi, \ F_u \to F, \ \lambda_u \to \lambda \ . \tag{2.94}$$

A typical order parameter equation then reads

$$\dot{\xi} = \lambda \xi - \beta \xi^3 + F(t) \ , \tag{2.95}$$

as is shown in synergetics.

The fluctuating force $F(t)$ obeys the relation

$$\langle F(t) F(t') \rangle = Q \delta(t - t') \ . \tag{2.96}$$

If we assume in addition that $F$ is Gaussian distributed we may establish a Fokker-Planck equation belonging to (2.95) in the form

$$\dot{f}(\xi;t) = -\frac{\partial}{\partial \xi}[(\lambda \xi - \beta \xi^3) f] + \frac{Q}{2} \frac{\partial^2}{\partial \xi^2} f \ . \tag{2.97}$$

Let us consider the steady state for which

$$\dot{f} = 0 \; . \tag{2.98}$$

Then (2.97) can easily be integrated,

$$f = N \exp[Q^{-1}(\lambda\xi^2 - \tfrac{1}{2}\beta\xi^4)] \; . \tag{2.99}$$

Equation (2.99) provides us with an explicit and typical example for the distribution function $f$ occurring in (2.87). Distribution functions for order parameters can be found also for several order parameters explicitly, provided the principle of detailed balance holds. Let us assume that $\lambda_u$ is real. A special case of the principle of detailed balance is the following: The Langevin equation has the form

$$\dot{\xi}_u = \lambda_u \xi_u + \frac{\partial V}{\partial \xi_u} + F_u(t) \tag{2.100}$$

where $V$ is a nonlinear function of $\xi_u$. We assume that

$$\langle F_u(t) F_{u'}(t') \rangle = \delta_{uu'} Q \delta(t - t') \tag{2.101}$$

holds. Then the solution of the Fokker-Planck equation belonging to the Langevin equation (2.100) can be written in the general form

$$f(\xi_u) = N \exp\left[ -\frac{2V(\xi_u)}{Q} \right] . \tag{2.102}$$

Even if the principle of detailed balance is not valid, but the distribution function is singly connected, it may always be written in the form

$$f(\xi_u) = N \exp[-\Phi(\xi_u)] \tag{2.103}$$

where $\Phi$ plays the role of a generalized thermodynamic potential. Now let us assume that a soft transition occurs when we change the control parameter so that the system becomes unstable and enters a new region. In such a case we may assume that the order parameters are still small, for instance that

$$\xi_u \propto \lambda_u^{\kappa} \; , \quad \kappa > 0 \tag{2.104}$$

In such a case we may expand $\Phi$ as a power series with respect to $\xi_u$

$$\begin{aligned} \Phi(\xi_u) = & \sum_u \lambda_u \xi_u^2 + \sum_{uu'} c_{uu'} \xi_u \xi_{u'} + \sum_{uu'u''} c_{uu'u''} \xi_u \xi_{u'} \xi_{u''} \\ & + \sum_{uu'u''u'''} c_{uu'u''u'''} \xi_u \xi_{u'} \xi_{u''} \xi_{u'''} \; . \end{aligned} \tag{2.105}$$

In most cases of non-equilibrium phase transitions, it is sufficient to retain the first 4 powers, though in exceptional cases higher powers are also needed. In general, systems possess internal symmetries that give rise to specific relations between the coefficients $c$ of each order.

## 2.8 Pattern Formation

We now wish to show how our above formalism can describe the formation of patterns. If we are dealing with continuously extended media which are described by a space coordinate $\boldsymbol{x}$, then in general, the operator $L$ in the linearization (2.72) contains derivatives with respect to the spatial coordinate. In such a case $\boldsymbol{w}$ and thus $\boldsymbol{v}$ which occurs in (2.73) become functions of the space coordinate $\boldsymbol{x}$, (2.79).

Quite generally the solution $\boldsymbol{q}$ of the nonlinear equation (2.66) can be written in the form

$$\boldsymbol{q} = \boldsymbol{q}_0 + \sum_u \xi_u(t)\boldsymbol{v}_u(\boldsymbol{x}) + \sum_s \xi_s(t)\boldsymbol{v}_s(\boldsymbol{x}) \ . \tag{2.106}$$

which is just the same as our previous formula (2.78).

As it turns out, $\xi_u$ is, in general, an order of magnitude larger than $\xi_s$, i.e. the evolving pattern is mainly determined by the first sum over $u$ in (2.106) which we therefore call the mode skeleton. If only a single order parameter is present and $\boldsymbol{v}_u$ has the form

$$\boldsymbol{v}_u = L^{-1/2} \sin kx \ , \tag{2.107}$$

then (2.106) is essentially given by

$$\boldsymbol{q} = \boldsymbol{q}_0 + \xi_u(t) L^{-1/2} \sin kx \ . \tag{2.108}$$

In many cases $\xi_u$ obeys an equation of the form (2.95) which describes the growth of $\xi_u$ out of an initial fluctuation to its final size (Fig.2.3,4). Quite evidently, with more order parameters and/or more complicated functions $\boldsymbol{v}_u$, far more complicated patterns than that described by (2.107), and (2.108) can be obtained. Thus, this theory is capable of deriving the emergent spatial structures of complex systems.

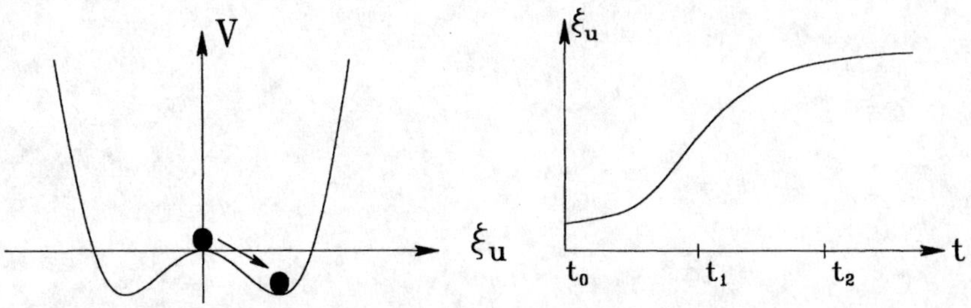


**Fig. 2.3.** Illustration of the behavior of the order parameter $\xi_u$ as a function of time. Left-hand side: The potential function $V$ in which a fictitious particle with coordinate $\xi_u$ may move. Right-hand side: The temporal evolution of $\xi_u$

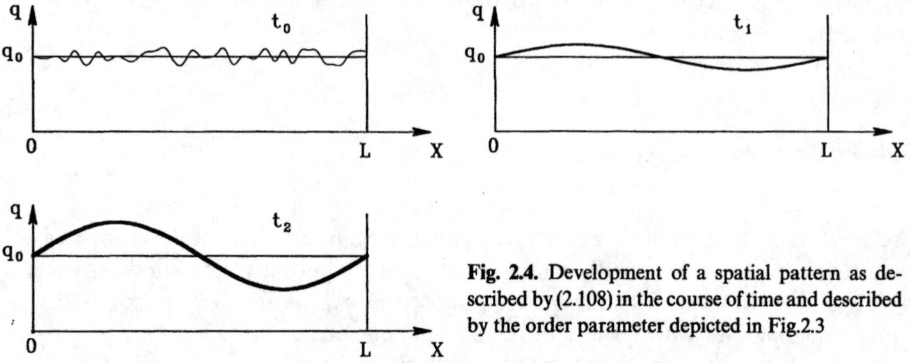


**Fig. 2.4.** Development of a spatial pattern as described by (2.108) in the course of time and described by the order parameter depicted in Fig.2.3

Our example here is just a brief reminder of what has been presented in my books *Synergetics* and *Advanced Synergetics.* I just wish to remind the reader that considerably more general cases have been treated there, such as evolving time-dependent patterns, namely limit cycles, quasi periodic motion, or chaos. But for what follows this brief reminder will, in most cases, be sufficient.

# 3. ... and Back Again: The Maximum Information Principle (MIP)

## 3.1 Some Basic Ideas

In this chapter we address the following question: Let some *macroscopic quantities* of a system be given. We then wish to devise a procedure by which we can derive the probability distribution of *macroscopic* or even *microscopic* variables. In other words, we start from the macroscopic world and wish to draw conclusions about the microscopic world. Depending on the kind of systems we are treating, the adequate macroscopic quantities may be quite different. In closed physical systems, to which thermodynamics applies, these quantities are energy, particle numbers etc., and we shall illustrate the general procedure by this example in Chap. 4. In open systems, e.g. in physics or biology, the adequate macroscopic quantities will turn out to be, for instance, intensities and intensity fluctuations. Indeed, it will be the main purpose of the following chapters, to deal with *open systems*.

Since the starting point of our approach is the concept of information, we shall derive this concept in this Sect. 3.1.

By some sort of new interpretation of probability theory we get an insight into a seemingly quite different discipline, namely information theory. Consider the sequence of tossing a coin with outcomes 0 and 1. Now interpret 0 and 1 as a dash and dot of a Morse alphabet. We all know that by means of a Morse alphabet we can transmit messages so that we may ascribe a certain meaning to a certain sequence of symbols. Or, in other words, a certain sequence of symbols carries information. In information theory we try to find a measure for the amount of information.

Let us consider a simple example and consider $R_0$ different possible events ("realizations") which are equally probable a priori. Thus when tossing a coin we have the events 1 and 0 and $R_0 = 2$. In the case of a die we have 6 different outcomes, therefore $R_0 = 6$. Thus the outcome of tossing a coin or throwing a die is interpreted as the receipt of a message and only one out of the possible $R_0$ outcomes is actually realized. Apparently the greater $R_0$, the greater is the uncertainty before the message is received and the larger will be the amount of information after the message is received. Thus we may interpret the whole procedure in the following manner: In the initial situation we have no information $I_0$, i.e., $I_0 = 0$ with $R_0$ equally probable outcomes.

In the final situation we have an information $I_1 \neq 0$ with $R_1 = 1$, i.e., a single outcome. We now want to introduce a measure for the amount of information, $I$, which apparently must be connected with $R_0$. To get an idea how the connection between $R_0$ and $I$ must appear we require that $I$ is additive for independent events.

Thus when we have two such sets with $R_{01}$ or $R_{02}$ outcomes so that the total number of outcomes is

$$R_0 = R_{01} R_{02} \tag{3.1}$$

we require

$$I(R_{01} R_{02}) = I(R_{01}) + I(R_{02}) \ . \tag{3.2}$$

This relation can be fulfilled by choosing

$$I = K \ln R_0 \tag{3.3}$$

where $K$ is a constant. It can even be shown that (3.3) is the only solution to (3.2). The constant $K$ is still arbitrary and can be fixed by some definition. Usually the following definition is used. We consider a so-called "binary" system which has only two symbols (or letters). These may be the head and the tail of a coin, or answers yes and no, or numbers 0 and 1 in a binomial system. When we form all possible "words" (or sequences) of length $n$, we find $R = 2^n$ realizations. We now want to identify $I$ with $n$ in such a binary system. We therefore require

$$I = K \ln R \equiv K n \ln 2 = n \tag{3.4}$$

which is fulfilled by

$$K = \frac{1}{\ln 2} = \log_2 \mathrm{e} \tag{3.5}$$

With this choice of $K$, another form of (3.4) reads

$$I = \log_2 R \ . \tag{3.4a}$$

Since a single position in a sequence of symbols (signs) in a binary system is called "bit", the information $I$ is now directly given in bits. Thus if $R = 8 = 2^3$ we find $I = 3$ bits and generally for $R = 2^n$, $I = n$ bits. The definition of information for (3.3) can be easily generalized to the case where we initially have $R_0$ equally probable cases and finally $R_1$ equally probable cases. In this case the information is

$$I = K \ln R_0 - K \ln R_1 \tag{3.6}$$

which reduces to the earlier definition (3.3), if $R_1 = 1$. A simple example for this is given by a die. Let us define a game in which the even numbers mean gain and the odd numbers mean loss. Then $R_0 = 6$ and $R_1 = 3$. In this case the information content is the same as that of a coin with originally just two possibilities.

We now derive a more convenient expression for the information: We first consider the following example of a simplified Morse alphabet with dash and dot (in the real Morse alphabet, the intermission is a third symbol). We consider a word of length $G$ which contains $N_1$ dashes and $N_2$ dots, with

$$N_1 + N_2 = N \ . \tag{3.7}$$

We ask for the information which is obtained by the receipt of such a word. In the spirit of information theory we must calculate the total number of words which can be constructed out of these two symbols for fixed $N_1, N_2$. The analysis is quite simple. According to the ways in which we can distribute the dashes and dots over the $N$ positions, there are

$$R = \frac{N!}{N_1!N_2!} \tag{3.8}$$

possibilities. Or, in other words, $R$ is the number of messages which can be transmitted by $N_1$ dashes and $N_2$ dots. We now want to derive the information per symbol, i.e. $i = I/N$. Inserting (3.8) into (3.3) we obtain

$$I = K \ln R = K[\ln N! - \ln N_1! - \ln N_2!] \ . \tag{3.9}$$

Using Stirling's formula in the approximation

$$\ln N! \approx N(\ln N - 1) \ , \tag{3.10}$$

which is good for $N > 100$, we readily find

$$I \approx K[N(\ln N - 1) - N_1(\ln N_1 - 1) - N_2(\ln N_2 - 1)] \ , \tag{3.11}$$

and from (3.7) we then have

$$i \equiv \frac{I}{N} \approx -K\left[\frac{N_1}{N}\ln\frac{N_1}{N} + \frac{N_2}{N}\ln\frac{N_2}{N}\right] . \tag{3.12}$$

We now introduce a quantity which may be interpreted as the probability of finding the sign "dash" or "dot". The probability is identical to the relative frequency with which dash or dot are found

$$p_j = \frac{N_j}{N} \ , \qquad j = 1, 2 \ . \tag{3.13}$$

With this, our final formula takes the form

$$i = \frac{I}{N} = -K(p_1 \ln p_1 + p_2 \ln p_2) \ . \tag{3.14}$$

This expression can be easily generalized to the case where we do not simply have two symbols but several, such as letters in the alphabet. Then we obtain, in an exactly analogous manner, an expression for the information per symbol which is given by

$$i = -K \sum_j p_j \ln p_j \ . \tag{3.15}$$

$p_j$ is the relative frequency of the occurrence of the symbols. From this interpretation it is evident that $i$ may be used in the context of transmission of information, etc.

Before continuing we should say a word about information used in the sense here. It should be noted that "useful" or "useless" or "meaningful" or "meaningless" are not contained in the theory; e.g., in the Morse alphabet defined above quite a number of words might be meaningless. Information in the sense used here rather refers to the scarcity of an event. Though this may seem to restrict the theory considerably, the theory will in fact turn out to be extremely useful.

The expression for the information can be viewed in two completely different ways. On the one hand we may assume that the $p_i$'s are given by their numerical values, and then we may write down a number for $I$ by use of formula (3.3). Of still greater importance, however, is a second interpretation; namely, to consider $I$ as a function of the $p_i$'s such that if we change the values of the $p_i$'s, the value of $I$ changes correspondingly. To make this interpretation clear we anticipate an application which we will treat later in much greater detail. Consider a gas of atoms moving freely in a box. It is then of interest to know about the spatial distribution of the gas atoms. We divide the container into $M$ cells of equal size and denote the number of particles in cell $k$ by $N_k$. The total number of particles is $N$. The relative frequency of a particle to be found in cell $k$ is then given by

$$\frac{N_k}{N} = p_k \;, \qquad k = 1, 2, \ldots, M \;. \tag{3.16}$$

$p_k$ may be considered as the distribution function of the particles over the cells $k$. Because the cells have equal size and do not differ in their physical properties, we expect that the particles will be found with equal probability in each cell, i.e.,

$$p_k = \frac{1}{M} \;. \tag{3.17}$$

We now want to derive this result (3.17) from the properties of information. Indeed the information may be as follows: Before we make a measurement or obtain a message, there are $R$ possibilities or, in other words, $K \ln R$ is a measure of our ignorance. Another way of looking at this is the following: $R$ gives us the number of realizations which are possible in principle.

Now let us look at an ensemble of $C$ containers, each with $N$ gas atoms. We assume that in each container the particles are distributed according to different distribution functions $p_k$, i.e.,

$$p_k^{(1)}, p_k^{(2)}, p_k^{(3)}, \ldots \;.$$

Accordingly, we obtain different numbers of realizations, i.e., different information. For example, if $N_1 = N$, $N_2 = N_3 = \cdots = 0$, we have $p_1^{(1)} = 1$, $p_2^{(1)} = p_3^{(1)} = \cdots = 0$ and thus $I^{(1)} = 0$. On the other hand, if $N_1 = N_2 = N_3 = \cdots = N/M$, we have $p_1^{(2)} = 1/M$, $p_2^{(2)} = 1/M, \ldots$, so that $I^{(2)} = -M \log_2 (1/M) = M \log_2 M$, which is a very large number if the number of boxes is large.

Thus when we consider any container with gas atoms, the probability that it is one with the second distribution function is much greater than one with the first distribution function. That means there is an overwhelming probability of finding that probability distribution $p_k$ realized which has the greatest number of possibilities

$R$ and thus the greatest information. Hence we are led to require that

$$-\sum p_i \ln p_i = \mathit{Extremum} \tag{3.18}$$

under the constraint that the total sum of the probabilities $p_i$ equals unity

$$\sum_{i=1}^{M} p_i = 1 \ . \tag{3.19}$$

This principle will turn out to be fundamental for application to realistic systems in physics, chemistry, and biology and we shall come back to it later.

The problem (3.18) with (3.19) can be solved using the method of Lagrange multipliers. This method consists in multiplying (3.19) by a still unknown parameter $\lambda$ and adding it to the left-hand side of (3.18) now requiring that the total expression becomes an extremum. Here we are now allowed to vary the $p_i$'s independently of each other, not taking into account the constraint (3.19). Varying the left-hand side of

$$-\sum p_i \ln p_i + \lambda \sum p_i = \mathit{Extr} \ . \tag{3.20}$$

means taking the derivative of it with respect to $p_i$ which leads to

$$-\ln p_i - 1 + \lambda = 0 \ . \tag{3.21}$$

Equation (3.21) has the solution

$$p_i = \mathrm{e}^{\lambda - 1} \tag{3.22}$$

which is independent of the index $i$, i.e., the $p_i$'s are all equal. Inserting them into (3.19) we may readily determine $\lambda$ so that

$$M \mathrm{e}^{\lambda - 1} = 1 \ , \tag{3.23}$$

or, in other words, we find

$$p_i = \frac{1}{M} \tag{3.24}$$

in agreement with (3.17) as expected.

## 3.2 Information Gain

The expression (3.15) for the information can be interpreted as an average over $f_j$,

$$i = \sum p_j f_j \ , \qquad \text{where} \tag{3.25}$$

$$f_j = -K \ln p_j \ , \qquad p_j \neq 0 \tag{3.26}$$

and the weight is $p_j$.

This suggests that $f_j$ in (3.26) be interpreted as the information content of the symbol with index $j$ and $p_j$ as the probability (or relative frequency).

Now let us assume that a set of measurements has led us to the relative frequency $p_j$ for the symbol with index $j$. Then let us assume that – possibly under different conditions – we determine a new relative frequency $p'_j$. What is the corresponding change of information, $\Delta_j$?

Adopting the interpretation of (3.26), we are immediately led to define it by

$$\Delta_j = K \ln p'_j - K \ln p_j \ . \tag{3.27}$$

To obtain the mean change of information, we average (3.27) over the new distribution function ("relative frequency") $p'_j$. We thus obtain the so-called information gain (or "Kullback information").

$$K(p', p) = \sum_j p'_j \Delta_j = K \sum_j p'_j \ln \frac{p'_j}{p_j} \ , \tag{3.28}$$

where, of course,

$$\sum_j p_j = 1 \quad \text{and} \tag{3.29}$$

$$\sum_j p'_j = 1 \ . \tag{3.30}$$

The information gain $K(p', p)$ has the following important property.

$$K(p', p) \geq 0 \ . \tag{3.31}$$

The equality sign holds if and only if

$$p' \equiv p, \text{ i.e. } p'_k = p_k \qquad \text{for all } k\text{'s} \ .$$

## 3.3 Information Entropy and Constraints

In this section and in the next chapter we will be concerned, in particular, with applications of the information concept to physics and shall thus follow the convention of denoting the information by $S$, and identifying the constant $K$ in (3.3) with Boltzmann's constant $k_B$. For reasons which will appear later, $S$ will be called information entropy. Because chemical and biological systems can be viewed as physical systems, our considerations apply equally well to these systems too. The general formalism of this chapter is also applicable to other sciences, such as information processing, etc. We start from the basic expression

$$S = -k_B \sum_i p_i \ln p_i \ . \tag{3.32}$$

The indices $i$ may be considered as describing individual features of the particles or subsystems. Let us explain this in some detail. The index $i$ may describe, for instance, the position of a gas particle or it may describe its velocity or both properties. In our previous examples the index $i$ referred to boxes filled with balls. In a more general interpretation the index $i$ represents the values that a random variable may acquire. In this section we assume for simplicity that the index $i$ is discrete.

A central task to be solved in this book consists in finding ways to determine the $p_i$'s (compare for example the gas molecules in a container where one wants to know their location). The problem we are confronted with in many disciplines is to make unbiased estimates leading to $p_i$'s which are in agreement with all the possible knowledge available about the system. Consider an ideal gas in one dimension. What we could measure, for instance, is the center of gravity. In this case we would have as constraint an expression of the form

$$\sum_i p_i q_i = M \tag{3.33}$$

where $q_i$ measures the position of the cell $i$. $M$ is a fixed quantity equal to $Q/N$, where $Q$ is the coordinate of the center of gravity, and $N$ the particle number. There are, of course, very many sets of $p_i$'s, which fulfill the relation (3.33). Thus we could choose a set $\{p_i\}$ rather arbitrarily, i.e., we would favor one set against another one. Similar to ordinary life, this is a biased action. How may it be unbiased? When we look again at the example of the gas atoms, then we can invoke the principle stated in Sect. 3.1. With an overwhelming probability we will find those distributions realized for which (3.32) is a maximum. However, due to (3.33) not all distributions can be taken into account. Instead we have to seek the maximum of (3.32) under the constraint (3.33). This principle can be generalized if we have a set of constraints. Let, for example, the variable $i$ distinguish between different velocities. Then we may have the constraint that the total kinetic energy $E_{\text{kin}}^{\text{tot}}$ of the particles is fixed. Denoting the kinetic energy of a particle with mass $m$ and velocity $v_i$ by $f_i$ $[f_i = (m/2)v_i^2]$ the mean kinetic energy per particle is given by

$$\sum_i p_i f_i = E_{\text{kin}} \; . \tag{3.33a}$$

In general the single system $i$ may be characterized by quantities $f_i^{(k)}$, $k = 1, 2, \ldots, M$ (position, kinetic energy or other typical features). If these features are additive, and the corresponding sums are kept fixed at values $f_k$, then the constraints take the form

$$\sum_i p_i f_i^{(k)} = f_k \; . \tag{3.34}$$

We further add the usual constraint that the probability distribution is normalized

$$\sum_i p_i = 1 \; . \tag{3.35}$$

The problem of finding the extremum of (3.32) under the constraints (3.34) and (3.35) can be solved by using the method of Lagrange multipliers $\lambda_k$, $k = 1, 2, \ldots, M$ (cf. Sect. 3.1). We multiply the left-hand side of (3.34) by $\lambda_k$ and the left-hand side of

(3.35) by $(\lambda - 1)$ and take the sum of the resulting expressions. We then subtract this sum from $(1/k_B)S$. The factor $1/k_B$ amounts to a certain normalization of $\lambda$, $\lambda_k$. We then have to vary the total sum with respect to the $p_i$'s

$$\delta\left[\frac{1}{k_B}S - (\lambda - 1)\sum_i p_i - \sum_k \lambda_k \sum_i p_i f_i^{(k)}\right] = 0 . \tag{3.36}$$

Differentiating with respect to $p_i$ and setting the resulting expression equal to zero, we obtain

$$-\ln p_i - 1 - (\lambda - 1) - \sum_k \lambda_k f_i^{(k)} = 0 , \tag{3.37}$$

which can be readily solved for $p_i$ yielding

$$p_i = \exp\left(-\lambda - \sum_k \lambda_k f_i^{(k)}\right) . \tag{3.38}$$

Inserting (3.38) into (3.35) yields

$$e^{-\lambda}\sum_i \exp\left(-\sum_k \lambda_k f_i^{(k)}\right) = 1 . \tag{3.39}$$

It is now convenient to abbreviate the sum over $i$, $\sum_i$ in (3.39) by

$$\sum_i \exp\left(-\sum_k \lambda_k f_i^{(k)}\right) = Z(\lambda_1, \ldots, \lambda_M) , \tag{3.40}$$

which we shall call the partition function. Inserting (3.40) into (3.39) yields

$$e^{\lambda} = Z \quad \text{or} \tag{3.41}$$

$$\lambda = \ln Z , \tag{3.42}$$

which allows us to determine $\lambda$ once the $\lambda_k$'s are determined. To find equations for the $\lambda_k$'s we insert (3.38) into the equations of the constraints (3.34) which lead immediately to

$$\langle f_i^{(k)} \rangle = \sum_i p_i f_i^{(k)} = e^{-\lambda}\sum_i \exp\left(-\sum_l \lambda_l f_i^{(l)}\right) f_i^{(k)} . \tag{3.43}$$

Equation (3.43) has a rather similar structure to (3.40). The difference between these two expressions arises because in (3.43) each exponential function is still multiplied by $f_i^{(k)}$. However, we may easily derive the sum occurring in (3.43) from (3.40) by differentiating (3.40) with respect to $\lambda_k$. Expressing the first factor in (3.43) by $Z$ according to (3.41) we thus obtain

$$\langle f_i^{(k)} \rangle = \frac{1}{Z}\left(-\frac{\partial}{\partial \lambda_k}\right)\underbrace{\sum_i \exp\left(-\sum_l \lambda_l f_i^{(l)}\right)}_{Z} \tag{3.44}$$

or in still shorter form

$$f_k \equiv \langle f_i^{(k)} \rangle = -\frac{\partial \ln Z}{\partial \lambda_k} . \tag{3.45}$$

Because the $f_k$ on the left-hand side are prescribed [compare (3.34)] and $Z$ is given by (3.40) which is a function of the $\lambda_k$'s in a special form, (3.45) is a concise form for a set of equations for the $\lambda_k$'s.

We further quote a formula which will become useful later on. Inserting (3.38) into (3.32) yields

$$\frac{1}{k_B} S_{\max} = \lambda \sum_i p_i + \sum_k \lambda_k \sum_i p_i f_i^{(k)} \tag{3.46}$$

which can be written using (3.34) and (3.35) as

$$\frac{1}{k_B} S_{\max} = \lambda + \sum_k \lambda_k f_k . \tag{3.47}$$

The maximum of the information entropy may thus be represented by the mean values $f_k$ and the Lagrange multipliers $\lambda_k$. Those readers who are acquainted with the Lagrange equations of the first kind in mechanics will remember that the Lagrange multipliers have a physical meaning, in that case, of forces. In a similar way we shall see later on that the Lagrange multipliers $\lambda_k$ have physical (or chemical or biological, etc) interpretations. In deriving the above formulas [i.e., (3.38,42) with (3.32,45,47)] we have completed our original task of finding the $p$'s and $S_{\max}$.

We now derive some further useful relations. We first investigate how the information $S_{\max}$ is changed if we change the functions $f_i^{(k)}$ and $f_k$ in (3.34). Because $S$ depends, according to (3.47), not only on the $f$'s but also on $\lambda$ and the $\lambda_k$'s which are functions of the $f$'s, we must exercise some care in taking the derivatives with respect to the $f$'s. We therefore first calculate the change of $\lambda$ (3.42)

$$\delta\lambda \equiv \delta \ln Z = \frac{1}{Z} \delta Z .$$

Inserting (3.40) for $Z$ yields

$$\delta\lambda = \mathrm{e}^{-\lambda} \sum_i \sum_k \{-\delta\lambda_k f_i^{(k)} - \lambda_k \delta f_i^{(k)}\} \exp\left(-\sum_l \lambda_l f_i^{(l)}\right)$$

which, from the definition of $p_i$ (3.38) transforms to

$$\delta\lambda = -\sum_k \left[\delta\lambda_k \sum_i p_i f_i^{(k)} + \lambda_k \sum_i p_i \delta f_i^{(k)}\right] .$$

Equation (3.43) and an analogous definition of $\langle \delta f_i^{(k)} \rangle$ allow us to write the last line as

$$-\sum_k [\delta\lambda_k \langle f_i^{(k)} \rangle + \lambda_k \langle \delta f_i^{(k)} \rangle] . \tag{3.48}$$

Inserting this into $\delta S_{\max}$ from (3.47), we find that the variation of the $\lambda_k$'s drops out and we are left with

$$\delta S_{\max} = k_B \sum_k \lambda_k [\delta \langle f_i^{(k)} \rangle - \langle \delta f_i^{(k)} \rangle] \ . \tag{3.49}$$

We write this in the form

$$\delta S_{\max} = k_B \sum_k \lambda_k \delta Q_k \tag{3.50}$$

where we define a "generalized heat" by means of

$$\delta Q_k = \delta \langle f_i^{(k)} \rangle - \langle \delta f_i^{(k)} \rangle \ . \tag{3.51}$$

The notation "generalized heat" will become clearer below when contact with thermodynamics is made. In analogy to (3.45), a simple expression for the variance of $f_i^{(k)}$ may be derived:

$$\langle f_i^{(k)2} \rangle - \langle f_i^{(k)} \rangle^2 = \frac{\partial^2 \ln Z}{\partial \lambda_k^2} \ . \tag{3.52}$$

In many practical applications, $f_i^{(k)}$ depends on a further quantity $\alpha$ (on a set of such quantities $\alpha_1, \alpha_2, \ldots$). Then we want to express the change of the mean value (3.34), when $\alpha$ is changed. Taking the derivative of $f_{i,\alpha}^{(k)}$ with respect to $\alpha$ and taking the average value, we find

$$\left\langle \frac{\partial f_{i,\alpha}^{(k)}}{\partial \alpha} \right\rangle = \sum_i p_i \frac{\partial f_{i,\alpha}^{(k)}}{\partial \alpha} \ . \tag{3.53}$$

Using the $p_i$'s in the form (3.38) and using (3.41), the right-hand side of (3.53) may be written in the form

$$\frac{1}{Z} \sum_i \frac{\partial f_{i,\alpha}^{(k)}}{\partial \alpha} \exp\left( -\sum_j \lambda_j f_{i,\alpha}^{(j)} \right) , \tag{3.54}$$

which may easily be expressed as a derivative of $Z$ with respect to $\alpha$:

$$(3.54) = -\frac{1}{Z} \frac{1}{\lambda_k} \frac{\partial Z}{\partial \alpha} \ . \tag{3.55}$$

Thus we are lead to the final formula

$$-\frac{1}{\lambda_k} \frac{\partial \ln Z}{\partial \alpha} = \left\langle \frac{\partial f_{i,\alpha}^{(k)}}{\partial \alpha} \right\rangle \ . \tag{3.56}$$

If there are several parameters $\alpha_l$ present, this formula can be readily generalized by writing $\alpha_l$ in place of $\alpha$ in (3.56).

As we have seen several times, the quantity $Z$, (3.40), or its logarithm, is very useful [see e.g., (3.45,52,56). We want to convince ourselves that $\ln Z \equiv \lambda$ [cf. (3.42)] may be directly determined by a variational principle. A glance at (3.36) reveals that (3.36) can also be interpreted in the following way: Seek the extremum of

$$\frac{1}{k_B} S - \sum_k \lambda_k \sum_i p_i f_i^{(k)} \tag{3.57}$$

under the only constraint

$$\sum p_i = 1 \ . \tag{3.58}$$

Now, by virtue of (3.34,47,42) the extremum of (3.57) is indeed identical with $\ln Z$. Note that the spirit of the variational principle for $\ln Z$ is different from that for $S$. In the former case, we had to seek the maximum of $S$ under the constraints (3.34,35) with $f_k$ fixed and $\lambda_k$ unknown. Here, only one constraint, (3.58), applies and the $\lambda_k$'s are assumed as given quantities. How such a switching from one set of fixed quantities to another one can be done will become more evident by the example from physics given in Chap. 4, which will also elucidate many other aspects of the aforegoing discussion.

## 3.4 Continuous Variables

In most applications that we have in mind, the variables $\xi$ are not discrete but continuous. One may then easily convince oneself that in such a case, at least in general, the information diverges. This is due to the fact that we have continuously many states. Therefore we have to discuss briefly how we can define information for continuous variables. We start from the definition of the probability density given by

$$\mathrm{Prob}(\chi \le \xi \le \chi + \Delta\xi) = P(\xi)\Delta\xi \ . \tag{3.59}$$

We now invoke the idea that measurements can be made only with finite accuracy. Therefore we introduce an interval of accuracy and define a new probability distribution by

$$P_\varepsilon(j) = \int_{\xi_j - \varepsilon/2}^{\xi_j + \varepsilon/2} P(\xi)\, d\xi \ . \tag{3.60}$$

Assuming that $P(\xi)$ is continuous we may approximate (3.60) by

$$P_\varepsilon(j) \approx P(\xi_j)\varepsilon \ . \tag{3.61}$$

We define the information with respect to the interval of accuracy $\varepsilon$ by means of

$$I_\varepsilon = -\sum_j P_\varepsilon(j) \ln P_\varepsilon(j) \ . \tag{3.62}$$

Inserting (3.61) into (3.62) we obtain

$$I_\varepsilon = -\sum_j \varepsilon P(\xi_j) \ln P(\xi_j) - \underbrace{\sum_j \varepsilon P(\xi_j)}_{=1} \ln \varepsilon \tag{3.63}$$

so that our final result reads

$$I_\varepsilon = -\int d\xi\, P(\xi) \ln P(\xi) - \ln \varepsilon \ . \tag{3.64}$$

In the following we shall drop the constant and uninteresting term $-\ln \varepsilon$. The extension to several variables is obvious.

# 4. An Example from Physics: Thermodynamics

To visualize the meaning of the index $i$, let us identify it with the velocity of a particle. In a more advanced theory $p_i$ is the occupation probability of a quantum state $i$ of a many-particle system. Further, we identify $f_{i,\alpha}^{(k)}$ with energy $E$ and the parameter $\alpha$ with the volume. Thus we put

$$f_{i,\alpha}^{(k)} = E_i(V) \; ; \qquad k = 1 \; , \tag{4.1}$$

and have the identifications

$$f_1 \leftrightarrow U \equiv \langle E_i \rangle \; ; \qquad \alpha \leftrightarrow V \; ; \qquad \lambda_1 = \beta \; . \tag{4.2}$$

We have, in particular, set $\lambda_1 = \beta$. With this, we may write a number of the previous formulas in a way which can be immediately identified with relations well known in thermodynamics and statistical mechanics. Instead of (3.38) we find

$$p_i = \exp[-\lambda - \beta E_i(V)] \tag{4.3}$$

which is the famous Boltzmann distribution function.

Equation (3.47) acquires the form

$$\frac{1}{k_B} S_{max} = \ln Z + \beta U \tag{4.4}$$

or, after a slight rearrangement of this equation

$$U - \frac{1}{k_B \beta} S_{max} = -\frac{1}{\beta} \ln Z \; . \tag{4.5}$$

This equation is well known in thermodynamics and statistical physics. The first term may be interpreted as the internal energy $U$, $1/\beta$ as the absolute temperature $T$ multiplied by Boltzmann's constant $k_B$. $S_{max}$ is the entropy. The right-hand side represents the free energy, $\mathcal{F}$, so that in thermodynamic notation (4.5) reads

$$U - TS = \mathcal{F} \; . \tag{4.6}$$

By comparison we find

$$\mathcal{F} = -k_B T \ln Z \; , \tag{4.7}$$

and $S = S_{max}$. Therefore we will henceforth drop the suffix "max". Equation (3.40) now reads

$$Z = \sum_i e^{-\beta E_i} \tag{4.8}$$

and is nothing but the usual partition function. A number of further identities of thermodynamics can easily be checked by applying the above formulas.

The only problem requiring some thought is the identification of independent and dependent variables. Let us begin with the information entropy, $S_{\max}$. In (3.47) it appears as a function of $\lambda$, $\lambda_\mu$ and the $f_k$. However, $\lambda$ and the $\lambda_\mu$ are themselves determined by equations which contain the $f_k$ and $f_i^{(k)}$ as given quantities [cf. (3.40,42,43)]. Therefore, the independent variables are $f_k$ and $f_i^{(k)}$, and the dependent variables are $\lambda$ and $\lambda_k$, and thus, by virtue of (3.47), $S_{\max}$. In practice the $f_i^{(k)}$ are fixed functions of $i$ (e.g., the energy of state "$i$"), but still depending on parameters $\alpha$ [e.g., the volume, cf. (4.1)]. Thus the truly independent variables in our approach are the $f_k$ (as above) and the $\alpha$'s. In conclusion we thus find $S = S(f_k, \alpha)$. In our example, $f_1 = E \equiv U$, $\alpha = V$, and therefore

$$S = S(U, V) \ . \tag{4.9}$$

Now let us apply the general relation (3.49) to our specific model. If we vary only the internal energy, $U$, but leave $V$ unchanged, then

$$\delta \langle f_i^{(1)} \rangle \equiv \delta f_1 \equiv \delta U \neq 0 \ , \qquad \text{and} \tag{4.10}$$

$$\delta f_{i,\alpha}^{(1)} \equiv \delta E_i(V) = \frac{\delta E_i(V)}{\delta V} \delta V = 0 \ , \qquad \text{and therefore} \tag{4.11}$$

$$\delta S = k_B \lambda_1 \delta U \qquad \text{or}$$

$$\frac{\delta S}{\delta U} = k_B \lambda_1 (\equiv k_B \beta) \ . \tag{4.12}$$

According to thermodynamics, the left-hand side of (4.12) defines the inverse of the absolute temperature

$$\frac{\delta S}{\delta U} = \frac{1}{T} \ . \tag{4.13}$$

This yields $\beta = 1/(k_B T)$ as anticipated above. On the other hand, varying $V$ but leaving $U$ fixed, i.e.,

$$\delta \langle f_i^{(1)} \rangle = 0 \ , \qquad \text{but} \tag{4.14}$$

$$\langle \delta f_i^{(1)} \rangle = \left\langle \frac{\delta E_i(V)}{\delta V} \right\rangle \delta V \neq 0 \tag{4.15}$$

yields in (3.49)

$$\delta S = k_{\rm B}(-\lambda_1)\left\langle \frac{\delta E_i(V)}{\delta V} \right\rangle \delta V \qquad \text{or}$$

$$\frac{\delta S}{\delta V} = -\frac{1}{T}\left\langle \frac{\delta E_i(V)}{\delta V} \right\rangle . \tag{4.16}$$

Since thermodynamics teaches us that

$$\frac{\delta S}{\delta V} = \frac{P}{T} \tag{4.17}$$

where $P$ is the pressure, we obtain by comparison with (4.16)

$$\left\langle \frac{\delta E_i(V)}{\delta V} \right\rangle = -P . \tag{4.18}$$

Inserting (4.13,17) into (3.49) yields

$$\delta S = \frac{1}{T}\delta U + \frac{1}{T}P\delta V . \tag{4.19}$$

In thermodynamics the right-hand side is equal to $dQ/T$ where $dQ$ is heat. This explains the notation "generalized heat" used after (3.51). These considerations may be generalized to different kinds of particles whose average numbers $N_k$; $k = 1, \ldots, m$, are prescribed quantities. We therefore identify $f_1$ with $E$, but $f_{k'+1}$ with $N_{k'}$; $k' = 1, \ldots, m$ (note the shift of index!). Since each kind of particle, $l$, may be present with different numbers $N_l$ we replace the index $i$ by $i, N_1, \ldots, N_m$ and put

$$f_i^{(k+1)} \to f_{i,N_1,\ldots,N_m}^{(k+1)} = N_k .$$

To be in accordance with thermodynamics, we put

$$\lambda_{k+1} = -\frac{1}{k_{\rm B}T}\mu_k , \tag{4.20}$$

where $\mu_k$ is called the chemical potential.

Equation (3.47) with (3.42) acquires (after multiplying both sides by $k_{\rm B}T$) the form

$$TS = \underbrace{k_{\rm B}T\ln Z}_{-\mathcal{F}} + U - \mu_1\bar{N}_1 - \mu_2\bar{N}_2 - \cdots - \mu_m\bar{N}_m . \tag{4.21}$$

Equation (3.49) permits us to identify

$$\frac{\delta S}{\delta \bar{N}_k} = -k_{\rm B}\lambda_{k+1} = \frac{1}{T}\mu_k . \tag{4.22}$$

The partition function reads

$$Z = \sum_{N_1 N_2 \dots N_m} \sum_i \exp\left\{-\frac{1}{k_B T}[E_i(V) - \mu_1 N_1 - \dots - \mu_m N_m]\right\} . \tag{4.23}$$

While the above considerations are most useful for irreversible thermodynamics, in thermodynamics the role played by independent and dependent variables is, to some extent, exchanged. It is not our task to treat these transformations which give rise to the different thermodynamic potentials (Gibbs, Helmholtz, etc). We just mention one important case: Instead of $U$, $V$ (and $N_1, \dots, N_m$), as independent variables, one may introduce $V$ and $T = (\partial S/\partial U)^{-1}$ (and $N_1, \dots, N_m$) as new independent variables. As an example we treat the $U$-$V$ case (putting formally $\mu_1, \mu_2, \dots = 0$). According to (4.7) the free energy, $\mathscr{F}$, is there directly given as a function of $T$. The differentiation $\partial \mathscr{F}/\partial T$ yields

$$-\frac{\partial \mathscr{F}}{\partial T} = k_B \ln Z + \frac{1}{T}\frac{1}{Z}\sum_i E_i e^{-\beta E_i} .$$

The second term on the right-hand side is just $U$, so that

$$-\frac{\partial \mathscr{F}}{\partial T} = k_B \ln Z + \frac{1}{T} U . \tag{4.24}$$

Comparing this relation with (4.5), where $1/\beta = k_B T$, yields the important relation

$$-\frac{\partial \mathscr{F}}{\partial T} = S \tag{4.25}$$

where we have dropped the suffix "max".

Readers who are interested in the application of the above formalism to irreversible thermodynamics, i.e. to *relaxation phenomena*, are referred to my book *Synergetics. An Introduction.* In the present book we shall be concerned with quite a different field, namely nonequilibrium phase transitions in physical and biological systems far from thermal equilibrium and in nonphysical systems (e.g. economy).

# 5. Application of the Maximum Information Principle to Self-Organizing Systems

## 5.1 Introduction

According to Chap.1, self-organizing systems are systems which can acquire macroscopic spatial, temporal, or spatio-temporal structures by means of internal processes without specific interference from the outside. Hitherto, the distribution functions of the order parameters governing the macroscopic structures could only be calculated by microscopic theories (cf. Chap.2). In the present section we derive them from macroscopic quantities, and we demonstrate this procedure explicitly by means of the single and multimode laser close to the lasing threshold.

The maximum information entropy principle allows one to make unbiased estimates on the probability distribution of microscopic states of systems of which otherwise only certain averages, corresponding to macroscopic observations, are known. As we have seen in the preceding section this principle provides one with a very elegant access to many of the basic relations and concepts of thermodynamics, i.e. it can be applied very nicely to systems in thermal equilibrium. On the other hand no successful attempts are known of a general application of this principle to systems far from thermal equilibrium.

In this section I wish to show how the maximum information entropy principle can indeed be very successfully applied to nonequilibrium systems provided they acquire macroscopic structures through self-organization. In this way we shall recover well-known distribution functions of such systems. These functions have been previously derived from microscopic theories (cf. Sects. 2.4,7). At the same time generalizations of these functions now become available, too. In order to illustrate our procedure we first focus our attention on lasers.

## 5.2 Application to Self-Organizing Systems: Single Mode Laser

Self-organizing systems are characterized by the occurrence of macroscopic structures which can be described by adequate order parameters. Instead of introducing abstract considerations we prefer to illustrate our procedure by means of explicit examples. It will be important to strictly stick to the basic notion of the maximum information entropy principle, namely to consider the macroscopically observed quantities.

The quantities observed experimentally of a single mode laser are its intensity and the second moment of the intensity in the steady state case. It is well known from laser theory that the basic difference between the light from lamps and that from lasers becomes apparent only when the second moments of the intensity are measured in addition to the first moment. Further measured quantities are intensity correlations, but because we have a time-independent theory in mind we shall ignore this information here.

The space- and time-dependent electric field strength of a single mode laser can be written in the form

$$E(x,t) = E(t) \sin kx \ , \tag{5.1}$$

where the amplitude $E(t)$ can be decomposed into its positive and negative frequency part according to

$$E(t) = \underbrace{B\,\mathrm{e}^{-\mathrm{i}\omega t}}_{b} + \underbrace{B^*\,\mathrm{e}^{\mathrm{i}\omega t}}_{b^*} \ . \tag{5.2}$$

If we measure the intensity of the light field over time intervals large compared to an oscillation period, but small compared to the fluctuation times of $B(t)$, the output intensity is proportional to $B^*B$ and to the loss rate, $2\kappa$, of the laser. For the sake of simplicity we drop all other constants and put

$$I = 2\kappa B^*B \ . \tag{5.3}$$

Similarly, the intensity squared, if averaged over the same time interval, turns out to be

$$I^2 = 4\kappa^2 B^{*2} B^2 \ . \tag{5.4}$$

Because of the fluctuations of the laser, $B^*$ and $B$ are random variables which belong to a stationary process. This leads us to identify $B$, $B^*$ with the indices $i$ of $p_i$ in (3.32), where we put $k_\mathrm{B} = 1$ and interpret the right-hand side as information, $i$. Because the random variables $B$ are no longer discrete but continuous we must replace the summation over the indices $i$ by an integration

$$i = -\int p(B,B^*) \ln p(B,B^*)\, d^2B \ . \tag{5.5}$$

Equation (3.34) may be interpreted as integrals over $d^2B$ with the probability $p(B,B^*)$ as weight functions. Denoting these averages by brackets we are led to consider the following two constraints

$$f_1 = \langle 2\kappa B^*B \rangle \ , \tag{5.6}$$

$$f_2 = \langle 4\kappa^2 B^{*2} B^2 \rangle \ . \tag{5.7}$$

Furthermore by the same analogy we are led to define $f_i^{(k)}$ by means of

$$f_{B,B^*}^{(1)} = 2\kappa B^*B \ , \tag{5.8}$$

$$f_{B,B^*}^{(2)} = 4\kappa^2 B^{*2} B^2 \ . \tag{5.9}$$

We are now in a position to apply formula (3.38) immediately and find

$$p(B, B^*) = \exp[-\lambda - \lambda_1 2\kappa B^* B - \lambda_2 4\kappa^2 (B^* B)^2] \ , \tag{5.10}$$

which in a somewhat different notation reads

$$p(B, B^*) = N \exp(-\alpha |B|^2 - \beta |B|^4) \ . \tag{5.11}$$

This function is well known is laser physics. It was derived by Risken by solving the Fokker-Planck equation belonging to the laser Langevin equation derived previously by the present author.

We note that in the laser case $\alpha$ must be negative. But close to threshold $\alpha$ can take both negative and positive values.

## 5.3 Multimode Laser Without Phase Relations

In this case the field strength is decomposed into its modes according to

$$E(x, t) = \sum_l E_l(t) \sin k_l x \ , \tag{5.12}$$

where for simplicity we consider only axial modes. Again the mode amplitudes can be decomposed into their positive and negative frequency parts

$$E_l(t) = B_l(t) \mathrm{e}^{-\mathrm{i}\omega_l t} + B_l^*(t)^{\mathrm{i}\omega_l t} \ . \tag{5.13}$$

The intensity averaged over time intervals long compared to an oscillation period and short to fluctuation periods is given by

$$I_l \sim 2\kappa_l B_l^* B_l \sim n_l \ . \tag{5.14}$$

The extension of the results of Sect. 5.2 is straightforward provided we now consider either $n_l$ or equivalently $B_l^*$ and $B_l$ as stochastic variables. We obtain

$$f_l = \langle n_l \rangle \ , \tag{5.15}$$

$$f_{l,l'} = \langle n_l, n_{l'} \rangle \ , \tag{5.16}$$

and identifying $k$ and $i$ according to

$$k \leftrightarrow \begin{cases} l \\ l, l' \end{cases}$$

$$i \leftrightarrow (n_1, n_2, \ldots, n_M) = \boldsymbol{n} \tag{5.17}$$

we have

$$f_{\boldsymbol{n}}^{(l)} = n_l \ , \tag{5.18}$$

$$f_n^{(l,l')} = n_l n_{l'} \ , \tag{5.19}$$

$$p_i \rightarrow p(\boldsymbol{n}) \ . \tag{5.20}$$

The application of (3.38) is now straightforward and yields

$$p(\boldsymbol{n}) \equiv p(n_1, \ldots, n_M) = \exp\left(-\lambda - \sum_l \lambda_l n_l - \sum_{ll'} \lambda_{l,l'} n_l n_{l'}\right) \tag{5.21}$$

as final result. Equivalently (5.21) can be written as

$$N \exp\left(-\sum_l \alpha_l n_l - \sum_{ll'} \beta_{ll'} n_l n_{l'}\right) \ . \tag{5.22}$$

This form can be derived from a multimode Fokker-Planck equation in special cases in which the solution can be explicitly constructed by means of the principle of detailed balance.

## 5.4 Processes Periodic in Order Parameters

We wish to show how guesses on distribution functions can be made if the processes considered are periodic in the order parameters.

Let us consider as a specific example the angle coordinate $\phi$ and let us consider moments which are periodic with $2\pi$. Then it would seem sensible, instead of the moments of $\phi$, to consider the corresponding moments of periodic functions i.e.

$$\langle \sin \phi \rangle \ , \qquad \langle \cos \phi \rangle \ , \ldots \tag{5.23}$$

or, more generally,

$$\langle \sin n\phi \rangle \ , \qquad \langle \cos n\phi \rangle \ , \tag{5.24}$$

where $n$ is a positive integer.

In order to illustrate our procedure let us consider the special case in which for symmetry reasons

$$\langle \sin n\phi \rangle = 0 \ , \qquad \text{for all } n \ . \tag{5.25}$$

In our treatment above we retained only the first few moments. If, in analogy to that, we keep only the first two terms of (5.24), we readily obtain

$$P(\phi) = \exp(\lambda + \lambda_1 \cos \phi + \lambda_2 \cos 2\phi) = \exp[V(\phi)] \ . \tag{5.26}$$

As we have shown explicitly, the maximum information entropy principle allows us to derive the general form of distribution functions of a nonequilibrium system, such as the laser, in a quite straightforward way. The results agree with distribution functions obtained from microscopic theories under certain restricting conditions.

It is now fairly obvious how one should proceed in other cases. The total state of the system $\boldsymbol{q}(\boldsymbol{x}, t)$ must be projected onto functions which describe the observed macroscopic spatial or temporal pattern. In this way amplitudes are obtained of which moments can be measured and thus may serve as the functions defining $f_k$ as well as $f_i^{(k)}$.

In spite of the success of our application of the maximum information entropy principle we must bear in mind that only little can be said about the Lagrange multipliers which can, of course, now be determined experimentally. On the other hand it has been the advantage of the microscopic theory that these constants can be determined from first principles and therefore in particular it could be predicted that $\alpha_l$ changes sign at instability points. However, it may also become possible to deduce such properties from a macroscopic theory.

Our approach can be applied to a number of problems such as convection in fluids, pattern formation in chemical reactions, and growth of morphogenetic fields.

In spite of the formal resemblance of our results to some of thermodynamics, there are still basic differences. First of all we realize that the constants have a physical meaning very different from those for systems in thermal equilibrium. For instance in nonequilibrium systems, such as lasers, we have to deal with output intensities whereas in equilibrium systems we deal with e.g. energies. This is also clearly reflected by a treatment on the microscopic level. While in the microscopic treatment of equilibrium systems energies play a decisive role, in nonequilibrium systems rate constants and growth rates determine the evolving patterns.

We may draw a number of rather far reaching conclusions. Until now, the maximum information (entropy) principle had been applied to thermodynamics and irreversible thermodynamics, but not to nonequilibrium phase transitions. Among the constraints used in the former two fields is the energy. In the present case we deal with the *output intensity* $I$. But what is still more important, we now have to include the intensity correlation in the form of the second moment, i.e. $\langle I^2 \rangle$. This is never done in equilibrium thermodynamics in the context of the maximum entropy principle. But now we see that the inclusion of $\langle I^2 \rangle$ is quite obvious and necessary for nonequilibrium phase transitions. At or above threshold, $\lambda_1 \geq 0$, and the integral over $\exp(\lambda_1 |b|^2)$ will diverge, reflecting the effect of critical fluctuations. Because of these, and in order to take care of their limitation due to saturation, $\langle I^2 \rangle$ must be taken into account. In other words, close to nonequilibrium phase transitions fluctuations become "observables" and must be taken into account by adequate constraints. It appears to me a safe bet that the same is true for phase transitions of systems in thermal equilibrium and that an extension of the maximum information entropy principle is also required there.

# 6. The Maximum Information Principle for Nonequilibrium Phase Transitions: Determination of Order Parameters, Enslaved Modes, and Emerging Patterns

## 6.1 Introduction

In the preceding chapter we formulated adequate constraints for the derivation of distribution functions of systems far from thermal equilibrium close to points of a nonequilibrium phase transition. Whenever a comparison was possible with previous results from a microscopic theory, perfect agreement was found. On the other hand our formulation was restricted to *order parameters*, i.e. we could derive the appropriate distribution functions for the order parameters only. It was assumed that the order parameters could be identified experimentally. In this section we wish to show that our treatment can be generalized in such a way that no prior knowledge of order parameters is required. By invoking adequate correlation functions instead, we shall be able to determine the order parameters as well as the enslaved modes and the emerging patterns by means of the formalism.

## 6.2 General Approach

We assume that the system to be studied is described by a state vector

$$\boldsymbol{q} = (q_1, \dots, q_N) , \tag{6.1}$$

whose components are accessible to measurements. The index $i$ of $q_i$ may stand for a cell, or for different kinds of physical or other quantities. We shall further assume that the statistical averages over the $q_i$ and their moments up to fourth order are known. We introduce the following quantities $f$ as constraints:

$$f_i = \langle q_i \rangle ; \qquad f_i^{(1)} = q_i , \tag{6.2}$$

$$f_{ij} = \langle q_i q_j \rangle ; \qquad f_{ij}^{(2)} = q_i q_j , \tag{6.3}$$

$$f_{ijk} = \langle q_i q_j q_k \rangle ; \qquad f_{ijk}^{(3)} = q_i q_j q_k , \tag{6.4}$$

$$f_{ijkl} = \langle q_i q_j q_k q_l \rangle ; \qquad f_{ijkl}^{(4)} = q_i q_j q_k q_l . \tag{6.5}$$

Maximizing the information entropy by use of Lagrange multipliers $\lambda$, which take care of the constraints (6.2)–(6.5), we obtain for the probability distribution

$$P = \exp[V(\boldsymbol{\lambda}, \boldsymbol{q})]. \tag{6.6}$$

In this $V$ is defined by

$$V(\boldsymbol{\lambda}, \boldsymbol{q}) = \lambda + \sum_i \lambda_i q_i + \sum_{ij} \lambda_{ij} q_i q_j + \sum_{ijk} \lambda_{ijk} q_i q_j q_k + \sum_{ijkl} \lambda_{ijkl} q_i q_j q_k q_l \ . \tag{6.7}$$

To make contact with nonequilibrium phase transitions we seek the extremum of $V$

$$\frac{\partial V}{\partial q_i} = 0 \qquad i = 1, \ldots, N \ . \tag{6.8}$$

In general we expect several extrema whose position we shall denote by $\boldsymbol{q}^0$. Having nonequilibrium phase transitions in mind we choose $\boldsymbol{q}^0$ so that $V(\boldsymbol{q}^0 + \boldsymbol{w})$ has the highest symmetry with respect to $\boldsymbol{w}$. In accordance with the spirit of the maximum information entropy principle this choice means that it is one without bias. Only lower symmetry would prefer a specific pattern and thus would introduce a bias. Another way of defining $\boldsymbol{q}^0$ appropriately as the position of the adequate extremum is given by following up $\boldsymbol{q}^0$ from the structureless state by changing a control parameter. Putting

$$\boldsymbol{q} = \boldsymbol{q}^0 + \boldsymbol{w} \tag{6.9}$$

we may rewrite $V$ in the form

$$V(\boldsymbol{\lambda}, \boldsymbol{q}) = \tilde{V}(\tilde{\boldsymbol{\lambda}}, \boldsymbol{w}) \qquad \text{with} \tag{6.10}$$

$$\tilde{V}(\tilde{\boldsymbol{\lambda}}, \boldsymbol{w}) = \tilde{\lambda} + O + \sum_{ij} \tilde{\lambda}_{ij} w_i w_j + \sum_{ijk} \tilde{\lambda}_{ijk} w_i w_j w_k + \sum_{ijkl} \tilde{\lambda}_{ijkl} w_i w_j w_k w_l \ . \tag{6.11}$$

Here, $\tilde{\lambda}_{ij}$ for instance, is given by

$$\tilde{\lambda}_{ij} = \frac{1}{2} \frac{\partial^2 V}{\partial q_i \partial q_j}\bigg|_{q^0} \ . \tag{6.12}$$

Simultaneously the old constraints (6.2)–(6.5), which can be written in the form

$$f_i = \left\langle \frac{\partial}{\partial \lambda_i} V \right\rangle \tag{6.13}$$

$$f_{ij} = \left\langle \frac{\partial}{\partial \lambda_{ij}} V \right\rangle \quad \text{etc.} \tag{6.14}$$

can be transformed into the new constraints

$$\tilde{f}_{ij} = \left\langle \frac{\partial \tilde{V}(\tilde{\boldsymbol{\lambda}}, \boldsymbol{w})}{\partial \tilde{\lambda}_{ij}} \right\rangle \quad \text{etc.} \tag{6.15}$$

## 6.3 Determination of Order Parameters, Enslaved Modes, and Emerging Patterns

Since the constraints (6.14) and (6.15) are symmetric in the indices $i$, $j$, so are the Lagrange multipliers

$$\tilde{\lambda}_{ij} = \tilde{\lambda}_{ji} \; . \tag{6.16}$$

Therefore we may diagonalize the matrix

$$\Delta = (\tilde{\lambda}_{ij}) \tag{6.17}$$

with real eigenvalues $\hat{\lambda}_\kappa$. This diagonalization is achieved by the transformation

$$w_i = \sum_\kappa a_{i\kappa}\xi_\kappa \qquad \text{where} \tag{6.18}$$

$$\tilde{V}(\tilde{\lambda}, w) = \hat{V}(\hat{\lambda}, \xi) \tag{6.19}$$

and the $a_{i\kappa}$ are orthogonal.

By means of (6.18) the expression (6.11) is transformed according to (6.19), where the r.h.s. reads more explicitly

$$\hat{V}(\hat{\lambda}, \xi) = \tilde{\lambda} + \sum_\kappa \hat{\lambda}_\kappa \xi_\kappa^2 + \sum_{\kappa\lambda\mu} \hat{\lambda}_{\kappa\lambda\mu}\xi_\kappa\xi_\lambda\xi_\mu + \sum_{\kappa\lambda\mu\nu} \hat{\lambda}_{\kappa\lambda\mu\nu}\xi_\kappa\xi_\lambda\xi_\mu\xi_\nu \; . \tag{6.20}$$

In general $\hat{V}$ represents the saddle point close to $\xi = 0$. Accordingly we shall distinguish between positive and negative $\lambda$'s and write

$$\begin{aligned} &\hat{\lambda}_\kappa \geq 0 \; ; \qquad \kappa \to u, \text{ total number } N_u \\ &\hat{\lambda}_\kappa < 0 \; ; \qquad \kappa \to s, \text{ total number } N_s \; . \end{aligned} \tag{6.21}$$

By a comparison with the results of the microscopic theory we may adopt the parlance of nonequilibrium phase transitions. We identify $\kappa$ belonging to $\hat{\lambda} \geq 0$ with the index $u$ (unstable) and denote $\xi_u$ accordingly as order parameters. Furthermore we identify $\kappa$ with the index $s$ for $\hat{\lambda} < 0$ and call $\xi_s$ the amplitude of the enslaved mode $s$. In correspondence to this decomposition we rewrite $\hat{V}$ in the form

$$\hat{V}(\hat{\lambda}, \xi) = \tilde{\lambda} + \hat{V}_u(\hat{\lambda}_u, \xi_u) + \hat{V}_s(\hat{\lambda}_u, \hat{\lambda}_s; \xi_s, \xi_u) \tag{6.22}$$

where the first part refers to the order parameters alone

$$\hat{V}_u = \sum_u \hat{\lambda}_u \xi_u^2 + \sum_{uu'u''} \hat{\lambda}_{uu'u''}\xi_u\xi_{u'}\xi_{u''} + \sum_{uu'u''u'''} \hat{\lambda}_{uu'u''u'''}\xi_u\xi_{u'}\xi_{u''}\xi_{u'''} \; . \tag{6.23}$$

More explicitly $\hat{V}_s$ reads

$$\hat{V}_s = \sum_s (-|\lambda_s|\xi_s^2) + \sum_{suu'} 3\hat{\lambda}_{suu'}\xi_s\xi_u\xi_{u'} + \sum_{suu'u''} 4\hat{\lambda}_{suu'u''}\xi_s\xi_u\xi_{u'}\xi_{u''} \tag{6.23a}$$

$$+ \text{ sums over products of } \xi_s\xi_{s'}\xi_u,\ \xi_s\xi_{s'}\xi_u\xi_{u'},\ \xi_s\xi_{s'}\xi_{s''},\ \xi_s\xi_{s'}\xi_{s''}\xi_u,\ \xi_s\xi_{s'}\xi_{s''}\xi_{s'''} \; . \tag{6.24}$$

The integral

$$\int e^{\hat{V}_s} d^{N_s}\xi_s = g(\xi_u) > 0 \tag{6.25}$$

defines a function of the order parameters $\xi_u$ alone. We introduce the function $h$ by means of

$$g(\xi_u) = e^{-h(\xi_u)} \tag{6.26}$$

and introduce a new function $W_s$ via

$$h(\xi_u) + \hat{V}_s = W_s(\xi_s|\xi_u) \ . \tag{6.27}$$

This definition guarantees that

$$P(\xi_s|\xi_u) = \exp[W_s(\xi_s|\xi_u)] \tag{6.28}$$

is normalized over the space of the enslaved modes for any $\xi_u$. In order that (6.22) remains unchanged by the introduction of $h$ we introduce the new function $W_u$ via

$$\tilde{\lambda} + \hat{V}_u(\hat{\lambda}_u, \xi_u) - h(\xi_u) = W_u(\xi_u) \ . \tag{6.29}$$

In conclusion we may thus rewrite (6.22) in the form

$$\hat{V}(\hat{\lambda}, \xi) = W_u(\xi_u) + W_s(\xi_s|\xi_u) \ . \tag{6.30}$$

This allows us to write

$$e^{\hat{V}} = P(\xi_u)P(\xi_s|\xi_u) \qquad \text{with} \tag{6.31}$$

$$P(\xi_u) = e^{W_u} \tag{6.32}$$

and $P(\xi_s|\xi_u)$ defined by (6.28).

Clearly $P(\xi_s|\xi_u)$ is a conditional probability, whereas $P(\xi_u)$ is the distribution function of the order parameters alone. So far our approach has been quite general. Our approach allows us to determine the distribution function for the order parameters as well as the conditional probability distribution of the enslaved modes. In particular (6.31), when written out explicitly, represents a special case of the slaving principle (cf. Sect. 2.6).

## 6.4 Approximations

In order to make contact with the results of the microscopic theory and the form of the slaving principle in its lowest approximation, we introduce the following approximations. We assume that $\xi_u$ and $\xi_s$ are small quantities so that the order of magnitude is $\xi_s \sim \xi_u^2$. In this spirit we shall neglect terms of order $\xi_u^5$, $\xi_u^6$ and higher in $\hat{V}_s$. Because of the normalization condition it readily follows that

$$\hat{V}_s \approx V_{s,\text{appr}} = -\sum_s |\lambda_s| [\xi_s - f_s(\xi_u)]^2 \quad \text{with} \tag{6.33}$$

$$f_s(\xi_u) = \frac{1}{2|\lambda_s|} \left( \sum_{uu'} 3\hat{\lambda}_{suu'} \xi_u \xi_{u'} + \sum_{uu'u''} 4\hat{\lambda}_{suu'u''} \xi_u \xi_{u'} \xi_{u''} \right) . \tag{6.34}$$

Note that $h$ occurring in (6.26) is given by

$$h(\xi_u) = \sum_s |\lambda_s| f_s^2(\xi_u) . \tag{6.35}$$

We quote another approach to approximate $\hat{V}_s$ which is still more in the spirit of the maximum information entropy principle. To this end we approximate $P(\xi_s|\xi_u)$ by

$$\bar{P} = \exp\left[ \bar{\lambda}(\xi_u) + \sum_s \bar{\lambda}_s(\xi_u)\xi_s + \sum_s \bar{\lambda}_{ss}(\xi_u)\xi_s^2 \right] \tag{6.36}$$

so that the constraints

$$\int \bar{P} d^{N_s}\xi_s = 1 , \tag{6.37}$$

$$\int \bar{P} \xi_{s'} d^{N_s}\xi_s = \int P(\xi_s|\xi_u)\xi_{s'} d^{N_s}\xi_s , \tag{6.38}$$

$$\int \bar{P} \xi_{s'}^2 d^{N_s}\xi_s = \int P(\xi_s|\xi_u)\xi_{s'}^2 d^{N_s}\xi_s \tag{6.39}$$

are fulfilled, i.e. we are now satisfied with the fulfillment of the constraints containing the first two moments instead of all four moments. In this way we obtain a quadratic expression in the exponential of (6.36) as it occurs in the lowest approximation of the slaving principle.

## 6.5 Spatial Patterns

The above approach allows us to determine the order parameters, the enslaved modes, and their distribution functions. When we identify the index $i$ with a lattice point, the coefficients $a_{i\kappa}$ in (6.18) determine the spatial mode pattern belonging to the order parameter with index $\kappa$. Thus the superposition of $a_{i\kappa}$, with $\kappa$ ranging over the indices of the order parameters, determines the mode skeleton (cf. Sect. 2.8). The total sum (6.18) determines the resulting patterns for a given set of order parameters and their corresponding enslaved modes where we assume that the symmetry of $\tilde{V}$ is broken by a specific set of order parameters belonging to a minimum of $\tilde{V}$. From the formulation it follows that our procedure is not only good for the determination of evolving patterns governed by order parameters in physical systems, but it also serves as a tool for the recognition of the dominant structures in general patterns and it is hoped that our procedure will find useful applications in pattern recognition by machines and in theories of pattern recognition by man and animals. (See also Chap. 12.)

## 6.6 Relation to the Landau Theory of Phase Transitions. Guessing of Fokker-Planck Equations

Our procedure casts new light on the Landau theory of phase transitions of systems in thermal equilibrium, where the starting point is the thermal distribution function in which in the exponent the free energy occurs. This free energy is then expanded into a power series of the order parameters. Quite clearly we can arrive at precisely the same expressions by invoking moments of the order parameters up to fourth order. Thus the maximum information principle gives us a new approach to the Landau theory of phase transitions with an entirely new interpretation.

An interesting question is whether we can determine the Fokker-Planck equation which has as its stationary solution just the distribution function that we have determined by the maximum information principle. In order to elucidate this problem, let us consider the case of a single variable $\xi_u \equiv \xi$. The general form of the Fokker-Planck equation with variable-dependent diffusion "constant" is chosen as

$$\frac{dP}{dt} = -\frac{d}{d\xi}[K(\xi)P] + \frac{d}{d\xi}D(\xi)\frac{d}{d\xi}P \ . \tag{6.40}$$

By comparing its stationary solution with the distribution function we have determined above, (6.32), we readily obtain

$$W_u(\xi_u) = -\int^{\xi_u} \frac{K(\xi)}{D(\xi)} d\xi \ . \tag{6.41}$$

In other words the still unknown drift coefficient $K$ and the diffusion coefficient $D$ have to obey the relation

$$-\frac{K}{D} = W_u' \tag{6.42}$$

which can be fulfilled in many ways.

The maximum information entropy principle is of no help in arriving at an appropriate choice of $K$ and $D$. All guesses obeying (6.42) are actually equally probable. Therefore we must look for another criterion and the one we choose is simplicity. To this end we put

$$D = Q = \text{const.} \tag{6.43}$$

The Fokker-Planck equation then reduces to

$$\frac{dP}{dt} = -\frac{d}{d\xi}[K(\xi)P] + Q\frac{d^2P}{d\xi^2} \tag{6.44}$$

where $Q$ still plays the role of a fixed, but unknown, parameter. The Langevin equation belonging to (6.44) reads

$$\dot{\xi} = K(\xi) + F(t) \tag{6.45}$$

where the drift coefficient is given by

$$K \propto -\partial W_u/\partial \xi \tag{6.46}$$

[on account of (6.42)], and $F$ is a Gaussian white-noise force. Inspecting (6.45,46) and the form of $W_u$ (6.41) with $D = Q =$ const. more closely, we discover that we have just found the Landau equation belonging to the Landau functional of the Landau theory of second-order phase transitions. This allows us to interpret the Landau theory anew. It represents a specific guess at the form of the Langevin equation close to a phase transition point. Quite clearly, the approach which we exemplified for a single variable is not quite satisfactory, especially when we proceed to multivariable Fokker-Planck equations because here, the variety of possibilities of the proper choice of drift and diffusion coefficients, is quite large.

Quite evidently new criteria or constraints are needed in order to fix the drift and diffusion coefficients. This can in fact be achieved if we take into account time-dependent correlation functions as will be shown in Chap. 9.

# 7. Information, Information Gain, and Efficiency of Self-Organizing Systems Close to Their Instability Points

## 7.1 Introduction

In this section we wish to further elaborate on the results of the preceding chapter. There we have shown that we may split the joint probability distribution function $P(\xi_u, \xi_s)$ which refers to the order parameters, $\xi_u$, and enslaved-mode amplitudes, $\xi_s$, into a product of the form

$$P(\xi_u, \xi_s) = \prod_s P_s(\xi_s|\xi_u) f(\xi_u) . \tag{7.1}$$

This decomposition is in accordance with the slaving principle of the microscopic theory. We now want to show that the form (7.1) allows us to decompose information and information gain into a part which refers to the order parameters alone and a second part which is a sum over the information of the enslaved modes averaged over the distribution of the order parameters. As we shall see, close to instability points the information of the order parameters changes dramatically whereas the information of the enslaved modes does not. Therefore close to these points it is sufficient to study the behavior of the order parameter information and information gain which is done explicitly here for a large class of systems undergoing nonequilibrium phaśe transitions. It will be shown how information and information gain as well as efficiency (in the sense defined in this section) can be measured directly.

The information is defined by

$$i = -\sum_j p_j \ln p_j , \tag{7.2}$$

and the information gain by

$$K = \sum_j p_j \ln \frac{p_j}{p_j'} . \tag{7.3}$$

We shall interpret $p_j$ as probability distribution of states characterized by the index $j$. In (7.3) $p_j$ and $p_j'$ are two different probability distributions. In order to make closer contact with the results obtained above, we shall identify the index $j$ with the values which the stochastic variables $\xi_u$, $\xi_s$ can take. As usual, we shall denote these values also by $\xi_u$, $\xi_s$. Note that via (2.78) the state vector $\boldsymbol{q}$ is determined once $\xi_u$ and $\xi_s$ are known.

## 7.2 The Slaving Principle and Its Application to Information

We wish to show that the information (7.2) can be cast into a specific form by means of (7.1). To this end we insert (7.1) into (7.2). Replacing the logarithm of a product by a sum of logarithms we obtain

$$i = -\sum_{\xi_u, \xi_s} \prod_s P_s(\xi_s|\xi_u) f(\xi_u) [\ln f(\xi_u) + \sum_s \ln P_s(\xi_s|\xi_u)] \ . \tag{7.4}$$

By use of the normalization condition

$$\sum_{\xi_s} P_s(\xi_s|\xi_u) = 1 \tag{7.5}$$

we may cast (7.4) into the form

$$i = -\sum_{\xi_u} f(\xi_u) \ln f(\xi_u) - \sum_{\xi_u, \xi_s, s} f(\xi_u) P_s(\xi_s|\xi_u) \ln P_s(\xi_s|\xi_u) \ . \tag{7.6}$$

This relation can be written as

$$i = I_f + \sum_{s, \xi_u} f(\xi_u) I_s(\xi_u) \ , \qquad \text{where} \tag{7.7}$$

$$I_f = -\sum_{\xi_u} f(\xi_u) \ln f(\xi_u) \tag{7.8}$$

is the information of the order parameters, whereas

$$I_s = -\sum_{\xi_s} P_s(\xi_s|\xi_u) \ln P_s(\xi_s|\xi_u) \tag{7.9}$$

is the information of the enslaved subsystem or enslaved mode with index $s$. As we shall show below, close to instability points $I_f$ changes dramatically, whereas $I_s$ changes only weakly.

The information $I_s$ is clearly an information under the hypothesis that $\xi_u$ has acquired a specific value. In our context it means that the order parameter enslaves the modes in a specific fashion which in turn guarantees a macroscopic structure via self-organization.

## 7.3 Information Gain

We distinguish the distributions corresponding to $p$ and $p'$ by the indices $n$ and $a$. According to the definition (7.3), the information gain is then given by

$$K = \sum_{\xi_u, \xi_s} P_n(\xi_u, \xi_s) \ln \frac{P_n(\xi_u, \xi_s)}{P_a(\xi_u, \xi_s)} \ . \tag{7.10}$$

Making use of (7.1), we can cast (7.10) into the form

$$K = \sum_{\xi_u, \xi_s} f_n \prod_s P_{s,n} \left[ \ln f_n + \sum_{s'} \ln P_{s',n} - \ln f_a - \sum_{s'} \ln P_{s',a} \right] , \tag{7.11}$$

where again the logarithm of a product has been decomposed into a sum of logarithms. Making use of (7.5) we find after a slight rearrangement of terms

$$K = \sum_{\xi_u} f_n \ln \frac{f_n}{f_a} + \sum_s \sum_{\xi_u} f_n \sum_{\xi_s} P_{s,n} \ln \frac{P_{s,n}}{P_{s,a}} , \tag{7.12}$$

which can be cast into the final form

$$K = K_f + \sum_s \sum_{\xi_u} f_n K_s , \tag{7.13}$$

where $K_f$ is defined by

$$K_f = \sum_{\xi_u} f_n \ln \frac{f_n}{f_a} , \tag{7.14}$$

and $K_s$ is defined by

$$K_s = \sum_{\xi_s} P_{s,n}(\xi_s|\xi_u) \ln \frac{P_{s,n}(\xi_s|\xi_u)}{P_{s,a}(\xi_s|\xi_u)} . \tag{7.15}$$

## 7.4 An Example: Nonequilibrium Phase Transitions

In this section we wish to illustrate the usefulness of the formulas derived above by applying them to nonequilibrium phase transitions, (cf. Sect. 2.7). Here we shall assume that $\lambda_u$ is real. To be more explicit we write down typical evolution equations for systems undergoing nonequilibrium phase transitions. In terms of order parameters $\xi_u$ and enslaved-mode amplitudes $\xi_s$, these equations can be written in the form

$$\dot{\xi}_u = \lambda_u \xi_u + h_u(\xi_u, \xi_s) + F_u(t) , \tag{7.16}$$

$$\dot{\xi}_s = \lambda_s \xi_s + g_s(\xi_u) + \xi_s k_s(\xi_u) + \cdots + F_s(t) . \tag{7.17}$$

If the system is controlled externally by a control parameter, all the quantities on the right hand sides of (7.16,17) depend on that control parameter, but in different ways.

When we normalize the control parameter such that the instability occurs at $\alpha = 0$, then $\lambda_u$, and $\lambda_s$ depend on $\alpha$ in the following manner.

$$\lambda_u \propto \alpha^\kappa , \qquad \lambda_s = \lambda_s(0) + 0(\alpha) \approx \lambda_s(0) , \tag{7.18}$$

where $\kappa$ is some positive number. Clearly, $\lambda_u$ depends very sensitively on $\alpha$, whereas $\lambda_s$ depends only weakly on it because the leading term is a nonvanishing constant. Similarly the functions $h_u$, $g_s$, $k_s$, $F_u$ and $F_s$ depend only weakly on $\alpha$. In order to make our example as explicit as possible we apply the slaving principle in its leading approximation, which gives rise to terms that are obtained by the adiabatic elimination technique by using

$$\dot{\xi}_s \approx 0 \, . \tag{7.19}$$

This allows us to solve (7.17) to leading order giving

$$\xi_s \approx -\frac{1}{\lambda_s(0)} g_s(\xi_u) - \frac{1}{\lambda_s(0)} F_s(t) \, . \tag{7.20}$$

In Sect. 2.6, Eq. (2.91), we derived the corresponding conditional probability distribution, $P_s(\xi_s|\xi_u)$. It reads

$$P_s(\xi_s|\xi_u) = N \exp\left\{ -\left[ \xi_s + \frac{g_s(\xi_u)}{\lambda_s} \right]^2 Q^{-1} \right\} \, . \tag{7.21}$$

We may now insert this explicit result into $I_s$ (7.9). However, by introducing the new variable

$$\xi_s' = \xi_s - \frac{g_s(\xi_u)}{|\lambda_s|} \tag{7.22}$$

we can eliminate the dependence of the probability distribution of the enslaved variables on $\xi_u$ so that $I_s$ becomes independent of $\xi_u$. Therefore in (7.7) we may perform the integration over $\xi_u$ in the second term. We thus obtain

$$I = I_f + \sum_s I_s \tag{7.23}$$

where the second part does not depend on $\alpha$, at least in the present approximation. Therefore the information change close to the instability point is governed by that of the order parameters alone

$$I(\alpha_1) - I(\alpha_2) \approx I_f(\alpha_1) - I_f(\alpha_2) \, . \tag{7.24}$$

Using the same kind of approximation we may cast the information gain into the form

$$K = K_f = \int d^n \xi_u f(\xi_u, \alpha_2) \ln \frac{f(\xi_u, \alpha_2)}{f(\xi_u, \alpha_1)} \, . \tag{7.25}$$

In the following we shall study $I_f$ in detail. In Sect. 7.11 we shall then investigate the dependence of $I_s$ on $\alpha$. As we shall see, $I_s$ changes monotonically with $\lambda$, whereas $I_f$ shows some kind of "singular" behavior.

## 7.5 Soft Single-Mode Instabilities

As we have seen in the last section, the change of information and information gain of the total system close to instability points is practically identical to the change of information and information gain, respectively, of the order parameters. In this section we wish to calculate $i$ and $K$ using the explicit form of the order parameter distribution function where we take the example of a single order parameter undergoing a second order nonequilibrium phase transition. As we have seen in Sect. 2.7, the distribution function $f$ has the form

$$f(\xi) = N \exp(\alpha\xi^2 - \beta\xi^4) \ . \tag{7.26}$$

Inserting (7.26) into the definition of the information (7.8) we readily obtain

$$I_f = -\ln N - \alpha\langle\xi^2\rangle + \beta\langle\xi^4\rangle \ , \tag{7.27}$$

where we have used the abbreviation

$$\langle\xi^n\rangle = \int_{-\infty}^{+\infty} f(\xi)\xi^n \, d\xi \ . \tag{7.28}$$

Similarly the information gain (7.14) acquires the explicit form

$$K(f_\alpha, f_{\alpha_0}) = \int f_\alpha \ln\frac{f_\alpha}{f_{\alpha_0}} d\xi = \ln N(\alpha) - \ln N(\alpha_0) + (\alpha - \alpha_0)\langle\xi^2\rangle_\alpha \ . \tag{7.29}$$

## 7.6 Can We Measure the Information and the Information Gain?

### 7.6.1 Efficiency

The explicit study of the behavior of self-organizing systems, e.g. of the laser, reveals that a measure for the macroscopic action of such a system is provided by the square of the order parameter.

For instance in the laser the field mode acts as an order parameter and laser action can be measured by the field mode amplitude squared. This suggests that we introduce the quantity

$$\Omega(\alpha) = \langle\xi_u^2\rangle \tag{7.30}$$

as a measure for the macroscopic action, or perhaps more precisely, the work of the system. The average is defined by

$$\langle\xi^2\rangle = \int \xi^2 f(\xi) \, d\xi \ , \tag{7.31}$$

where we shall use the explicit distribution function

$$f(\xi) = N \exp(\alpha\xi^2 - \beta\xi^4) \tag{7.32}$$

with the normalization factor defined by

$$N^{-1} = \int_{-\infty}^{+\infty} \exp(\alpha\xi^2 - \beta\xi^4) \, d\xi \ . \tag{7.33}$$

We define the efficiency, $W$, by the rate of change of $\Omega$ when the control parameter $\alpha$ is changed because in a number of self-organizing systems $\alpha$ is connected with the power input. Note that for the time being we shall interpret $\alpha$ directly as power input.

$$W = \frac{d\Omega}{d\alpha} . \tag{7.34}$$

Using (7.31–33) we readily find

$$W = \frac{\int \xi^4 \exp(\alpha\xi^2 - \beta\xi^4)\, d\xi}{\int \exp(\alpha\xi^2 - \beta\xi^4)\, d\xi} - \frac{[\int \exp(\alpha\xi^2 - \beta\xi^4)\xi^2\, d\xi]^2}{[\int \exp(\alpha\xi^2 - \beta\xi^4)\, d\xi]^2} \tag{7.35}$$

which can be written

$$W = \langle \xi_u^4 \rangle - \langle \xi_u^2 \rangle^2 \tag{7.36}$$

(where we write $\xi_u$ instead of $\xi$). If the control parameter $\alpha$ enters the distribution function (7.32) in an implicit fashion,

$$\exp[\lambda(\alpha)\xi_u^2 - \beta\xi_u^4] , \tag{7.37}$$

we have to replace (7.36) by

$$W = \frac{d\lambda}{d\alpha}(\langle \xi_u^4 \rangle - \langle \xi_u^2 \rangle^2) . \tag{7.38}$$

It is not difficult the evaluate $W$ well below and well above the instability point and we shall present the corresponding results below.

### 7.6.2 Information and Information Gain

Let us now turn to the evaluation of the information $I_f$. Using the explicit form (7.32) with (7.33) we readily obtain

$$-I_f = \ln N + \alpha\langle \xi^2 \rangle - \beta\langle \xi^4 \rangle . \tag{7.39}$$

While the second and fourth moments can be directly measured, the logarithm of $N$ is still to be connected with measurable quantities. In a first step towards this we differentiate $\ln N$ with respect to $\alpha$ and obtain

$$\frac{d \ln N}{d\alpha} = - \frac{\int \exp(\alpha\xi^2 - \beta\xi^4)\xi^2\, d\xi}{\int \exp(\alpha\xi^2 - \beta\xi^4)\, d\xi} \tag{7.40}$$

which can clearly be cast into the form

$$\frac{d \ln N}{d\alpha} = - \langle \xi^2 \rangle . \tag{7.41}$$

Integrating (7.41) we obtain

$$\ln N = -\int_{\alpha_0}^{\alpha} \langle \xi^2 \rangle \, d\alpha + \ln N(\alpha_0) \; . \tag{7.42}$$

Inserting this result into (7.39) we obtain

$$-I_f = -\int_{\alpha_0}^{\alpha} \langle \xi^2 \rangle \, d\alpha + \ln N(\alpha_0) + \alpha \langle \xi^2 \rangle - \beta \langle \xi^4 \rangle \; . \tag{7.43}$$

In order to eliminate $\ln N$, we calculate the corresponding information for $\alpha_0$ and obtain our final result

$$-I_f(\alpha) + I_f(\alpha_0) = -\int_{\alpha_0}^{\alpha} \langle \xi^2 \rangle \, d\alpha + \alpha \langle \xi^2 \rangle_\alpha - \beta \langle \xi^4 \rangle_\alpha - \alpha_0 \langle \xi^2 \rangle_{\alpha_0} + \beta \langle \xi^4 \rangle_{\alpha_0} \tag{7.44}$$

or in differential form

$$-\frac{dI_f}{d\alpha} = -\alpha \langle \xi^2 \rangle^2 + \alpha \langle \xi^4 \rangle + \beta \langle \xi^2 \rangle \langle \xi^4 \rangle - \beta \langle \xi^6 \rangle \; . \tag{7.45}$$

Similarly the information gain acquires the form

$$K = -\int_{\alpha_0}^{\alpha} \langle \xi^2 \rangle_{\alpha'} \, d\alpha' + (\alpha - \alpha_0) \langle \xi^2 \rangle_\alpha \; . \tag{7.46}$$

## 7.7 Several Order Parameters

We now wish to generalize the results of Sect. 7.6 to the case of several order parameters. We adopt the following form for the distribution function $f$

$$f(\xi_u) = N \exp\left[ \sum_k \Lambda_k(\alpha) \xi_k^2 + h(\xi_u) \right] , \tag{7.47}$$

which can be explicitly derived in a number of cases from microscopic models (cf. Sect. 2.4) or according to Chap. 6 by the maximum information principle.

Here, we shall present the results for the information gain which turns out to be insensitive to the specific form of the nonlinear functions $h$ provided they do not depend on the control parameters. In reality they will depend on $\alpha$ but in general only in a weak fashion. Inserting (7.47) into the definition of $K_f$ we readily obtain

$$K_f = \int f \ln \frac{f}{f_0} d^n\xi = \int N \exp\left[ \sum_k \Lambda_k(\alpha) \xi_k^2 + h(\xi_u) \right] \times \left\{ \ln N - \ln N_0 + \sum_k [\Lambda_k(\alpha) - \Lambda_k(\alpha_0)] \xi_k^2 \right\} d^n\xi \tag{7.48}$$

which can be written in the form

$$K = \ln N - \ln N_0 + \sum_k [\Lambda_k(\alpha) - \Lambda_k(\alpha_0)] \langle \xi_k^2 \rangle_\alpha . \tag{7.49}$$

It remains to calculate $\ln N$. To this end we differentiate $\ln N$ and obtain

$$\frac{d \ln N}{d\alpha} = -\frac{\sum_k (\partial \Lambda_k / \partial \alpha) \int \xi_k^2 \, d\xi^n \exp(\dots)}{\int d^n \xi \exp(\dots)} \tag{7.50}$$

which can be reexpressed as

$$\frac{d \ln N}{d\alpha} = -\sum_k \frac{\partial \Lambda_k}{\partial \alpha} \langle \xi_k^2 \rangle . \tag{7.51}$$

Integrating (7.51) with respect to $\alpha$ and inserting the result into (7.49) provides us with the final result

$$K = -\int_{\alpha_0}^{\alpha} \sum_k \frac{\partial \Lambda_k}{\partial \alpha} \langle \xi_k^2 \rangle_\alpha \, d\alpha + \sum_k [\Lambda_k(\alpha) - \Lambda_k(\alpha_0)] \langle \xi_k^2 \rangle_\alpha . \tag{7.52}$$

In conclusion we wish to calculate the efficiency with respect to the order parameter $\xi_l$. Its output is defined by

$$\Omega(\alpha) = \langle \xi_l^2 \rangle . \tag{7.53}$$

A little calculation shows that the efficiency is given by

$$W = \frac{d\Omega}{d\alpha} = \sum_k \frac{\partial \Lambda_k}{\partial \alpha} (\langle \xi_l^2 \xi_k^2 \rangle - \langle \xi_l^2 \rangle \langle \xi_k^2 \rangle) . \tag{7.54}$$

In this section we have shown that the information change, the information gain and the efficiency, are quantities which can be expressed in terms of the order parameters. This sheds new light on the behavior of self-organizing systems which, with respect to the just-mentioned quantities, also behave as if they were governed by only a few degrees of freedom. In particular it was shown that information, information gain, and efficiency are measurable quantities; it can be expected that these quantities can be measured in a similar way as the entropy of a system in thermal equilibrium is a measurable quantity. However, the internal mechanism by which the order in nonequilibrium systems is produced, is quite different from the mechanism by which order in systems in thermal equilibrium is established. Similarly the physical meaning of thermodynamic entropy and information is different in the two cases.

## 7.8 Explicit Calculation of the Information of a Single Order Parameter

We now explicitly calculate the information of a single order parameter close to a nonequilibrium phase transition, i.e. (7.27). With reference to this result we then discuss why the relevant quantities are interpreted as information rather than as entropy.

We recall that the order parameter is continuous. In such a case we must use the definition

$$I_f = -\int d\xi\, f(\xi) \ln f(\xi) - \ln \varepsilon \,, \tag{7.55}$$

where $\varepsilon$ is an interval (in dimensionless units) which represents the accuracy with which measurements are done.

Using (7.26), we wish to evaluate (7.27)

$$I_f = -\ln N - \alpha\langle\xi^2\rangle + \beta\langle\xi^4\rangle - \ln \varepsilon \,. \tag{7.56}$$

In the regions well below and well above threshold (7.56) can be evaluated explicitly in a simple fashion.

### 7.8.1 The Region Well Below Threshold

In this region $\alpha$ is negative and its absolute value fairly large so that in (7.26) only the term quadratic in $\xi$ is important. So we have to evaluate (7.56) with respect to the Gaussian

$$f \approx N \exp(-|\alpha|\xi^2) \,. \tag{7.57}$$

The normalization factor can easily be determined and is given by

$$N = \sqrt{\frac{|\alpha|}{\pi}} \,. \tag{7.58}$$

The integrals which occur in (7.56) can be easily evaluated and we obtain in particular

$$|\alpha|\langle\xi^2\rangle = \tfrac{1}{2} \quad \text{and} \tag{7.59}$$

$$\beta\langle\xi^4\rangle = \frac{3\beta}{(2|\alpha|)^2} \,, \tag{7.60}$$

which for large enough $|\alpha|$ can be neglected against (7.59). Therefore we are left with the final result

$$I_f = -\tfrac{1}{2}\ln|\alpha| + \tfrac{1}{2} + \tfrac{1}{2}\ln \pi - \ln \varepsilon \,. \tag{7.61}$$

Let us now discuss the behavior of $\ln|\alpha|$ and $\ln \varepsilon$. Since $\varepsilon$ is a measure of the degree of accuracy, we may assume that it is a small quantity so that

$$\varepsilon \to 0 \quad \text{and} \quad -\ln \varepsilon \to +\infty \,. \tag{7.62}$$

On the other hand, because the system is well below threshold, we may assume

$$|\alpha| \to \infty \,, \qquad -\tfrac{1}{2}\ln|\alpha| \to -\infty \,. \tag{7.63}$$

When we keep $\varepsilon$ fixed and let $|\alpha|$ grow, $I_f$ will become 0 and eventually negative which is forbidden for mathematical reasons because the information is a positive quantity. But what is still more important, there is a physical reason why $|\alpha|$ cannot grow beyond a certain value, namely the distribution function (7.26) is then essentially only different from zero within the region of the accuracy of measurement, $\varepsilon$. This means that it has become meaningless to distinguish between different states of the system within this interval or, in other words, it is almost certain that the system is just in a specific state within that interval. At this moment there is still some ambiguity as to how we wish to fix $\alpha$, i.e. how to define what "almost certain" means. For instance we may require

$$I_f = 0 \tag{7.64}$$

from which a condition for $\alpha$ follows

$$-\tfrac{1}{2}\ln|\alpha| + \tfrac{1}{2} + \tfrac{1}{2}\ln\pi - \ln\varepsilon = 0 \ . \tag{7.65}$$

Since we shall be concerned only with changes of the information we may choose the zero of the information arbitrarily so that we may equally well require

$$I_f = \tfrac{1}{2} + \tfrac{1}{2}\ln\pi \tag{7.66}$$

which is chosen in such a way that the relation between $\alpha$ and $\varepsilon$ becomes particularly simple, namely

$$-\tfrac{1}{2}\ln|\alpha| - \ln\varepsilon = 0 \ . \tag{7.67}$$

From it we can readily deduce

$$|\alpha|^{1/2} = \frac{1}{\varepsilon} \ . \tag{7.68}$$

### 7.8.2 The Region Well Above Threshold

A little study of the behavior of the distribution function (7.26) reveals that it can be approximated by Gaussian functions provided $\alpha$ is big enough. To implement this we apply the method of steepest descent and write (7.26) in the form

$$f(\xi) = N\exp[g(\xi)] \ , \tag{7.69}$$

where, of course,

$$g(\xi) = \alpha\xi^2 - \beta\xi^4 \ . \tag{7.70}$$

The extremum of (7.70) and therefore of $f$ is determined by

$$g'(\xi) = 0 = 2\alpha\xi - 4\beta\xi^3 \tag{7.71}$$

and this possesses the solutions

$$\xi_0 = \pm\sqrt{\frac{\alpha}{2\beta}} , \tag{7.72}$$

where for now we will choose the plus sign. We introduce a new variable $\eta$ defined by

$$\xi = \xi_0 + \eta . \tag{7.73}$$

By means of the expansion

$$g(\xi) = g(\xi_0 + \eta) = g(\xi_0) + g'(\xi_0)\eta + \tfrac{1}{2}g''(\xi_0)\eta^2 \tag{7.74}$$

we may cast (7.69) into the form

$$f = N\exp\left(\frac{\alpha^2}{4\beta^2} - 2\alpha\eta^2\right) , \tag{7.75}$$

where the normalization is now to be taken in the interval $-\infty < \xi < +\infty$. The normalization factor is then given by

$$N = \frac{2\alpha}{\pi}\exp\left(-\frac{\alpha^2}{4\beta}\right) . \tag{7.76}$$

So far we have been dealing with only one maximum of $f$. Let us first evaluate the information for the case in which only one maximum can be realized, i.e. in which we break the symmetry artificially. In this case $I_f$ can be easily evaluated and is given by

$$I_f = \tfrac{1}{2}\ln\alpha - \tfrac{1}{2}\ln 2 + \tfrac{1}{2}\ln\pi + \tfrac{1}{2} - \ln\varepsilon . \tag{7.77}$$

We now turn to the main case of interest which takes into account the total distribution function which shows two maxima, i.e., roughly speaking, the system can be in one of two states (cf. Fig.7.1). Evidently we can store information in the system because we may identify one state with 0 and the other one with 1.

The normalization factor $N$ is determined by

$$\int_{-\infty}^{+\infty} N\exp(\alpha\xi^2 - \beta\xi^4)\,d\xi = 1 . \tag{7.78}$$

Because the maxima lie symmetrically with respect to the origin $\xi = 0$ we may replace the left-hand side of (7.78) by

$$2N\int_0^{\infty}\exp(\alpha\xi^2 - \beta\xi^4)\,d\xi . \tag{7.79}$$

Under the condition that we evaluate the integral only for one maximum we can replace the integral from 0 to $\infty$ by the integral from $-\infty$ to $+\infty$

$$\int_0^{\infty} \approx \int_{-\infty}^{+\infty}\exp(\alpha\xi^2 - \beta\xi^4)\,d\xi . \tag{7.80}$$

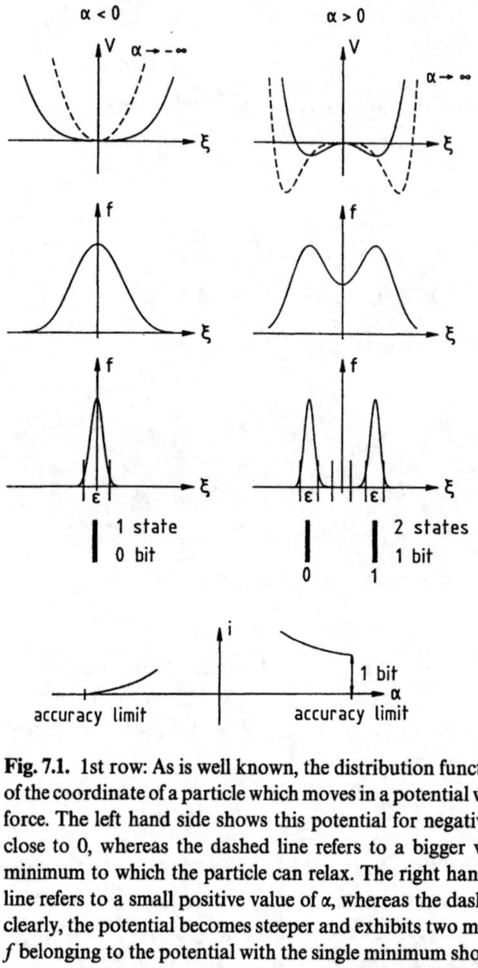


**Fig. 7.1.** 1st row: As is well known, the distribution function (7.69) with (7.70) can be interpreted as that of the coordinate of a particle which moves in a potential well and is subjected to an additional fluctuating force. The left hand side shows this potential for negative $\alpha$, where the solid line refers to a value of $\alpha$ close to 0, whereas the dashed line refers to a bigger value of $|\alpha|$. Quite evidently there is only one minimum to which the particle can relax. The right hand side shows the potential for $\alpha > 0$. The solid line refers to a small positive value of $\alpha$, whereas the dashed line refers to an increased value of $\alpha$. Quite clearly, the potential becomes steeper and exhibits two minima. 2nd row, l.h.s.: The distribution function $f$ belonging to the potential with the single minimum shows one maximum, r.h.s.: For $\alpha > 0$ two distinct maxima of $f$ have developed. 3rd row, l.h.s.: With increased value of $|\alpha|$ the distribution function becomes narrower and has one minimum. r.h.s.: The two-peaked distribution function has become narrower and each of its distributions falls into the limit of accuracy $\varepsilon$. Note that the scale of $f$ is different from the row above. 4th row: Sketch of the information versus $\alpha$. The behavior of the information is shown close to the accuracy limits

According to the above calculation this integral can be approximated by

$$\exp\left(\frac{\alpha^2}{4\beta}\right)\int_{-\infty}^{+\infty}\exp(-2\alpha\eta^2)\,d\eta \tag{7.81}$$

where the integral in (7.81) can be evaluated to give

$$\sqrt{\frac{\pi}{2\alpha}}\;. \tag{7.82}$$

Using (7.79–82) we may calculate the normalization factor $N$ by means of (7.78) and obtain

$$N = \exp\left(-\frac{\alpha^2}{4\beta}\right)\sqrt{\frac{\alpha}{2\pi}} \ . \tag{7.83}$$

In order to evaluate $I_f$ we repeat these steps in an analogous fashion and obtain the final result

$$I_f = -\tfrac{1}{2}\ln 2\alpha + \tfrac{1}{2}\ln\pi + \ln 2 + \tfrac{1}{2} - \ln\varepsilon \ . \tag{7.84}$$

Now let us compare (7.84), which holds well above threshold, with (7.61) which holds below threshold.

Let us first discuss the role of the accuracy limit $\varepsilon$. In Sect. 7.8.1 we put $\varepsilon$ in relation with $\alpha$, which obviously describes how quickly the Gaussian decreases. As is evident, e.g. from (7.81), the decay constant in the case well above threshold now reads $2\alpha$ instead of $|\alpha|$ below threshold. By the same physical argument as before we therefore have to require

$$-\tfrac{1}{2}\ln 2\alpha - \ln\varepsilon = 0 \ . \tag{7.85}$$

Taking now the difference between (7.84) and (7.61), using (7.85) and (7.67) we find

$$\Delta I_f = \ln 2 \ . \tag{7.86}$$

If we recall that for a proper definition of the information we should have used the logarithm to the base 2 instead of the natural logarithm, we find

$$\Delta I_f = \log_2 2 \equiv 1 \ . \tag{7.87}$$

This means that the system well above threshold can store one bit reliably whereas below threshold no bit can be stored. We shall come back to a discussion of this interpretation in Sect. 7.8.4 below.

### 7.8.3 Numerical Results

In Fig. 7.2 we have evaluated $I_f$ for $\varepsilon = 1$ and $\beta = 1$. With respect to the results of Sects. 7.8.1,2 it is interesting to note that the information proceeds from negative to positive values of $\alpha$ via a hump. The origin of this is clear for physical reasons. Close

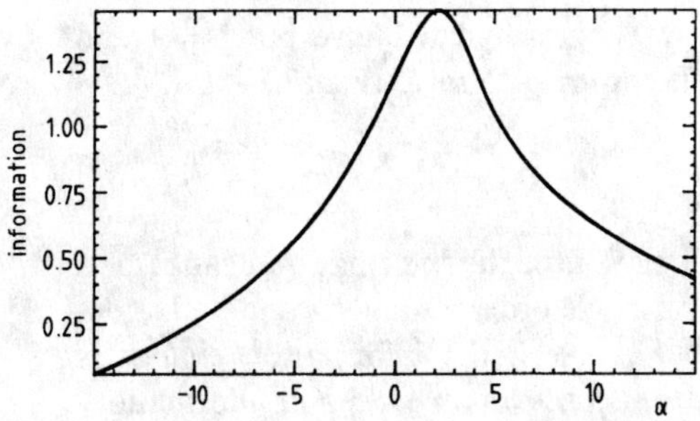


**Fig. 7.2.** The information (1.5) versus $\alpha$ for $\beta = \varepsilon = 1$

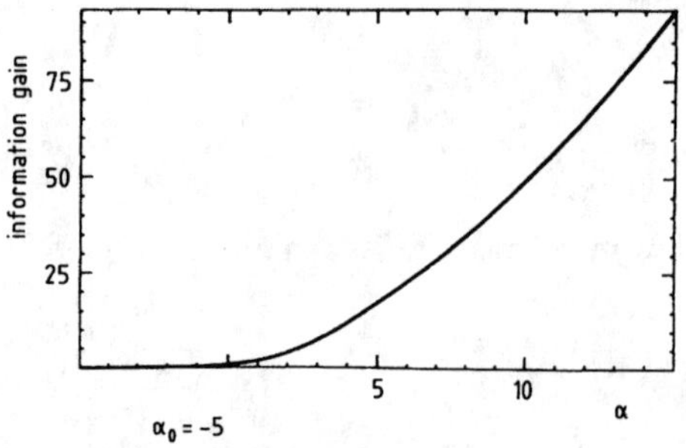


Fig. 7.3. The information gain for $\alpha_0 = -5.00$ and $\alpha$ varying between $-5$ and $+15$

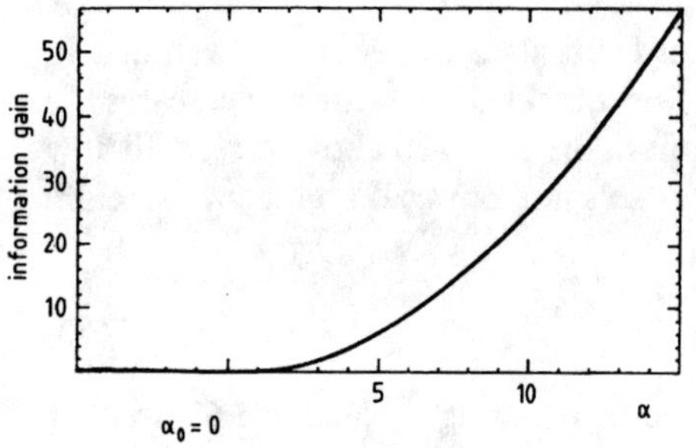


Fig. 7.4. Same as Fig.7.3 but for $\alpha_0 = 0$

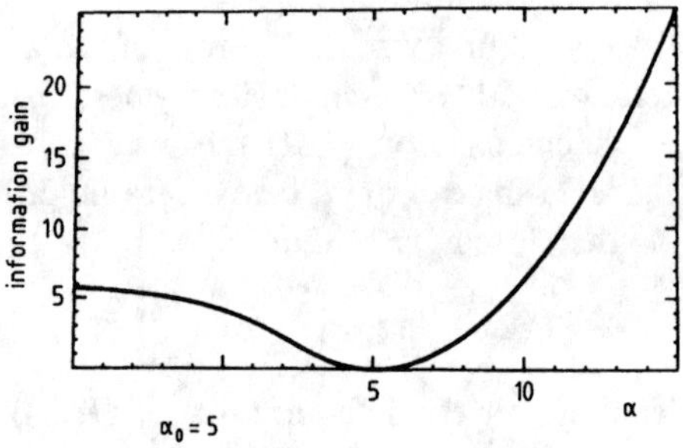


Fig. 7.5. Same as Fig.7.3 but for $\alpha_0 = 5$

to $\alpha = 0$ critical fluctuations, well known from phase transitions, occur which cause a spread of $f$ and therefore an increase of $I_f$. When $\alpha$ goes to negative values, $f$ centers around one maximum (cf. Fig.7.1) so that $I_f$ drops, whereas for growing positive values of $\alpha$, $f$ acquires two distinct maxima and the information drops again though to a higher level than for equivalent negative $\alpha$'s. For sake of completeness we have also calculated the information gain which is defined by

$$K = \int_{-\infty}^{+\infty} d\xi \, f_\alpha \ln \frac{f_\alpha}{f_{\alpha_0}} \tag{7.88}$$

where $f_\alpha$ is given by

$$f_\alpha = N(\alpha) \exp(\alpha \xi^2 - \beta \xi^4) \; . \tag{7.89}$$

In order to calculate (7.88) in the vicinity of threshold we have put $\beta = 1$ and calculated $K$ for $\alpha_0$ fixed and $\alpha$ taking values from $-15$ to 15 (Figs.7.3–5).

### 7.8.4 Discussion

Above we have presented an explicit calculation of the information change of a self-organizing system that is described by a single order parameter and undergoes a second-order phase transition. As we have seen the information change is quite small but represents precisely what we expect on physical and information-theoretical grounds. The system has become able to store one bit of information provided we may realize either one of the two possible states by means of symmetry breaking. This is one reason why we prefer the notion of information to that of entropy when we apply (7.55) to systems far from thermal equilibrium. On the other hand, the behavior close to threshold shows that here the information can increase considerably due to critical fluctuations. In this case one may be inclined to prefer the term "entropy". However, there is yet another reason which makes me reluctant to apply the term "entropy". As we saw in Chap. 5, the constraints under which the information entropy (7.55) is maximized are quite different for systems in thermal and away from thermal equilibrium. Among the constraints of systems in thermal equilibrium there is always energy and when the maximum entropy principle is used the corresponding Lagrange parameter is the inverse temperature. A further remark may be in order:

As is well known from nonequilibrium phase transitions, e.g. in lasers and fluid dynamics, in the region $\alpha > 0$ *ordered* structures appear. Our result exhibited in Fig.7.2 tells us that in this ordered state the information entropy is *higher* than in the disordered state. This result is counterintuitive and indeed some authors have claimed that $dS < 0$ for the transition from the disordered to the ordered state, *both* in equilibrium and nonequilibrium situations. But such conclusions are not justified. This becomes evident when we look more closely at thermodynamics. According to thermodynamics, entropy and thus disorder may increase but not decrease in a *closed* system. Thus, when we keep the total energy of a system constant, and compare two states, then the one with higher entropy is connected with greater disorder. But in the situation of an *open system* treated above, we do not compare states of a system with the same energy. Clearly, when we cut down the energy in- and outputs of the system, e.g. of a laser, and keep it now at a constant energy, the system will rearrange, increase its entropy and abandon its ordered state. In general, we may not draw conclusions on disorder or order in open systems if we consider the (information) entropy alone. From a more formal point of view, the difference in the interpretation of entropy and information entropy rests on the different constraints to be used for closed or open systems.

Let us assume that we have found the information entropy under the constraints for an *open* system. Then let us calculate the mean energy for the distribution function. When we take this energy as a new constraint, the entropy is *not* maximized by the distribution function of the open system but rather by a function for which just energy is the given constraint. We shall elucidate this result from still another point of view in Sect. 7.10.

## 7.9 Exact Analytical Results on Information, Information Gain, and Efficiency of a Single Order Parameter

In this section we shall present exact analytical expressions for the information, $I_f$, the information gain, $K_f$, and the efficiency $W$, all of which will be evaluated for

$$f(\xi) = N \exp(\alpha\xi^2 - \beta\xi^4) . \tag{7.90}$$

We briefly repeat these quantities:

$$I_f = -\ln N - \alpha\langle\xi^2\rangle + \beta\langle\xi^4\rangle \,, \tag{7.91}$$

where the $n$th moment of the distribution function $f(\xi)$ is given by

$$\langle\xi^n\rangle = \int_{-\infty}^{+\infty} f(\xi)\,\xi^n\,d\xi \,. \tag{7.92}$$

$$K_f = K_{ff_0} = \ln N - \ln N_0 + (\alpha - \alpha_0)\langle\xi^2\rangle_\alpha \,. \tag{7.93}$$

Finally, since the second moment $\langle\xi^2\rangle_\alpha$ is a measure of the intensity of radiation and hence of the work done, the efficiency of the system can be established to be

$$W = \frac{d\langle\xi^2\rangle_\alpha}{d\alpha} \,. \tag{7.94}$$

Using (7.90) to evaluate (7.94) results in [cf. (7.36)]

$$W = \langle\xi^4\rangle_\alpha - \langle\xi^2\rangle_\alpha^2 \,. \tag{7.95}$$

It is interesting to find out the relationship between $I_f$, $K_f$ and $W$ and the control parameters $\alpha$ and $\beta$. As we have seen above, depending on the magnitude of $\alpha$ there are various regimes, i.e. the instability region when $\alpha \approx 0$ and the stable regions when $|\alpha| \gg 0$.

The main mathematical tool we shall employ is the standard integral

$$\int_0^\infty x^n \exp(\alpha x^{2m} - \beta x^{4m})\,dx$$
$$= (2m)^{-1}(2\beta)^{-(n+1)/4m}\,\Gamma\!\left(\frac{n+1}{2m}\right) D_{-(n+1)/2m}(\lambda)\exp(\lambda^2/4) \tag{7.96}$$

where $\lambda = -\alpha/\sqrt{2\beta}$, $\Gamma$ is the gamma function and $D$ is the parabolic cylinder function. Since

$$N^{-1} = \int_{-\infty}^{+\infty} \exp(\alpha\xi^2 - \beta\xi^4)\,d\xi \tag{7.97}$$

simple application of (7.96) to (7.97) yields for the normalization constant $N$

$$N = (2\beta)^{1/4}\pi^{-1/2}\,e^{-\lambda^2/4}[D_{-1/2}(\lambda)]^{-1} \,. \tag{7.98}$$

Similarly, using (7.92) and (7.96) we can easily calculate the second and the fourth moments of $f(\xi)$ as

$$\langle\xi^2\rangle = (2\beta)^{-1/2}\frac{D_{-3/2}(\lambda)}{2D_{-1/2}(\lambda)} \quad \text{and} \tag{7.99}$$

$$\langle\xi^4\rangle = (8\beta)^{-1}\frac{3D_{-5/2}(\lambda)}{D_{-1/2}(\lambda)} \,, \tag{7.100}$$

respectively. Having found $N$, $\langle \xi^2 \rangle$ and $\langle \xi^4 \rangle$ in terms of the control parameters $\alpha$ and $\beta$ alone, we can simply substitute these formulas into the expressions for $I_f$, $K_f$ and $W$, i.e. (7.91,93,95). The information contained in the system with the distribution $f(\xi)$ is then obtain as

$$I_f = 0.3991 - \tfrac{1}{4}\ln\beta + \tfrac{1}{4}\lambda^2 + \ln D_{-1/2}(\lambda) + \frac{(\lambda/2)D_{-3/2}(\lambda) + (3/8)D_{-5/2}(\lambda)}{D_{-1/2}(\lambda)} \ . \tag{7.101}$$

Similarly, the information gain on changing the parameter $\alpha$ in $f(\xi)$ from $\alpha_0$ to $\alpha$ can be calculated to be

$$K_{ff_0} = \frac{\alpha_0^2 - \alpha^2}{8\beta} + \ln\frac{D_{-1/2}(\lambda_0)}{D_{-1/2}(\lambda)} - (\lambda - \lambda_0)\frac{D_{-3/2}(\lambda)}{2D_{-1/2}(\lambda)} \tag{7.102}$$

where $\lambda_0 = -\alpha_0/\sqrt{2\beta}$. Finally, the efficiency of the system is found from (7.95,99,100) to be

$$W = \frac{3D_{-5/2}(\lambda)D_{-1/2}(\lambda) - [D_{-3/2}(\lambda)]^2}{8\beta[D_{-1/2}(\lambda)]^2} \ . \tag{7.103}$$

It is worth emphasizing that all the formulas presented in this section, i.e. (7.98–103), are expressed in terms of analytic functions of the control parameters $\alpha$ and $\beta$, provided $\beta \neq 0$ and $\alpha < \infty$. The quantities of special interest to us, $I_f$, $K_f$ and $W$, depend in a critical way on $\lambda$. Since the parabolic cylinder function $D$ is a special function, our expressions (7.101–103) cannot, in general, be further simplified unless we tabulate the functions for selected values of $\beta$. However, there exist three special cases which are worth investigating. The first is when the instability point itself occurs when $\alpha = 0$, hence $\lambda = 0$. In the second, the immediate vicinity of the instability point is characterized by $\alpha \to 0$ and $\lambda \to 0$. In the third case, the stable region is characterized by $\lambda \gg 1$. We shall analyze each of these special cases in detail.

### 7.9.1 The Instability Point

The parabolic cylinder function is related to the Weber function via

$$D_{-p-1/2}(\lambda) = U(p,\lambda) \ . \tag{7.104}$$

At the instability point $\alpha = 0$, we use the property

$$U(p,0) = \frac{\sqrt{\pi}}{2^{p/2+1/4}}\Gamma\left(\frac{p}{2} + \frac{3}{4}\right)^{-1} \tag{7.105}$$

and find that the normalization constant $N$ achieves a maximum of

$$N_c = 0.5516\beta^{1/4} \ . \tag{7.106}$$

The second and the fourth moments are given by

$$\langle \xi^2 \rangle_c = 0.3380\beta^{-1/2} \quad \text{and} \tag{7.107}$$

$$\langle \xi^4 \rangle_c = 0.2500\beta^{-1} , \tag{7.108}$$

respectively. Consequently, we find that the information $I_f$ at the instability point is

$$I_c = 0.8449 - \tfrac{1}{4}\ln\beta . \tag{7.109}$$

Similarly, the information gain on going from the distribution $f_c$ ($\alpha = 0$) to $f$ is

$$K_{ff_c} = 0.5949 - \tfrac{1}{4}\ln\beta + \ln N + \alpha\langle \xi^2 \rangle_\alpha . \tag{7.110}$$

Finally, the efficiency at the instability point can be readily shown to be

$$W_c = 0.1358\beta^{-1} . \tag{7.111}$$

Obviously, all these values depend only on $\beta$ (except for the information gain) and it will be shown that they represent the maxima of their functions with respect to $\alpha$. It is interesting to explore the approach to these maxima as $\alpha \to 0$.

### 7.9.2 The Approach to Instability

In order to study the immediate vicinity of the instability point we use the following asymptotic expansion of the Weber function

$$U(p, \mp|\lambda|) \approx \left[\frac{\sqrt{\pi}}{2^{p/2+1/4}}\Gamma\left(\frac{p}{2}+\frac{3}{4}\right)^{-1}\right] e^{\mp\sqrt{p}\lambda} \tag{7.112}$$

where $\lambda \approx 0$ and thus higher powers of $\lambda$ have been neglected in the exponential function. Using this, we easily obtain the dominant behavior of $N$, $\langle \xi^2 \rangle$ and $\langle \xi^4 \rangle$ close to $\lambda = 0$:

$$N \simeq N_c e^{-\lambda^2/4} \tag{7.113}$$

$$\langle \xi^2 \rangle \simeq \langle \xi^2 \rangle_c e^{\mp\lambda} \quad \text{and} \tag{7.114}$$

$$\langle \xi^4 \rangle \simeq \langle \xi^4 \rangle_c e^{\mp\sqrt{2}\lambda} \tag{7.115}$$

where $N_c$, $\langle \xi^2 \rangle_c$ and $\langle \xi^4 \rangle_c$ are those of (7.106), (7.107) and (7.108) respectively. In (7.114) and (7.115) the negative sign corresponds to $\lambda > 0$ ($\alpha < 0$) and the positive sign to $\lambda < 0$ ($\alpha > 0$). Therefore, upon approaching $\alpha = 0$ the values of $N$, $\langle \xi^2 \rangle$ and $\langle \xi^4 \rangle$ increase exponentially and tend to their critical values $N_c$, $\langle \xi^2 \rangle_c$ and $\langle \xi^4 \rangle_c$, respectively. Applying these results to $I_f$ gives

$$I_f \simeq I_c + \lambda^2/4 + 0.478\lambda\exp(\mp\lambda) + 0.250[\exp(\mp\sqrt{2}\lambda) - 1] \tag{7.116}$$

which also approaches its critical value $I_c$ along a combination of exponential functions of $\lambda$. The information gain achieved on changing the distribution function from $f_0$ to $f$, both of which are close to $f_c$, can be calculated as

$$K_{ff_0} \simeq K_{ff_c} + \frac{\alpha_0^2}{8\beta} - \alpha_0 \langle \xi^2 \rangle_\alpha . \tag{7.117}$$

Thus, $K_{ff_0}$ approaches $K_{ff_c}$ via a combination of a linear term in $\alpha_0$ and a quadratic term in $\alpha_0$. Obviously, $\alpha_0 \to 0$. Allowing $\lambda \simeq 0$ we can further reduce (7.117) to

$$K_{ff_0} \simeq \lambda_0^2 - \lambda^2 + 0.478(\lambda - \lambda_0)\mathrm{e}^{\mp\lambda} . \tag{7.118}$$

Finally, we can calculate the efficiency function close to the instability point obtaining

$$W \simeq W_c \mathrm{e}^{-\sqrt{2\lambda}} . \tag{7.119}$$

We conclude by noting that all quantities of interest to us, except for $K$, approach their critical values via exponential functions or combinations of exponential functions. This will be shown to be in sharp contrast to their behavior in the stable region.

### 7.9.3 The Stable Region

The other asymptotic limit, $\lambda \to \infty$, can be achieved either by increasing $\alpha$ and keeping $\beta$ constant, or (as in the mean field approximation) by letting both $\alpha$ and $\beta$ tend to infinity at the same rate, e.g. by increasing the volume of the system. We shall therefore depart asymptotically from the instability point. The relevant expansion for the Weber function is

$$U(p, \lambda) \simeq \mathrm{e}^{-\lambda^2/4} \lambda^{-p-1/2} \tag{7.120}$$

where we have dropped higher powers of $(1/\lambda)$. Applying (7.120) to (7.98) yields for the normalization constant

$$N \simeq (2\beta)^{1/4} \left(\frac{\lambda}{\pi}\right)^{1/2} \tag{7.121}$$

and it obviously requires that $\alpha < 0$. The second moment is similarly obtained as

$$\langle \xi^2 \rangle \simeq \tfrac{1}{2}(2\beta)^{-1/2} \lambda^{-1} \tag{7.122}$$

and the fourth moment as

$$\langle \xi^4 \rangle \simeq \tfrac{3}{8} \beta^{-1} \lambda^{-2} . \tag{7.123}$$

Substituting these expressions into (7.91) gives the information

$$I_f \simeq 0.5724 - \tfrac{1}{2} \ln \lambda - \tfrac{1}{4} \ln(2\beta) + \tfrac{1}{2} + \tfrac{3}{8} \lambda^{-2} . \tag{7.124}$$

This formula for $I_f$ also applies only when $\alpha < 0$ and it represents a predominantly power-law decrease of $I_f$ to zero as $\lambda \to \infty$. Of course, there is a maximum value of

$\lambda$ beyond which (7.124) cannot be used since it would otherwise produce a negative information. Moreover, in such a case one needs to re-incorporate the effects due to the enslaved modes which become significant.

Next, we calculate the information gain on going from $f_{\alpha_0}$ to $f_\alpha$ and find

$$K \simeq \frac{1}{2}\ln\frac{\alpha}{\alpha_0} - \frac{1}{2}\left(1 - \frac{\alpha_0}{\alpha}\right) . \tag{7.125}$$

Finally, the efficiency of the system is found to be

$$W \simeq \tfrac{1}{4}\beta^{-1}\lambda^{-2} \tag{7.126}$$

which decreases monotonically as $\lambda$ tends to infinity. We therefore see that the relevant quantities decrease with increasing $\lambda$ according to various power laws.

### 7.9.4 The Injected Signal

Without introducing an external field that couples with $\xi$ via a term $-\sigma\xi$, the most likely value of $\xi$ is doubly degenerate, i.e. both $\xi = +(\alpha/2\beta)^{1/2}$ and $\xi = -(\alpha/2\beta)^{1/2}$ are equally probable. Injecting a signal with amplitude $\sigma$ will remove such degeneracy and cause the value of $\xi$ with the same sign as $\sigma$ to be preferred. In such a case we should modify our distribution function to

$$f_\sigma(\xi) = N\exp(-\sigma\xi + \alpha\xi^2 - \beta\xi^4) . \tag{7.127}$$

We shall henceforth denote the properties in the presence of the injected signal by a subscript "$\sigma$" and in the absence of it by a subscript "0". With $f_\sigma(\xi)$ given by (7.127) we can no longer afford exact formulas since (7.96) does not apply to it. Instead, we shall expand $f_\sigma(\xi)$ in a series about $f_0(\xi)$ of (7.90) in powers of $\sigma$. Since $\sigma$ is assumed to be small compared to $\alpha$ and $\beta$, we shall include only the lowest terms in our analysis and approximate $N_\sigma$ at the outset by $N_0$ of (7.98). Simple application of (7.96) to the series expansion of $f_\sigma(\xi)$ yields to the lowest order of approximation:

$$\langle\xi^2\rangle_\sigma \simeq \langle\xi^2\rangle_0\left[\frac{1 + 3}{4\sqrt{2\beta}}\frac{D_{-5/2}(\lambda)}{D_{3/2}(\lambda)}\sigma^2 + \cdots\right] \tag{7.128}$$

and

$$\langle\xi^4\rangle_\sigma \simeq \langle\xi^4\rangle_0\left[\frac{1 + 5}{4\sqrt{2\beta}}\frac{D_{-7/2}(\lambda)}{D_{-5/2}(\lambda)}\sigma^2 + \cdots\right] , \tag{7.129}$$

where $\langle\xi^2\rangle_0$ and $\langle\xi^4\rangle_0$ are those of (7.99) and (7.100), respectively. In both of these expressions the amplitude, $\sigma$, of the injected signal enters via a correction term and appears squared in the lowest order. We can analyze these terms in more detail when we specify the regime of $\lambda$. In the vicinity of the instability point we obtain

$$\langle\xi^2\rangle_\sigma \simeq \langle\xi^2\rangle_0[1 + 0.3698\beta^{-1/2}\exp(\mp 0.4142\lambda)\sigma^2] \tag{7.130}$$

and

$$\langle\xi^4\rangle_\sigma \simeq \langle\xi^4\rangle_0[1 + 0.5070\beta^{-1/2}\exp(\mp 0.3178\lambda)\sigma^2] \ . \tag{7.131}$$

So the correction due to $\sigma$ is modified by an exponential function of $\lambda$ and thus it becomes more and more pronounced as we approach the instability point. We can carry it further to evaluate its effect on the information and efficiency. There,

$$I_\sigma \simeq I_0 - 0.1768\beta^{-1/2}\exp(\mp\sqrt{2\lambda})[\lambda - 0.7172\exp(\mp 0.9036\lambda)]\sigma^2 \tag{7.132}$$

and

$$W_\sigma \simeq W_0 + 0.0845\beta^{-1/2}\exp(\mp 2.4142\lambda)[1 + 1.5\exp(\pm 0.0964\lambda)]\sigma^2 \ . \tag{7.133}$$

Again, the correction terms are dominated by factors quadratic in $\sigma$ and modified by exponential terms in $\lambda$.

On the other hand, far away from the instability point we obtain for the second and the fourth moments

$$\langle\xi^2\rangle_\sigma \simeq \langle\xi^2\rangle_0(1 + 0.5303\lambda^{-1}\beta^{-1/2}\sigma^2) \tag{7.134}$$

$$\langle\xi^4\rangle_\sigma \simeq \langle\xi^4\rangle_0(1 + 0.8839\lambda^{-1}\beta^{-1/2}\sigma^2) \tag{7.135}$$

both of which have correction terms quadratic in $\sigma$ and decreasing linearly with $\lambda^{-1}$. As a result the information is given by

$$I_\sigma \simeq I_0 - 0.1326\beta^{-1/2}\lambda^{-2}(\alpha - 5\beta\lambda^{-1}) \tag{7.136}$$

and the efficiency is approximated by

$$W_\sigma \simeq W_0 + 0.5966\beta^{-1/2}\lambda^{-3}\sigma^2 \ . \tag{7.137}$$

### 7.9.5 Conclusions

In the neighbourhood of the instability point both information and efficiency of a self-organizing system increase exponentially with respect to the power input $\alpha$ and thus are likely to overshadow the contributions from the enslaved modes. On the other hand, far away from the instability point the information and efficiency of the system decrease according to a second power of the power input $\alpha$.

Therefore, the enslaved modes will reenter the scene in this region. It is interesting to note that the quantity $\langle\xi^2\rangle^2/\langle\xi^4\rangle$ has a constant value in both limits. In the neighbourhood of the instability it equals 0.4570 signalling a much higher intensity than that of the stable region where $\langle\xi^2\rangle^2/\langle\xi^4\rangle$ is only 0.083. Since we have also given general formulas for $\langle\xi^2\rangle$, $\langle\xi^4\rangle$, $I_f$, $K_{ff_0}$ and $W$, which apply to all values of $\alpha$ and $\beta$, and which are analytic functions, we expect to find all the values between 0.083 and 0.4570 in the intermediate region between $\lambda = \infty$ and $\lambda = 0$. The injected signal has been found to introduce correction terms to the unperturbed expressions. The dominant contributions are proportional to the square of the amplitude of the injected signal.

## 7.10 The $S$-Theorem of Klimontovich

In this section we wish to do two things. First we will show how the laser distribution function, which we derived in Sect. 5.2 by means of our macroscopic approach, can also be found by the microscopic theory. We then shall use this result in order to illustrate a theorem of Klimontovich which he called the $S$-theorem and which sheds new light on our result that in a system far from equilibrium the entropy may increase inspite of the fact that the system enters a state of higher order.

But let us start first with the microscopic theory. We consider a complex order parameter $\xi$. We shall consider the case in which it obeys the equation

$$\dot{\xi} = \lambda\xi - \beta\xi|\xi|^2 + F(t) . \tag{7.138}$$

In the following we shall decompose $\xi$ into its real and imaginary parts

$$\xi = q_1 + \mathrm{i}q_2 . \tag{7.139}$$

We decompose the fluctuating force $F$ in a similar fashion

$$F = F_1 + \mathrm{i}F_2 , \tag{7.140}$$

where $F_1$ and $F_2$ are assumed to be real. Then the order parameter equation (7.138) can be split into the equations

$$\dot{q}_j = \lambda q_j - \beta q_j(q_1^2 + q_2^2) + F_j , \qquad j = 1, 2 \tag{7.141}$$

for the real quantities $q_1$ and $q_2$. They can also be written in the abbreviated form

$$\dot{q}_j = K_j(q_1, q_2) + F_j . \tag{7.142}$$

As a more detailed analysis shows, the fluctuating forces have the following properties

$$\langle F_j(t)\rangle = 0 \tag{7.143}$$

$$\langle F_j(t)F_k(t')\rangle = \delta_{jk}Q\delta(t - t') . \tag{7.144}$$

The rules of Sect. 2.3 allow us immediately to establish the Fokker-Planck equation corresponding to (7.141) or (7.138); it reads

$$\dot{f} = -\frac{\partial}{\partial q_1}(K_1 f) - \frac{\partial}{\partial q_2}(K_2 f) + \frac{Q}{2}\left(\frac{\partial^2 f}{\partial q_1^2} + \frac{\partial^2 f}{\partial q_2^2}\right) . \tag{7.145}$$

One may readily convince oneself that the conditions for detailed balance are fulfilled. Therefore, for the stationary state with

$$\dot{f} = 0 \tag{7.146}$$

we immediately find

$$f(q_1, q_2) = \mathcal{N} \exp\left\{\left[\lambda(q_1^2 + q_2^2) - \frac{\beta}{2}(q_1^4 + q_2^4)\right] Q^{-1}\right\} . \tag{7.147}$$

For what follows, it is convenient to transform the distribution function to new coordinates. To this end we consider the probability function

$$f(q_1, q_2)\, dq_1\, dq_2 . \tag{7.148}$$

We now introduce new coordinates defined by

$$q_1 = r \cos\phi \tag{7.149}$$

$$q_2 = r \sin\phi \tag{7.150}$$

or equivalently

$$\xi = r \exp(\mathrm{i}\phi) . \tag{7.151}$$

In these new coordinates $r$ and $\phi$ the volume element can be written as

$$dV = dq_1\, dq_2 = r\, dr\, d\phi . \tag{7.152}$$

Therefore (7.148) can be written explicitly in the form

$$f(r) r\, dr\, d\phi = \mathcal{N} \exp[(\lambda r^2 - \tfrac{1}{2}\beta r^4) Q^{-1}] r\, dr\, d\phi \tag{7.153}$$

when we use the new coordinates $r$ and $\phi$. Now let us introduce the quantity $U$ according to

$$r^2 = U . \tag{7.154}$$

It may be interpreted in a number of cases in physics as potential energy. From (7.153) and (7.154) we immediately obtain

$$f(U)\, dU = \mathcal{N} \exp[(\lambda U - \tfrac{1}{2}\beta U^2) Q^{-1}]\, dU , \tag{7.155}$$

where we have omitted the common factor $d\phi$. For the following we shall require that $f(U)$ is normalized according to

$$\int_0^\infty f(U)\, dU = 1 . \tag{7.156}$$

In order to prepare ourselves for the explanation of the $S$-theorem, we shall consider the information entropy

$$i = -\int_0^\infty f(U) \ln f(U)\, dU , \tag{7.157}$$

and the average energy which is defined by

$$\langle U \rangle = \int_0^\infty U f(U)\, dU . \tag{7.158}$$

In order to be able to present explicit formulas for (7.157) and (7.158), we consider three limiting cases which are well known from laser physics, but which can also be found in other areas.

### 7.10.1 Region 1: Below Laser Threshold

In this region we have $\lambda < 0$. In this case the laser oscillator is excited only weakly so that its energy obeys the inequality

$$\frac{\beta}{2}U^2 \ll |\lambda| U \ . \tag{7.159}$$

Under this assumption we may neglect the quadratic term in comparison to the linear term in the distribution function (7.147). Then it is simple to calculate (7.158) explicitly which yields

$$\langle U \rangle = \frac{Q}{|\lambda|} \ . \tag{7.160}$$

Using (7.160) in (7.159) we may cast the inequality (7.159) into the form

$$\frac{\beta Q}{2|\lambda|^2} \ll 1 \ . \tag{7.161}$$

The distribution function used in this case reads explicitly

$$f_1(U) = \frac{|\lambda|}{Q} \exp\left(-\frac{|\lambda| U}{Q}\right) , \tag{7.162}$$

where we have included the exact normalization factor. It is now a simple task to calculate (7.157) and (7.158) to obtain

$$i_1 = \ln \frac{Q}{|\lambda|} + 1 \ , \qquad \text{and} \tag{7.163}$$

$$\langle U \rangle_1 = \frac{Q}{|\lambda|} \ . \tag{7.164}$$

Because we are calculating (7.157) and (7.158) in the regime 1, i.e. below theshold, we have added the index 1 to $i$ and $\langle U \rangle$. When we pump the laser higher, that is when we increase the pump power, $\lambda$ decreases. Equation (7.164) then tells us that the average energy increases. Let us now consider the region at which the laser is at threshold.

### 7.10.2 Region 2: At Threshold

In this case the control parameter $\lambda$ fulfills

$$\lambda = 0 \ . \tag{7.165}$$

The corresponding distribution function reads

$$f_2(U) = \sqrt{\frac{2\beta}{\pi Q}} \exp\left(-\frac{\beta U^2}{2Q}\right) . \tag{7.166}$$

We have added the index 2 to indicate the threshold region. It is again quite simple to calculate the information entropy and the mean energy:

$$i_2 = \frac{1}{2} \ln \frac{\pi Q}{2\beta} + \frac{1}{2} , \tag{7.167}$$

$$\langle U \rangle_2 = \sqrt{\frac{2Q}{\pi \beta}} . \tag{7.168}$$

### 7.10.3 Region 3: Well Above Threshold

In this region we have $\lambda > 0$. We shall consider the region where the inequality

$$\frac{Q\beta}{\lambda^2} \ll 1 \tag{7.169}$$

holds. In such a case the distribution function $f$ may be well approximated by

$$f_3(U) = \mathcal{N} \exp\left[-\frac{\beta}{2Q}\left(U - \frac{\lambda}{\beta}\right)^2\right] , \tag{7.170}$$

where the index 3 labels the region well above threshold. The information entropy and mean energy can easily be calculated yielding

$$i_3 = \frac{1}{2} \ln \frac{2\pi Q}{\beta} + \frac{1}{2} , \tag{7.171}$$

$$\langle U \rangle_3 = \frac{\lambda}{\beta} . \tag{7.172}$$

We may now compare the information entropies in the regions 1, 2 and 3, and also the mean energies in these regions. We can readily convince ourselves that they obey the inequalities

$$i_1 < i_2 < i_3 \quad \text{and} \tag{7.173}$$

$$\langle U \rangle_1 < \langle U \rangle_2 < \langle U \rangle_3 . \tag{7.174}$$

With increasing pumping the information entropy increases, a result which we have already found for a real order parameter in Sect. 7.9. In that section we already observed that an adequate comparison of the entropies can be made only when we take them at the same energies. So according to an idea of Klimontovich, we now choose the fluctuation strength $Q$ in such a way that the energies become equal. In other words we shall change the fluctuating forces in the cases 1 and 2 so that we now obtain

$$\langle U\rangle_{1,r} = \langle U\rangle_{2,r} = \langle U\rangle_{3,r} \,. \tag{7.175}$$

We want to study the consequences in detail and start in the region below threshold.

To obtain a constant energy with decreasing $\lambda$, we must decrease the noise level $Q$ in proportion to $\lambda$. It is interesting to discuss the corresponding results in terms of laser physics. When we decrease $\lambda$, i.e. when we increase the pump strength, the process of stimulated emission occurs which means that the individual wave tracks of the light field are enhanced. As a consequence the mean energy increases. On the other hand, when we decrease the noise level, then the spontaneous emission events become scarcer. When we decrease the noise level or the production rate of spontaneous emission according to the enhancement factor, we may readily find the conditions under which the energy remains constant. At the same time we find that the information entropy remains constant, i.e. in the so-called linear amplifier region the information entropy remains constant under the circumstances just described.

Klimontovich chose another comparison; he used the equality (7.175) for a fixed mean energy above threshold according to

$$\langle U\rangle_{3,r} = \frac{\lambda}{\beta} \,, \tag{7.176}$$

where $\lambda$ and $\beta$ are now fixed quantities. In the equation

$$\langle U\rangle_{2,r} = \sqrt{\frac{2Q}{\pi\beta}} \tag{7.177}$$

we now choose $Q = Q_2$ so that the relation

$$\langle U\rangle_{2,r} = \frac{\lambda}{\beta} \tag{7.178}$$

is fulfilled. For $Q_2$ we then obtain

$$Q_2 = \frac{\pi}{2}\frac{\lambda^2}{\beta} \,. \tag{7.179}$$

Inserting (7.179) in the expression for the information entropy (7.167), we readily obtain

$$i_{2,r} = \frac{1}{2}\ln\frac{\pi Q_2}{2\beta} + \frac{1}{2} = \ln\frac{\pi\lambda}{2\beta} + \frac{1}{2} \,. \tag{7.180}$$

We do the same thing for region 1. In the equation

$$\langle U\rangle_{1,r} = \frac{Q}{|\lambda_1|} \tag{7.181}$$

we choose $Q = Q_1$ such that

$$\langle U \rangle_{1,r} = \frac{\lambda}{\beta} \, . \tag{7.182}$$

We readily obtain

$$Q_1 = \frac{\lambda |\lambda_1|}{\beta} \, . \tag{7.183}$$

Inserting this result into the expression for the information entropy (7.163) we find

$$i_{1,r} = \ln \frac{Q_1}{|\lambda_1|} + 1 = \ln \frac{\lambda}{\beta} + 1 \, . \tag{7.184}$$

When we recall the inequality

$$\frac{Q\beta}{\lambda^2} \ll 1 \, , \tag{7.185}$$

we now obtain the following inequality for the entropies

$$i_{1,r} > i_{2,r} > i_{3,r} \, . \tag{7.186}$$

Thus when we keep the mean energy fixed by adjusting the noise level, we indeed obtain the result that the entropy decreases when we increase the control parameter and achieve a more ordered state of the system, e.g. laser action. This decrease of information entropy upon ordering, or, in other words, the occurrence of self-organization, was called the *S*-theorem by Klimontovich. He also considered the general case of regions intermediate between 1 and 3 considered here. His theorem was derived by considering the change of energy and information entropy with changing control parameter.

In this treatment Klimontovich considered only the change of energy with respect to the control parameter but not the simultaneous change of the noise level which is required to keep the energy constant. Therefore, the derivation of his theorem requires a slight change which we shall not present here.

## 7.11 The Contribution of the Enslaved Modes to the Information Close to Nonequilibrium Phase Transitions

In Sect. 7.2 we have shown that we may decompose the information of the total system into the information of the order parameters and that of the enslaved modes. In the previous sections we have calculated the information of the order parameters. We shall now turn to a study of the information of the enslaved modes. To this end we use the slaving principle in its leading approximation. As we have shown in

(7.21), the conditional probability $P$ is then given by

$$P(\xi_s|\xi_u) = \mathcal{N}_s \exp\left\{-\frac{[\lambda_s \xi_s + g_s(\xi_u)]}{Q_s}\right\} , \tag{7.187}$$

where the normalization factor $\mathcal{N}_s$ is given by

$$\mathcal{N}_s = |\lambda_s|(\pi Q_s)^{-1/2} . \tag{7.188}$$

In the following we shall allow $\lambda_s$, $Q_s$ and thus the normalization factor $\mathcal{N}_s$, to be functions of the control parameter $\alpha$.

$$\lambda_s = \lambda_s(\alpha) , \tag{7.189}$$

$$Q_s = Q_s(\alpha) , \tag{7.190}$$

$$\mathcal{N}_s = \mathcal{N}_s(\alpha) . \tag{7.191}$$

We may further assume that $g_s$ depends on this control parameter $\alpha$ and that

$$g_s(0) = 0 . \tag{7.192}$$

The information of the mode $s$ for a given fixed value of the control parameter $\xi_u$ is given by

$$I_s(\xi_u, \alpha) = -\int P(\xi_s|\xi_u) \ln P(\xi_s|\xi_u)\, d\xi_s . \tag{7.193}$$

Note that the integral is extended from minus infinity to plus infinity. We now introduce a new variable $\xi'$ instead of $\xi_s$ by means of the definition

$$\xi_s + \frac{g_s}{\lambda_s} = \xi' . \tag{7.194}$$

Using (7.187) and (7.194) in (7.193) we obtain

$$I_s = -\int \mathcal{N}_s(\alpha) \exp\left(-\frac{\lambda_s^2 \xi'^2}{Q_s}\right) \times \ln\left[\mathcal{N}_s(\alpha) \exp\left(-\frac{\lambda_s^2 \xi'^2}{Q_s}\right)\right] d\xi' . \tag{7.195}$$

In order to eliminate the dependence of $I_s$ on $\alpha$ as far as possible we make the following substitution

$$|\lambda_s(\alpha)| Q_s(\alpha)^{-1/2} \xi' = |\lambda_s(0)| Q_s(0)^{-1/2} \eta , \tag{7.196}$$

or solving for $\xi'$ we write

$$\xi' = \frac{|\lambda_s(0)|}{|\lambda_s(\alpha)|} \left(\frac{Q_s(0)}{Q_s(\alpha)}\right)^{-1/2} \eta . \tag{7.197}$$

Considered as a function of the order parameter $\xi_u$ and of the control parameter $\alpha$, $I_s$ then acquires the form

$$I_s(0,\alpha) = -\int \frac{\mathcal{N}_s(\alpha)}{\mathcal{N}_s(0)} \mathcal{N}_s(0) \exp\left[-\frac{\lambda_s(0)^2\eta^2}{Q_s(0)}\right]$$
$$\times \ln\left\{\frac{\mathcal{N}_s(\alpha)}{\mathcal{N}_s(0)} \mathcal{N}_s(0) \exp\left[-\frac{\lambda_s(0)^2\eta^2}{Q_s(0)}\right]\right\} \times \frac{|\lambda_s(0)|}{|\lambda_s(\alpha)|}\left(\frac{Q_s(0)}{Q_s(\alpha)}\right)^{-1/2} d\eta \,. \quad (7.198)$$

Note that this expression is now independent of the order parameter $\xi_u$ as indicated by the argument 0 in $I_s$. Using the relation (7.188) we observe that a number of factors in (7.198) cancel so that (7.198) can be written in the simpler form

$$I_s(0,\alpha) = -\mathcal{N}_s(0) \int \exp\left[-\frac{\lambda_s(0)^2\eta^2}{Q_s(0)}\right]\{A + B\}\, d\eta \,, \quad (7.199)$$

where $A$ is given by

$$A = \ln\frac{\mathcal{N}_s(\alpha)}{\mathcal{N}_s(0)} \,, \quad (7.200)$$

and $B$ by

$$B = \ln\left\{\mathcal{N}_s(0) \exp\left[-\frac{\lambda_s(0)^2\eta^2}{Q_s(0)}\right]\right\} \,. \quad (7.201)$$

Since $A$ is a constant, the integral over the Gaussian function can be evaluated immediately and yields, together with a normalization factor, unity. One can easily convince oneself that the integral which contains the term $B$ is just the definition of the information $I_s$ for the control parameter value $\alpha = 0$. Therefore, we may cast (7.199) into the form

$$I_s(0,\alpha) = -\ln\frac{\mathcal{N}_s(\alpha)}{\mathcal{N}_s(0)} + I_s(0,0) \,. \quad (7.202)$$

Using the explicit definition of the normalization factor $\mathcal{N}_s$ (7.188), (7.202) can also be represented in the form

$$I_s(0,\alpha) = -\ln\left[\frac{|\lambda_s(\alpha)|}{|\lambda_s(0)|}\left(\frac{Q_s(\alpha)}{Q_s(0)}\right)^{-1/2}\right] + I_s(0,0) \,. \quad (7.203)$$

We now recall the general decomposition of the information $i$ of the total system into the information of the order parameters $I_f$ and of the enslaved modes. Since $I_s$ is independent of $\xi_u$, the integration over the order parameters can be performed immediately and yields the factor 1. Therefore, we obtain as a final result

$$i = I_f + \sum_s I_s(0,0) - \sum_s \ln\frac{\mathcal{N}_s(\alpha)}{\mathcal{N}_s(0)} \,. \quad (7.204)$$

When we wish to compare the information for two different values of the control parameter $\alpha_2$ and $\alpha_1$, we may form

$$i(\alpha_2) - i(\alpha_1) = I_u(\alpha_2) - I_u(\alpha_1) - \sum_s \ln \frac{\mathcal{N}_s(\alpha_2)}{\mathcal{N}_s(\alpha_1)} . \tag{7.205}$$

Note that $\mathcal{N}^{-1}$ corresponds to the partition function of thermodynamics, and $\ln \mathcal{N}$ corresponds, according to

$$\ln \mathcal{N} = -\lambda \tag{7.206}$$

to the free energy. Clearly (7.204) then allows a generalization of the concepts of thermodynamics. When the control parameter $\alpha$ is chosen in such a way that at the critical point $\alpha = 0$, we may easily calculate the change of $I_s$ in the vicinity of the critical point by taking the derivative. This yields

$$\left.\frac{dI_s}{d\alpha}\right|_{\alpha=0} = -\frac{\mathcal{N}_s'(0)}{\mathcal{N}_s(0)} . \tag{7.207}$$

## An Explicit Example: The Single-Mode Laser

Let us now study the behavior of the information of the enslaved modes close to non-equilibrium phase transitions by means of an explicit example. One of the simplest examples in this respect is provided by the single-mode laser. Therefore we wish to remind the reader briefly of the corresponding theory. The laser is composed of a single field mode which has an amplitude $b(t)$ and which is made dimensionless. $|b|^2$ can be interpreted as the average number of photons present in the laser.

The laser is further composed of the laser-active atoms which are assumed to possess only two levels. The atomic dipole moment is described by a dimensionless variable $\alpha_\mu$. Finally the inversion of each atom, i.e. the difference between the occupation number of the upper and the lower level is given by $d_\mu$. The laser equations then have the following form

$$\dot{b} = (-\mathrm{i}\omega - \kappa)b - \mathrm{i}g \sum_\mu \alpha_\mu \tag{7.208}$$

$$\dot{\alpha}_\mu = (-\mathrm{i}\omega - \gamma)\alpha_\mu + \mathrm{i}gb\, d_\mu \tag{7.209}$$

$$\dot{d}_\mu = \gamma_\| (d_0 - d_\mu) + 2\mathrm{i}g(\alpha_\mu b^* - \alpha_\mu^* b) , \tag{7.210}$$

where $\omega$ is the optical transition frequency of the laser atoms, $\kappa$ is the decay constant of the field mode and $g$ is the coupling constant between the field mode and the atomic dipole moment. $\gamma$ is the decay constant of the dipole moment of the atom and $\gamma_\|$ is the relaxation constant of the atomic inversion. $d_0$ is the prescribed inversion given by the pump strength with which the atoms are excited and by incoherent relaxation processes.

In order to eliminate the constant $\omega$, we make the hypothesis

$$b = B\mathrm{e}^{-\mathrm{i}\omega t} , \qquad \alpha_\mu = A_\mu \mathrm{e}^{-\mathrm{i}\omega t} , \tag{7.211}$$

where $B$ and $A_\mu$ are time-dependent variables. Inserting (7.211) into (7.208–210) we readily obtain the equations

$$\dot{B} = -\kappa B - \mathrm{i}g \sum_\mu A_\mu \tag{7.212}$$

$$\dot{A}_\mu = -\gamma A_\mu + igb\, d_\mu \tag{7.213}$$

$$\dot{d}_\mu = \gamma_\parallel (d_0 - d_\mu) + 2ig(A_\mu B^* - A_\mu^* B) \ . \tag{7.214}$$

$d_0$ may be identified with the control parameter. As it is known that for weak inversion $d_0$, no laser action takes place, we first study this case. In it the field vanishes and also the atomic dipole moments are zero, whereas the inversion $d_\mu$ acquires a value identical with that of the inversion $d_0$ described by the pump process and incoherent relaxation processes. Therefore we obtain

$$B_0 = 0 \ , \qquad A_{\mu,0} = 0 \ , \qquad d_{\mu,0} = d_0 \ . \tag{7.215}$$

We now wish to study under which values of $d_0$ the solution (7.215) is stable and at which critical value $d_0$ it loses its stability. To this end we make the usual stability analysis described in Sect. 2.6, namely, we make the hypothesis

$$B = B_0 + \delta B \ , \qquad A_\mu = A_{\mu,0} + \delta A_\mu \ . \tag{7.216}$$

Inserting (7.216) into (7.212–214) and using (7.215) we obtain in the leading approximation, i.e. in the linear terms, the equations

$$\delta\dot{B} = -\kappa\delta B - ig\sum_\mu \delta A_\mu \tag{7.217}$$

$$\delta\dot{A}_\mu = -\gamma\delta A_\mu + ig\delta B\, d_0 \ . \tag{7.218}$$

In addition it turns out that (7.214) yields no first-order change of $d_\mu$. Note that the equations (7.217) and (7.218) are of the same form as equation (2.72) of Sect. 2.6 and illustrate the set of linear equations derived there. In order to solve the equations (7.217) and (7.218) we make the usual hypothesis

$$\delta B(t) = \delta B(0)\mathrm{e}^{\lambda t} \tag{7.219}$$

$$\delta A_\mu(t) = \delta A_\mu(0)\mathrm{e}^{\lambda t} \ , \tag{7.220}$$

where the time-independent amplitudes and $\lambda$ are still unknown quantities. Inserting (7.219) and (7.220) into (7.217) and (7.218) we obtain the linear algebraic equations

$$\delta B(0)(\lambda + \kappa) = -ig\sum_\mu \delta A_\mu(0) \tag{7.221}$$

$$\delta A_\mu(0)(\lambda + \gamma) = ig\delta B(0)\, d_0 \ . \tag{7.222}$$

The determinant belonging to (7.221,222) reads

$$\begin{vmatrix} -\kappa - \lambda & -ig & -ig & \dots & -ig \\ ig\, d_0 & -\gamma - \lambda & 0 & & 0 \\ ig\, d_0 & 0 & -\gamma - \lambda & & 0 \\ \vdots & \vdots & \vdots & & \vdots \\ ig\, d_0 & 0 & 0 & \dots & -\gamma - \lambda \end{vmatrix} \ .$$

It can easily be evaluated and we obtain

$$-(\kappa+\gamma)(\gamma+\lambda)^N + g^2 d_0 N(\gamma+\lambda)^{N-1} = 0 \ . \tag{7.223}$$

The eigenvalues $\lambda$ are:

a) for the stable modes, $\lambda_s$:

$$\lambda_1 = -\frac{\kappa+\gamma}{2} - \frac{1}{2}\sqrt{(\kappa-\gamma)^2 + 4g^2 D_0} \ ; \qquad D_0 = N d_0 \qquad \text{(simple)} \tag{7.224}$$

and for $s = 2, \dots, N$

$$\lambda_s = -\gamma \qquad \text{i.e. } (N-1)\text{-fold.} \tag{7.225}$$

b) for the unstable mode, $\lambda_u$:

$$\lambda_u = -\frac{\kappa+\gamma}{2} + \frac{1}{2}\sqrt{(\kappa-\gamma)^2 + 4g^2 D_0} \tag{7.226}$$

which can also be written in the form

$$\lambda = -\frac{\kappa+\gamma}{2} + \frac{1}{2}\sqrt{(\kappa-\gamma)^2 + 4g^2 D_0} \tag{7.227}$$

The unstability condition then yields the critical inversion $D_c$, namely

$$\lambda_u \geq 0 \ ; \qquad D_c = \frac{\kappa\gamma}{g^2} \ . \tag{7.228}$$

We now wish to study the change of the information of the enslaved modes close to the instability point. To this end we put

$$D_0 = D_c + \alpha' = \frac{\kappa\gamma}{g^2} + \alpha' \tag{7.229}$$

where $\alpha'$ plays the role of a control parameter.

Let us now assume, in accordance with usual laser theory, that the fluctuating forces do not depend on the control parameter. Under such an assumption, and because of (7.188), the relation (7.207) reduces to a calculation of

$$\left.\frac{dI_s}{d\alpha'}\right|_{\alpha'=0} = -\frac{\lambda_s'(\alpha'=0)}{\lambda_s(\alpha'=0)} \ , \qquad s = 1, \dots, N \ . \tag{7.230}$$

We readily obtain the result that (7.230) = 0 for $s = 2, \dots, N$ and

$$\frac{\lambda_1'(0)}{\lambda_1(0)} = \frac{g^2}{(\kappa+\gamma)^2} \approx \frac{g^2}{\gamma^2} = \frac{\kappa}{\gamma}\frac{1}{D_c} \tag{7.231}$$

where we have specialized the result to the case in which

$$\kappa \ll \gamma \tag{7.232}$$

which is valid for most lasers. Thus our final result is given by

$$\frac{dI_s}{d\alpha'}\alpha' = -\frac{\kappa}{\gamma}\frac{\alpha'}{D_c} \quad \text{for } s = 1 \tag{7.233}$$

$$= 0 \quad \text{otherwise} .$$

This tells us that with increasing pump rate represented by $\alpha'$, the information decreases. This result can be easily interpreted when we recall the general form of the probability distribution of the enslaved modes given by (7.187). If $Q_s$ is a constant and after $g_s$ has been transformed away according to (7.194), the information of the enslaved modes depends on a Gaussian distribution function, the width of which decreases when the absolute value of $\lambda_s$ increases which is the case for increasing $\alpha'$. A narrower distribution function, however, means a greater certainty and at the same time a decrease of information.

In the leading approximation of the slaving principle, therefore, we notice that the information entropy of the laser atoms decreases when the laser is pumped harder. Let us finally relate the control parameter $\alpha'$, which we introduced in (7.229), to the control parameter $\alpha$ which occurs in the distribution function of the order parameter in the equation

$$f = \mathcal{N} \exp(\alpha\xi^2 - \beta\xi^4) . \tag{7.234}$$

To achieve this we must recall some results of laser theory for the special case (7.232) and some basic equations of the microscopic theory.

If we assume for the time being, and just for illustration, that the field amplitude $b$ is a real variable, we may identify it with $\xi$ occurring in (7.234). Then in laser theory it is shown that $B \equiv \xi$ obeys a Langevin equation of the type (2.1) with (2.3). This Langevin equation may be transformed into a Fokker-Planck equation whose stationary solution is then just given by (7.234) where we may make the identification

$$\alpha = \frac{\lambda_u(\alpha')}{Q_{tot}} . \tag{7.235}$$

Here $Q_{tot}$ is the total fluctuating force acting on the laser field, whereas $\lambda_u(\alpha')$ is just given by

$$\lambda_u = -\frac{\kappa + \gamma}{2} + \frac{1}{2}\sqrt{(\kappa + \gamma)^2 + 4g^2\alpha'} \tag{7.236}$$

where we have used (7.226,229). Equation (7.236) can be easily expanded into a series of powers of $\alpha'$ where the leading term is given by

$$\lambda_u \approx \frac{g^2\alpha'}{\kappa + \gamma} . \tag{7.237}$$

Comparing (7.235) with (7.237) we obtain the relation between the two types of control parameters $\alpha$ and $\alpha'$:

$$\alpha = \frac{g^2}{(\kappa + \gamma)Q_{\mathrm{tot}}}\alpha' \ . \tag{7.238}$$

Using a result from laser theory according to which

$$Q_{\mathrm{tot}} \approx \frac{2\kappa N_{2s}}{D_{\mathrm{c}}} \tag{7.239}$$

where $N_{2s}$ is the number of excited atoms, we may cast (7.238) into the form

$$\alpha = \frac{\alpha'}{2N_{2s}} \tag{7.240}$$

where we have used the relations (7.229,232). To a good approximation we may put

$$N_{2s} \approx \frac{N}{2} \tag{7.241}$$

where $N$ is the total number of laser active atoms.

By means of (7.240,241) we may transform (7.231) into

$$\frac{d\lambda_1}{d\alpha}\Big/\lambda_1 = \frac{\kappa}{\gamma}\frac{N}{D_{\mathrm{c}}} \quad \text{and} \quad \frac{d\lambda_1}{d\alpha} = 0 \qquad \text{for } s \geq 2 \tag{7.242}$$

and thus the information change (7.230) into

$$\sum_{s=1}^{N} \frac{dI_s}{d\alpha}\bigg|_{\alpha=0} = -\frac{\kappa}{\gamma}\frac{N}{D_{\mathrm{c}}} \ . \tag{7.243}$$

$D_{\mathrm{c}}/N$ can be interpreted as the critical inversion per atom, $d_{\mathrm{c}}$. For typical data of the helium-neon laser we obtain for a complex laser amplitude

$$\sum_{s} \frac{dI_s}{d\alpha}\bigg|_{\alpha=0} \approx 10^6 \ ; \qquad \frac{dI_u}{d\alpha}\bigg|_{\alpha=0} \approx 10^8 \ . \tag{7.244}$$

# 8. Direct Determination of Lagrange Multipliers

## 8.1 Information Entropy of Systems Below and Above Their Critical Point

It is well known from the theory of nonequilibrium phase transitions that the distribution function of the order parameters simplifies when the system is brought to a region well below or well above the critical point. The critical fluctuations of the system close to a critical point then give way to normal fluctuations described by Gaussian distributions of the order parameters. In addition, below the critical point, i.e. in the disordered phase, the order parameters $\xi_u$ are relatively small so that in

$$P_s(\xi_s|\xi_u) = \mathcal{N} \exp\left[ -\frac{(\xi_s - g_s(\xi_u)/|\lambda_s|)^2}{Q_s} \right] \tag{8.1}$$

we may put $\xi_u = 0$.

Above threshold, the system settles down (by means of symmetry breaking) into a state $\xi_u = \xi_0 \neq 0$, but where again the fluctuations are small so that we may replace $\xi_u$ by $\xi_0$ in (8.1). Thus the joint distribution function $P(\xi_u, \xi_s)$ becomes a *Gaussian.* In this section we shall show how we can determine the Lagrange multipliers explicitly using only the first and second moment. The higher order moments are then uniquely determined. The comparison between the higher moments which are thus calculated, and the experimentally determined moments may then serve as a check of the validity of the Gaussian approximation.

Let us write the distribution function whose Lagrange multipliers we wish to determine in the usual form

$$f(\lambda, \boldsymbol{q}) = \exp\left[ -\lambda - \sum_j \lambda_j q_j - \sum_{jk} \lambda_{jk} q_j q_k \right] , \tag{8.2}$$

and let us compare it with the following distribution function

$$\tilde{f}(\boldsymbol{q}) = (2\pi)^{-n/2} |Q|^{-1/2} \exp[-\tfrac{1}{2}(\boldsymbol{q} - \boldsymbol{m})^T Q^{-1} (\boldsymbol{q} - \boldsymbol{m})] , \tag{8.3}$$

where we use the vector notation. The superscript $T$ means transposed vector, and $Q = (Q_{ik})$ is a matrix; $|Q|$ is the determinant of $Q$. The function (8.3) is normalized to one in $\boldsymbol{q}$-space as can readily be shown. One can also easily convince oneself that

$$m_j = \langle q_j \rangle \qquad \text{and that} \tag{8.4}$$

$$Q_{ij} = \langle (q_i - m_i)(q_j - m_j) \rangle \equiv \langle q_i q_j \rangle - \langle q_i \rangle \langle q_j \rangle \, . \tag{8.5}$$

This means that once the moments $\langle q_i \rangle$ and $\langle q_i q_j \rangle$ are known (or given), we may determine $\boldsymbol{m}$ and $Q$ occurring in (8.3).

Therefore, in order to determine the Lagrange multipliers in (8.2), we must establish a relationship between $\boldsymbol{m}$ and $Q$ on the one hand, and between $\lambda$, $\lambda_j$, $\lambda_{jk}$ on the other. The steps are trivial, and the result reads

$$\exp[-\lambda] = (2\pi)^{-n/2} |Q|^{-1/2} \exp[-\tfrac{1}{2} \boldsymbol{m}^T Q^{-1} \boldsymbol{m}] \tag{8.6}$$

$$\lambda_{jk} = \tfrac{1}{2} [Q^{-1}]_{jk} \tag{8.7}$$

$$\lambda_j = -\tfrac{1}{2} \sum_k m_k \{ [Q^{-1}]_{kj} + [Q^{-1}]_{jk} \} \, . \tag{8.8}$$

Now we may use formula (3.47), which we may cast in the form

$$i = \lambda + \sum_j \lambda_j \langle q_j \rangle + \sum_{jk} \lambda_{jk} \langle q_j q_k \rangle \, . \tag{8.9}$$

Inserting (8.6–8) into (8.9) and using (8.4,5) we obtain

$$i = \frac{n}{2} \ln 2\pi + \frac{1}{2} \ln |Q| + \frac{1}{2} \sum_{jk} [Q^{-1}]_{jk} \underbrace{(\langle q_j q_k \rangle - \langle q_j \rangle \langle q_k \rangle)}_{Q_{kj}} \, . \tag{8.10}$$

But the matrix relation

$$\sum_{jk} [Q^{-1}]_{jk} Q_{kj} = \sum_j (Q^{-1} Q)_{jj} = \sum_j 1 = n \tag{8.11}$$

allows us to obtain our final result for the information belonging to the distribution function (8.2) with the constraints (8.4,5):

$$i = \frac{n}{2} \ln 2\pi + \frac{1}{2} \ln |Q| + \frac{n}{2} \, . \tag{8.12}$$

Quite evidently, the first moments $\langle q_i \rangle$ drop out.

A comparison with (7.61) shows that (8.12) can be interpreted as a sum over $n$ information entropies of independent Gaussian distributions of widths $1/\alpha_k$, where

$$\ln |Q| = -\sum_k \ln 2\alpha_k \, . \tag{8.13}$$

**Exercise:** Verify the normalization of $\tilde{f}(\boldsymbol{q})$ in (8.3) and also the relations (8.4,5).

*Hint:* Diagonalize $Q$ by means of a transformation $\boldsymbol{q} = A\xi$.
Verify the relations (8.6–8).

*Hint:* Multiply out the brackets in (8.3) and compare (8.3) with (8.2) term by term in the exponent.

## 8.2 Direct Determination of Lagrange Multipliers Below, At and Above the Critical Point

In this section we wish to show how we can determine the Lagrange multipliers which we need in the maximum information entropy principle without the evaluation of integrals. The price we have to pay is the following: We shall have to assume that we are dealing with a specific kind of transition, for instance a soft mode transition as treated in Sects. 7.4–9. We shall proceed in two steps. Starting from the constraints of the form

$$f_j = \langle q_j \rangle \ , \qquad f_{jj'} = \langle q_j q_{j'} \rangle, \dots \tag{8.14}$$

where we consider moments up to the fourth order, we transform the variables $q_j$ to a new set of variables and determine the corresponding transformations. Then we specialize to the case of a soft mode instability and show how the unknown Lagrange multipliers can be determined explicitly. At the same time self-consistency relations for the constraints (8.14) are derived. By means of these conditions one may check whether the system in question undergoes a soft mode transition in accordance with our original hypothesis.

To facilitate the calculation we shall introduce the following notation: Taking for instance $f_{jj'}$ in the form (8.14) we may form a matrix and this matrix can be considered as a direct product of the vector $\boldsymbol{q}$ with itself

$$\langle \boldsymbol{q} \circ \boldsymbol{q} \rangle = (\langle q_j q_{j'} \rangle) \ . \tag{8.15}$$

In a similar fashion with respect to three indices $j, j', j''$, we may form

$$\langle \boldsymbol{q} \circ \boldsymbol{q} \circ \boldsymbol{q} \rangle = (\langle q_j q_{j'} q_{j''} \rangle) \ , \tag{8.16}$$

where on the right-hand side we have to take all the combinations of $j, j', j''$. The fourth order moments can be treated in a similar fashion. We now have a look at the microscopic approach where we have represented the state vector $\boldsymbol{q}$ as a superposition of the form

$$\boldsymbol{q} = \boldsymbol{q}_0 + \sum_k \xi_k \boldsymbol{v}_k \ , \tag{8.17}$$

where $\boldsymbol{q}_0$ is a constant state vector which we shall put equal to zero

$$\boldsymbol{q}_0 = 0 \ . \tag{8.18}$$

It is not difficult, however, to include the case where $\boldsymbol{q}_0 \neq 0$. $\xi_k$ are stochastic variables whereas $\boldsymbol{v}_k$ are constant vectors which remain to be determined. We introduce the adjoint vector by means of

$$\bar{\boldsymbol{v}}_k = (v_{k1}^*, v_{k2}^*, \dots, v_{kn}^*) \ , \tag{8.19}$$

and we assume the orthogonality relation

$$\bar{v}_k v_{k'} = \delta_{kk'} \, . \tag{8.20}$$

Taking the average over (8.17) we immediately obtain

$$\langle \boldsymbol{q} \rangle = \sum_k \langle \xi_k \rangle \boldsymbol{v}_k \, . \tag{8.21}$$

Multiplying (8.21) by means of the adjoint vector $\bar{\boldsymbol{v}}_k$, we obtain

$$\bar{\boldsymbol{v}}_k \langle \boldsymbol{q} \rangle = \langle \xi_k \rangle \tag{8.22}$$

as a result of (8.20). Introducing the abbreviation

$$C_k \equiv \bar{\boldsymbol{v}}_k \langle \boldsymbol{q} \rangle \tag{8.23}$$

we may write (8.22) in the form

$$\langle \xi_k \rangle = C_k \, . \tag{8.24}$$

Multiplying (8.15) by the adjoint vectors $\bar{\boldsymbol{v}}$ in an appropriate manner and using the relation (8.20) we immediately find

$$\langle \xi_k \xi_{k'} \rangle = \langle \bar{\boldsymbol{v}}_k \boldsymbol{q} \cdot \bar{\boldsymbol{v}}_{k'} \boldsymbol{q} \rangle \equiv C_{kk'} \, , \tag{8.25}$$

where the last equation is a definition of $C_{kk'}$. The middle part of (8.25) is defined more explicitly by

$$\langle \bar{\boldsymbol{v}}_k \boldsymbol{q} \cdot \bar{\boldsymbol{v}}_{k'} \boldsymbol{q} \rangle = \sum_l v_{kl}^* q_l \cdot \sum_{l'} v_{k'l'}^* q_{l'} \, . \tag{8.26}$$

In exactly the same manner we find

$$\langle \xi_k \xi_{k'} \xi_{k''} \rangle = \langle \bar{\boldsymbol{v}}_k \boldsymbol{q} \cdot \bar{\boldsymbol{v}}_{k'} \boldsymbol{q} \cdot \bar{\boldsymbol{v}}_{k''} \boldsymbol{q} \rangle = C_{kk'k''} \, , \tag{8.27}$$

and

$$\langle \xi_k \xi_{k'} \xi_{k''} \xi_{k'''} \rangle = \left\langle \prod_{\nu=1}^{4} \bar{\boldsymbol{v}}_{k_\nu} \boldsymbol{q} \right\rangle = C_{kk'k''k'''} \tag{8.28}$$

for the third and fourth order moments. So far we have transformed the old moments in terms of $\boldsymbol{q}$ into new moments in terms of the new variables $\xi_k$. In the following we shall assume that the old moments (8.14) are given experimentally, whereas the new moments

$$\langle \xi_k \xi_{k'} \ldots \rangle \tag{8.29}$$

together with the vectors $\boldsymbol{v}_k$ must still be determined. To this end we now explicitly assume that the phenomenon we are dealing with and which is described by the moments (8.14) is a soft mode instability. To proceed further we remind the reader of the approach at the microscopic level. Below threshold, the system is very well approximated by a Gaussian distribution in all the variables of the system. Then it is sufficient to deal with the first and second moments only. In order to find the appropriate mode vectors $\boldsymbol{v}_k$ we diagonalize

$$\langle \boldsymbol{q} \circ \boldsymbol{q} \rangle \; . \tag{8.30}$$

By means of the diagonalization we obtain real eigenvalues which represent the second moments of the modes. Now we know that within a soft mode instability one mode will show critical fluctuations when we approach the critical point. Thus we expect that one eigenvalue will become very large whereas all others remain small. The eigenvalue becoming large will then denote the unstable mode and the other eigenvalues must belong to the stable modes. In this way we can distinguish between the soft mode and the stable modes.

Note, however, that in contrast to the mathematical linearization in which the eigenvalue of the soft mode will diverge, in the present case the eigenvalue will remain bounded because we are dealing in (8.30) with experimental data where the nonlinearities will prevent the eigenvalue from becoming infinitely large.

We now make the distinction between unstable and stable modes:

$$\boldsymbol{v}_j \rightarrow \boldsymbol{v}_u, \boldsymbol{v}_s \; . \tag{8.31}$$

Furthermore we shall consider the typical soft mode transition in which, for to symmetry reasons,

$$\langle \xi_u \rangle = 0 \; . \tag{8.32}$$

In the following we shall additionally assume that

$$\langle \xi_u \xi_s \rangle = 0 \; . \tag{8.33}$$

To be quite explicit we shall further assume that the joint probability function over which the average is taken has the form

$$P(\xi_u, \xi_s) = f(\xi_u) \prod_s P_s(\xi_s | \xi_u) \; , \tag{8.34}$$

which we may take from a microscopic approach or from the macroscopic approach according to Chap. 6. Here $f$ is the distribution function of the order parameter and $P_s$ is the conditional probability for the enslaved mode amplitude $\xi_s$ given by

$$P_s = \mathcal{N}_s \exp\left\{ -\frac{[\xi_s - g_s(\xi_u)]^2}{Q_s} \right\} . \tag{8.35}$$

In the following we shall consider a special case, but one which is quite often fulfilled, namely that $g_s$ is an even function of the order parameter. By making the transformation

$$\xi_s = \bar{\xi}_s + g_s(\xi_u) \; , \tag{8.36}$$

average values can easily be calculated, for instance

$$\langle \xi_s \rangle = \langle \bar{\xi}_s + g_s(\xi_u) \rangle = \langle \bar{\xi}_s \rangle + \langle g_s(\xi_u) \rangle \ . \tag{8.37}$$

On account of (8.35) we have the condition

$$\langle \bar{\xi}_s \rangle = 0 \ . \tag{8.38}$$

In the following we shall abbreviate the variables $\xi_s$ and $\xi_u$ by the indices $s$ and $u$ alone:

$$\xi_s \rightarrow s \ , \qquad \xi_u \rightarrow u \ . \tag{8.39}$$

In this new notation (8.37) reads

$$\langle s \rangle = \langle \bar{s} + g_s \rangle = \langle \bar{s} \rangle + \langle g_s \rangle \ . \tag{8.40}$$

We now study the expressions (8.24,25,27,28) in the light of the specific form of the distribution function, namely (8.34,35). From (8.32,23,24) we readily find

$$\bar{v}_u \langle \boldsymbol{q} \rangle = 0 \ , \tag{8.41}$$

which is now a condition on the average value of $\boldsymbol{q}$, as well as

$$\langle s \rangle = \langle g_s(u) \rangle = C_s \ , \tag{8.42}$$

which, as we shall see, will fix $g_s$ under suitable assumptions. From (8.25), by identifying $k = u$, $k' = u$, we obtain

$$\langle u^2 \rangle = C_{uu} \ . \tag{8.43}$$

Identifying $k = u$, $k' = s$ and using the forms (8.34,35) we obtain

$$\langle us \rangle = 0 = C_{us} \ , \tag{8.44}$$

which is a condition on $C_{us}$. Finally identifying $k = s$, $k' = s'$, we obtain

$$\langle ss' \rangle = \langle \bar{s}^2 \rangle \delta_{ss'} + \langle g_s g_{s'} \rangle = C_{ss'} \ . \tag{8.45}$$

By using

$$\langle \bar{s}^2 \rangle = Q_s \tag{8.46}$$

we may cast (8.45) into the form

$$Q_s \delta_{ss'} + \langle g_s g_{s'} \rangle = C_{ss'} \ . \tag{8.47}$$

As we shall see later, this equation may serve to determine $Q_s$. We now study the consequences of the relation

$$\langle \xi_k \xi_{k'} \xi_{k''} \rangle = C_{kk'k''} \ . \tag{8.48}$$

1) We first make the identifications

$$k = u \ , \qquad k' = s' \ , \qquad k'' = s'' \tag{8.49}$$

or permutations thereof. We readily find

$$C_{us's''} = 0 \tag{8.50}$$

which results from the properties of the left-hand side of (8.48). This is again a condition that the experimentally found moments must satisfy.

2) We now make the identification

$$k = u \ , \qquad k' = u \ , \qquad k'' = s'' \tag{8.51}$$

or permutations thereof. We obtain

$$\langle \xi_u^2 g_{s''}(\xi_u) \rangle = C_{uus''} \ . \tag{8.52}$$

3) We consider the case

$$k = u \ , \qquad k' = u \ , \qquad k'' = u \tag{8.53}$$

and readily obtain

$$C_{uu'u''} = 0 \tag{8.54}$$

which again is a condition that must be satisfied by the experimentally found moments.

4) We identify

$$k = s \ , \qquad k' = s' \ , \qquad k'' = s'' \ . \tag{8.55}$$

We now use the decomposition (8.36) in the notation used in (8.40). We then obtain instead of (8.48)

$$\langle (\bar{s} + g_s)(\bar{s}' + g_{s'})(\bar{s}'' + g_{s''}) \rangle \ . \tag{8.56}$$

We now multiply out the factors in (8.56) and consider the individual terms thus resulting. We readily obtain

$$\langle \bar{s}\bar{s}'\bar{s}'' \rangle = 0 \ . \tag{8.57}$$

Furthermore we obtain

$$\langle \bar{s}\bar{s}' g_{s''} \rangle = \delta_{ss'} \langle \bar{s}^2 \rangle \langle g_{s''} \rangle \ , \tag{8.58}$$

$$\langle \bar{s} g_{s'} g_{s''} \rangle = 0 \ , \tag{8.59}$$

and finally the expression

$$\langle g_s g_{s'} g_{s''} \rangle \ , \tag{8.60}$$

which cannot be simplified further for the time being. Collecting the terms (8.57–60) together with their appropriate permutations, we obtain the following relation

$$Q_s \delta_{ss'} \langle g_{s''} \rangle + Q_s \delta_{ss''} \langle g_{s'} \rangle + Q_{s'} \delta_{s's''} \langle g_s \rangle + \langle g_s g_{s'} g_{s''} \rangle = C_{ss's''} \ , \tag{8.61}$$

which follows from (8.48). For the sake of completeness we now represent the results stemming from

$$\langle \xi_k \xi_{k'} \xi_{k''} \xi_{k'''} \rangle = Q_{kk'k''k'''} \ , \tag{8.62}$$

but the impatient reader may immediately proceed to the final result presented in formula (8.83) below.

In assigning the set of indices $k, k', k'', k'''$ we distinguish various cases.

$$1. \ u, u', u'', u''' \ . \tag{8.63}$$

We obtain

$$\langle u^4 \rangle = C_{uuuu} \ . \tag{8.64}$$

$$2. \ u, u, u, s'' \quad \text{for which} \tag{8.65}$$

$$C_{uuus''} = 0 \ . \tag{8.66}$$

$$3. \ u, u, s'', s''' \ . \tag{8.67}$$

We make the decomposition (8.36) which yields

$$\langle u^2(\bar{s}'' + g_{s''})(\bar{s}''' + g_{s'''}) \rangle \ , \tag{8.68}$$

and consider the individual terms for which we readily obtain

$$\langle u^2 \bar{s}'' \bar{s}''' \rangle = \langle u^2 \rangle Q_{s''} \delta_{s''s'''} \tag{8.69}$$

$$\langle u^2 \bar{s}'' g_{s'''} \rangle = 0 \quad \text{and} \tag{8.70}$$

$$\langle u^2 g_{s''} g_{s'''} \rangle \neq 0 \ . \tag{8.71}$$

$$4. \ u, s', s'', s''' \ . \tag{8.72}$$

For symmetry reasons as mentioned above we have

$$C_{us's''s'''} = 0 \ . \tag{8.73}$$

$$5. \ s, s', s'', s''' \ . \tag{8.74}$$

Again making the appropriate decomposition, we have to consider the expression

$$\langle (\bar{s} + g_s)(\bar{s}' + g_{s'})(\bar{s}'' + g_{s''})(\bar{s}''' + g_{s'''}) \rangle \ , \tag{8.75}$$

and may then treat the individual terms. In order to evaluate the term

$$\langle \bar{s} \bar{s}' \bar{s}'' \bar{s}''' \rangle \tag{8.76}$$

we have to consider those combinations of the indices $s$ which give rise to non-vanishing mean values. These are those combinations where two pairs of indices coincide, i.e. we must fulfill the relations

$$s = s' \qquad s'' = s'''$$
$$s = s'' \qquad s' = s''' \tag{8.77}$$
$$s = s''' \qquad s' = s'' \ .$$

Thus, instead of (8.76) we find the expression

$$\langle \bar{s}\bar{s}'\bar{s}''\bar{s}''' \rangle = \langle \bar{s}^2 \rangle \langle \bar{s}''^2 \rangle \delta_{ss'} \delta_{s''s'''} + \langle \bar{s}^2 \rangle \langle \bar{s}'^2 \rangle \delta_{ss''} \delta_{s's'''}$$
$$+ \langle \bar{s}^2 \rangle \langle \bar{s}'^2 \rangle \delta_{ss'''} \delta_{s's''} \ . \tag{8.78}$$

The next term to be considered is of the type

$$\langle \bar{s}\bar{s}'\bar{s}'' g_{s'''} \rangle = 0 \tag{8.79}$$

and its permutations in which $g_s$ replaces one of the other $s$-terms. A further term consists of

$$\langle \bar{s}\bar{s}' g_{s''} g_{s'''} \rangle = \langle \bar{s}^2 \rangle \delta_{ss'} \langle g_{s''} g_{s'''} \rangle$$
$$= Q_s \delta_{ss'} \langle g_{s''} g_{s'''} \rangle \ , \tag{8.80}$$

where we must choose $\bar{s} = \bar{s}'$ in order to find a non-vanishing result. We also find other terms of this type by exchanging the indices. For terms of the form

$$\langle \bar{s} g_{s'} g_{s''} g_{s'''} \rangle = 0 \tag{8.81}$$

we obtain the indicated result. Finally we have the term

$$\langle g_s g_{s'} g_{s''} g_{s'''} \rangle \ . \tag{8.82}$$

Collecting all the terms from (8.64–82) we obtain our final result in the form

$$Q_s Q_{s''} \delta_{ss'} \delta_{s''s'''} + Q_s Q_{s'} \delta_{ss''} \delta_{s's'''} + Q_s Q_{s'} \delta_{ss'''} \delta_{s's''}$$
$$+ Q_s [\delta_{ss'} \langle g_{s''} g_{s'''} \rangle + \delta_{ss''} \langle g_{s'} g_{s'''} \rangle + \delta_{ss'''} \langle g_{s'} g_{s''} \rangle]$$
$$+ Q_{s'} [\delta_{s's''} \langle g_s g_{s'''} \rangle + \delta_{s's'''} \langle g_s g_{s''} \rangle]$$
$$+ Q_{s''} \delta_{s''s'''} \langle g_s g_{s'} \rangle + \langle g_s g_{s'} g_{s''} g_{s'''} \rangle = C_{ss's''s'''} \tag{8.83}$$

which stems from (8.62) for the case in which all indices are of $s$-type. Let us now have a look at our results in order to fix the unknown quantities which occur as Lagrange multipliers in the distribution function (8.34,35). We confine our analysis to a region not too far above threshold in which we may approximate $g_s$ by the leading term, namely

$$g_s(u) \approx a_s u^2 \ . \tag{8.84}$$

We further make an explicit hypothesis about the form of the distribution function of the order parameters, namely

$$f(u) = \mathcal{N} \exp(\alpha u^2 - \beta u^4) \tag{8.85}$$

in accordance with our previous results from the microscopic and macroscopic theory. Thus, we now have the following constants to determine: $\alpha$, $\beta$, $Q_s$, $a_s$. The constants $\alpha$ and $\beta$ and the normalization factor $\mathcal{N}$ are determined by means of (8.43,64), i.e. by the relations

$$\langle u^2 \rangle = C_{uu} \tag{8.86}$$

$$\langle u^4 \rangle = C_{uuuu} \ . \tag{8.87}$$

$a_s$ can be determined by the relation (8.42), i.e. by

$$a_s \langle u^2 \rangle = C_s \ . \tag{8.88}$$

Finally $Q_s$ is fixed by (8.47) when we choose $s = s'$, and by the relation

$$\langle g_s^2 \rangle = a_s^2 \langle u^4 \rangle \ , \tag{8.89}$$

which follows from (8.84). Thus $Q_s$ is determined by

$$Q_s = C_{ss} - \frac{C_s^2 \langle u^2 \rangle^2}{\langle u^4 \rangle} \ . \tag{8.90}$$

In this way we are able to fix all unknown constants and one can easily convince oneself that in this way we have determined the Lagrange multipliers which occur in (3.38) or now equivalently in the specific form (8.34,35,84,85). It should be noted that the system is overdetermined, i.e., there are more equations than unknowns. Therefore, the remaining relations which have been established above and in which the measured quantities $C$ occur, are self-consistency conditions for the occurrence of a non-equilibrium phase transition with a single soft mode.

It is not difficult to generalize the above results to more complicated cases, for instance when (8.84) also contains odd and/or higher powers of $\xi_u$ or when several order parameters occur.

# 9. Unbiased Modeling of Stochastic Processes: How to Guess Path Integrals, Fokker-Planck Equations and Langevin-Îto Equations

## 9.1 One-Dimensional State Vector

In the previous chapters we have shown how the maximum information entropy principle allows us to derive the distribution functions of nonequilibrium systems by a suitable choice of constraints. While in these chapters we focussed our attention on the steady state distribution function, in the present chapter we wish to study time-dependent distribution functions. A suitable starting point is provided by the maximum calibre principle due to Jaynes, who extended his maximum entropy principle by maximizing an entropy-like functional over a space-time domain. Jaynes himself applied his principle to derive important formulas applicable to irreversible thermodynamics in order to study relaxation processes.

Here we wish to show that by choosing appropriate constraints we may treat arbitrary physical (or other) systems close to, or far away from, equilibrium. The formulation of our procedure becomes particularly elegant if we assume that the underlying process is continuous Markovian. In such a case we find a very quick access to the path integrals and the Fokker-Planck equation, whose drift and diffusion coefficients can be determined explicitly. Thus our method can be considered as a tool to derive the underlying deterministic and fluctuating forces from experimental data.

We consider a sequence of times $t_0, t_1, \ldots, t_N$ at which the system is measured, with measured values of the state vector $\boldsymbol{q}_i$ at time $t_i$. We wish to make an unbiased estimate on the joint distribution function

$$P_N = P(\boldsymbol{q}_N, t_N; \boldsymbol{q}_{N-1}, t_{N-1}; \ldots; \boldsymbol{q}_0, t_0) \; . \tag{9.1}$$

To this end we maximize the information $i$ or, in the formulation of Jaynes, the calibre

$$i = -\int \mathscr{D}q P \ln P \tag{9.2}$$

with respect to $P$ under given constraints. $\mathscr{D}q$ is the integration volume element over the space spanned by all vectors $\boldsymbol{q}_j$.

As is well known, in this type of problem the crucial task is the adequate choice of the constraints. This choice is greatly facilitated if we assume that the underlying process which leads to (9.1) is Markovian. Under this assumption (9.1) can be split into

$$P_N = \prod_{l=0}^{N-1} P(\boldsymbol{q}_{l+1}, t_{l+1} | \boldsymbol{q}_l, t_l) \cdot P_0(\boldsymbol{q}_0, t_0) \; . \tag{9.3}$$

In order to simplify the notation we shall drop the times $t_l$ so that the individual factors of (9.3) acquire the form

$$P(\boldsymbol{q}_{i+1}|\boldsymbol{q}_i) \ . \tag{9.4}$$

The expression in (9.4) is the conditional probability. In the first part of this chapter we shall assume that we are dealing with a stationary process.

In order to bring out the essentials we consider a single variable $q$ instead of a state vector. We now introduce the constraints

$$f_1 = \langle q_{i+1} \rangle_{q_i} \quad \text{and} \tag{9.5}$$

$$f_2 = \langle q_{i+1}^2 \rangle_{q_i} \tag{9.6}$$

which are defined by

$$f_k = \int P(q_{i+1}|q_i) q_{i+1}^k \, dq_{i+1} \ . \tag{9.7}$$

In this way we are considering the first and second moment of $q$ at time $t_{i+1}$ under the condition that $q_i$ was measured at time $t_i$. Applying the maximum information entropy principle to (9.4) with the constraints (9.5) and (9.6) we immediately obtain

$$P(q_{i+1}|q_i) = \exp(\lambda + \lambda_1 q_{i+1} + \lambda_2 q_{i+1}^2) \ . \tag{9.8}$$

Note that $f_1$ and $f_2$ are functions of $q_i$ so that $\lambda$, $\lambda_1$, $\lambda_2$ will also become functions of $q_i$, at least in general. Since we shall have the limit $\tau \to 0$ in mind, in what follows we write $q(i + \tau)$ instead of $q(i + 1)$. We now impose the usual requirements on (9.8) namely that

$$\int P(q_{i+\tau}|q_i) \, dq_{i+\tau} = 1 \ . \tag{9.9}$$

Furthermore we have to require

$$\tau \to 0: P(q_{i+\tau}|q_i) \to \delta(q_{i+\tau} - q_i) \ . \tag{9.10}$$

The $\delta$-function of the r.h.s. is to be understood in a specific manner, namely that $P$ behaves similarly to a $\delta$-function provided we integrate over $q_{i+\tau}$.

To exploit the conditions (9.9) and (9.10) further we cast the r.h.s. of (9.8) into the form

$$\exp\left[-|\lambda_2|\left(q_{i+\tau} - \frac{\lambda_1}{2|\lambda_2|}\right)^2 + \frac{\lambda_1^2}{4|\lambda_2|} + \lambda\right]. \tag{9.11}$$

Since

$$\int \exp(-|\lambda_2| q^2) \, dq = \sqrt{\pi/|\lambda_2|} \tag{9.12}$$

the normalization condition (9.9) can be written as

$$\exp\left(\lambda + \frac{\lambda_1^2}{4|\lambda_2|}\right) = \sqrt{\frac{|\lambda_2|}{\pi}} \tag{9.13}$$

so that $P$ acquires the form

$$P(q_{i+\tau}|q_i) = \sqrt{|\lambda_2(q_i)|/\pi}\exp\left[-|\lambda_2|\left(q_{i+\tau} - \frac{\lambda_1}{2|\lambda_2|}\right)^2\right] . \tag{9.14}$$

In order that (9.14) becomes a $\delta$-function for $\tau \to 0$ we must require the behavior

$$|\lambda_2| = \frac{G}{\tau} \quad \text{for } \tau \to 0 . \tag{9.15}$$

(Another function of $\tau$ which vanishes for $\tau \to 0$ would just mean another scaling of time, the scaling function being a function of time itself.)

From (9.10) we further deduce that for $\tau \to 0$

$$q_{i+\tau} - \frac{\lambda_1}{2|\lambda_2|} \to q_{i+\tau} - q_i \tag{9.16}$$

or, in other words,

$$\tau \to 0: \frac{\lambda_1}{2|\lambda_2|} \to q_i . \tag{9.17}$$

For finite but small times $\tau$ (9.17) clearly generalizes to

$$\frac{\lambda_1}{2|\lambda_2|} = q_i + \tau K(q_i) + \tau^2 H(q_i) + \dots . \tag{9.18}$$

Evidently we can state that the l.h.s. of (9.18) is required to behave as an analytic function of $\tau$ for $\tau$ small enough. With (9.14) and (9.18) we obtain the main result of this section, namely the conditional probability in the short time limit.

$$P(q_{i+\tau}|q_i) = \sqrt{G/(\tau\pi)}\exp\left\{-\frac{G}{\tau}[q_{i+\tau} - q_i - \tau K(q_i)]^2\right\} . \tag{9.19}$$

Note that $G$ may be still a function of $q_i$.

## 9.2 Generalization to a Multidimensional State Vector

In this section we want to make an unbiased guess of the conditional probability occurring in (9.3) under the assumption of a stationary continuous Markov process. We denote the components of the state vectors at times $i + \tau$ and $i$ by

$$q_l(i + \tau), \qquad q_k(i) . \tag{9.20}$$

As a generalization of the first and second moment of a single variable (9.5,6), we introduce the following first and second conditional moments

$$f_{1,l} = \langle q_l(i+\tau)\rangle_{q(i)} \tag{9.21}$$

$$f_{2,l,k} = \langle q_l(i+\tau) q_k(i+\tau)\rangle_{q(i)} \ . \tag{9.22}$$

Applying the maximum information entropy principle to the conditional probability we obtain

$$P(q(i+\tau)|q(i)) = \exp\left[\lambda + \sum_l \lambda_l q_l(i+\tau) + \sum_{kl} \lambda_{kl} q_k(i+\tau) q_l(i+\tau)\right] \tag{9.23}$$

where the Lagrange multipliers may still be functions of $q(i)$. In the following we shall assume that the problem is nondegenerate, which means that

$$\operatorname{Det} \lambda_{kl} \neq 0 \ . \tag{9.24}$$

A simple algebraic manipulation shows that (9.23) can be cast into the form

$$P = \exp\left\{\tilde{\lambda} + \sum_{kl} [q_k(i+\tau) - h_k] \lambda_{kl} [q_l(i+\tau) - h_l]\right\} \tag{9.25}$$

where $h_l$ is defined by

$$h_l = \sum_m A_{lm} \lambda_m [q(i)] \ . \tag{9.26}$$

We denote the matrix with elements $\lambda_{kl}$ by $\Lambda$. $A$ in (9.26) is defined by

$$A = \tfrac{1}{2}\Lambda^{-1} \ . \tag{9.27}$$

Note that the matrix $\Lambda$ is symmetric. Furthermore $\tilde{\lambda}$ occurring in (9.25) is given by

$$\tilde{\lambda} = \lambda - \sum_{kl} h_k \lambda_{kl} h_l \ . \tag{9.28}$$

In generalization of condition (9.10) we require that for $\tau \to 0$ $P$ behaves as a $\delta$-function provided we integrate over $q(i+\tau)$

$$P \to \delta[q(i+\tau) - q(i)] \ . \tag{9.29}$$

The singular behavior of the $\delta$-function is achieved by putting

$$\lambda_{kl} = -\frac{1}{\tau} G_{kl}[q(i)] \ . \tag{9.30}$$

Generalizing (9.18) we expand $h_k$ into a power series in $\tau$ and keep the first term

$$h_k = q_k(i) + \tau K_k[q(i)] + \dots \tag{9.31}$$

Taking the results (9.25,28,30,31) together, we can cast the conditional probability into the form

$$P = \exp[\tilde{\lambda}(\boldsymbol{q}_i)]\exp\left\{-\frac{1}{\tau}\sum_{kl}[q_k(i+\tau) - q_k(i) - \tau K_k]\right.$$

$$\left.\times\, G_{kl}[q_l(i+\tau) - q_l(i) - \tau K_l]\right\} . \tag{9.32}$$

Abbreviating the first factor in (9.32) by $\mathscr{N}(\boldsymbol{q}_i)$ we may write the result of (9.32) also in the form

$$P = \mathscr{N}(\boldsymbol{q}_i)\exp(\mathscr{L}_i) . \tag{9.33}$$

Inserting (9.32) into (9.3), where $P_0$ is a prescribed distribution function on which guesses can be made if required by prescribing a certain number of moments of $\boldsymbol{q}_0$, we obtain our desired final result, namely the joint distribution function (9.1).

In practical applications one is mainly interested in the conditional probability which is the distribution function of $\boldsymbol{q}$ at time $T$ provided the state vector $\boldsymbol{q}$ was known at time $T = 0$. This probability can be obtained by integrating out over all $\boldsymbol{q}$'s at intermediate times. Setting

$$N\tau = T \tag{9.34}$$

and identifying $\mathscr{L}_i$ by means of (9.33,32) we write the desired result as

$$P(\boldsymbol{q}(T)|\boldsymbol{q}(0)) = \int \mathscr{D}q \exp\left(\int_{i=0}^{N-1} \mathscr{L}_i\right) \tag{9.35}$$

where $\mathscr{N}(\boldsymbol{q}_i)$ is contained in $\mathscr{D}q$.

In the limit $\tau \to 0$ the sum over $i$ is converted into an integral and (9.35) is just the familiar path integral formulation of $P$. It is obvious from Sect. 2.5 on the path integral solution of the Fokker-Planck equation that (9.35) with (9.33) and (9.32) is the solution of a Fokker-Planck equation with drift coefficients $K_l$ and the diffusion matrix given by the inverse matrix with the components $G_{kl}$.

Within our context it suffices to know that (9.35) is the solution of a Fokker-Planck equation with the just mentioned drift and diffusion coefficients (compare Sect. 9.4).

To conclude this section we mention that our approach can easily be extended to a time-dependent continuous Markov process. In such a case, in general, the substitutions

$$\boldsymbol{K}[\boldsymbol{q}(i)] \to \boldsymbol{K}[\boldsymbol{q}(i); t_i] , \tag{9.36}$$

$$G_{kl}[\boldsymbol{q}(i)] \to G_{kl}[\boldsymbol{q}(i); t_i] \tag{9.37}$$

must be made.

## 9.3 Correlation Functions as Constraints

In the preceding sections we have used conditional moments, i.e. moments of the variables $q_{i+\tau}$ provided that $q_i$ has been measured. In a number of practical applications we may not know $q_i$ but rather we may measure correlation functions of the type

$$\langle q_{i+\tau} q_i \rangle \tag{9.38}$$

which have to be interpreted as averages over joint distribution functions according to

$$\langle q_{i+\tau} q_i \rangle = \int\int dq_{i+\tau} dq_i q_{i+\tau} q_i P(q_{i+\tau}, q_i) \ . \tag{9.39}$$

It will then be our task to make unbiased estimates of the joint distribution functions occurring in (9.39) instead of the conditional probabilities studied previously. To facilitate our task we assume that the measurements are made under steady-state conditions. In this case we can express the joint probability by a product of the conditional probability and the steady state probability distribution

$$P(q_{i+\tau}, q_i) = P(q_{i+\tau} | q_i) P_{st}(q_i) \ . \tag{9.40}$$

Because we have no precise information on $q_i$ but only an information via correlation functions we now need more constraints. The constraints we are going to use are

$$\langle q_i^\mu \rangle \ , \qquad \mu = 1, \ldots, m \ , \tag{9.41}$$

$$\langle q_{i+\tau} q_i^\nu \rangle \ , \qquad \nu = 0, \ldots, n \ , \tag{9.42}$$

$$\langle q_{i+\tau}^2 q_i^\kappa \rangle \ , \qquad \kappa = 0, \ldots, k \tag{9.43}$$

where the numbers $m$, $n$, and $k$ obey certain self-consistency conditions discussed below and must be chosen such that a reasonable convergence of data is achieved. Close to nonequilibrium phase transitions we expect that $n$ and $k$ may be confined to $n = k = 3$. There is a fundamental difference between the present section and the foregoing one. Note that now the multipliers $\lambda_i$ are independent of $q_i$. Because the maximum information entropy principle or, more precisely speaking, the maximum calibre principle refers to two variables $q_{i+\tau}$ and $q_i$ the maximum information entropy principle yields

$$P_{st} = \exp\left( \lambda_{st} + \sum_{\mu=1}^{m} \lambda_{\mu,st} q_i^\mu \right) \tag{9.44}$$

while the maximum calibre principle provides us with the following expression for the joint probability

$$P(q_{i+\tau}, q_i) = \exp[\lambda_0 + a(q_i) + q_{i+\tau} b(q_i) + q_{i+\tau}^2 C(q_i)] \ . \tag{9.45}$$

Here we have used the abbreviations

$$a(q_i) = \sum_{\mu=1}^{m} \lambda_\mu^{(1)} q_i^\mu , \tag{9.46}$$

$$b(q_i) = \sum_{\nu=0}^{n} \lambda_\nu^{(2)} q_i^\nu , \tag{9.47}$$

$$C(q_i) = \sum_{\kappa=0}^{k} \lambda_\kappa^{(3)} q_i^\kappa . \tag{9.48}$$

Using (9.40) we may determine the conditional probability occurring in that formula to be

$$P(q_{i+\tau}|q_i) = \exp[\hat{\lambda} + \hat{a}(q_i) + q_{i+\tau} b(q_i) + q_{i+\tau}^2 C(q_i)] \tag{9.49}$$

with the abbreviations

$$\hat{\lambda} = \lambda_0 - \lambda_{st} \qquad \text{and} \tag{9.50}$$

$$\hat{a} = a - \sum_{\mu=1}^{m} \lambda_{\mu, st} q_i^\mu . \tag{9.51}$$

In this way we have found a form for the conditional probability which can be put in direct analogy to the form (9.8). From now on we may proceed in precisely the same way as in Sect. 9.1 by making the following identifications.

$$\lambda(q_i) \Leftrightarrow \hat{\lambda} + \hat{a}(q_i) , \tag{9.52}$$

$$\lambda_1(q_i) \Leftrightarrow b(q_i) , \tag{9.53}$$

$$\lambda_2(q_i) \Leftrightarrow C(q_i) . \tag{9.54}$$

Note that some self-consistency requirements with respect to the r.h.s. of (9.52–54) must be fulfilled because $\lambda$ depends on $\lambda_1$ and $\lambda_2$ [cf. (9.13)]. This dependence becomes somewhat complicated if $\lambda_2$ is a function of $q_i$ because the relation (9.13) requires, at least in principle, infinitely high powers of $q_i$ or, in other words, infinitely many moments as constraints.

In practical cases, however, we may expect that a few moments will still be sufficient. On the other hand if $\lambda_2$ is dependent of $q_i$, it is a polynomial of order $m = 2n$. It is now quite obvious how the results of this section, which refer to a single variable at given time, can be extended to a state vector at given time. In order to save space we shall not write down the corresponding results explicitly.

In this section we have shown how simple moments or correlation functions, when used as constraints, allow us to derive basic expressions of statistical physics, namely the joint probabilities and path integrals which in turn may be interpreted as solutions of an underlying Fokker-Planck equation. Thus our approach allows us to make guesses about underlying mechanisms of observed experimental data; we may determine the drift coefficients, which are the deterministic forces, and diffusion coefficients, which are a measure of the fluctuating forces. In practical cases we may not always assume that the underlying process is Markovian or continuous

Markovian. In such a situation, however, our approach represents a guess of the underlying deterministic and fluctuating forces. Clearly our approach can be generalized in various directions, for instance taking into account higher order moments or correlation functions. We may make guesses of joint distribution functions or conditional probabilities for longer time intervals. On the other hand in such cases at least the elegance of our approach may be lost.

## 9.4 The Fokker-Planck Equation Belonging to the Short-Time Propagator

The derivation of the Fokker-Planck equation belonging to (9.35) is a well-known procedure. In order to show explicitly which type of Fokker-Planck equation (Îto or Stratonovich) belongs to (9.32) we briefly indicate the main steps. Since the Fokker-Planck equation refers to infinitesimal time steps it is sufficient to consider two subsequent times $i$ and $(i+\tau)$.

The corresponding probability distribution functions are connected by

$$P(\boldsymbol{q}_{i+\tau}) = \int P(\boldsymbol{q}_{i+\tau}|\boldsymbol{q}_i)P(\boldsymbol{q}_i)\,d^N q_i \tag{9.55}$$

where the conditional probability has been given by (9.32). We wish to derive a differential equation for $P(\boldsymbol{q}_{i+\tau})$. To this end we multiply (9.55) by an arbitrary function $g(\boldsymbol{q}_{i+\tau})$ which yields

$$\begin{aligned} A &\equiv \int g(\boldsymbol{q}_{i+\tau})P(\boldsymbol{q}_{i+\tau})\,d^N q_{i+\tau} \\ &= \int d^N q_i P(\boldsymbol{q}_i) \underbrace{\int d^N q_{i+\tau} g(\boldsymbol{q}_{i+\tau})P(\boldsymbol{q}_{i+\tau}|\boldsymbol{q}_i)}_{F(\boldsymbol{q}_i)} \,. \end{aligned} \tag{9.56}$$

To evaluate $F(\boldsymbol{q}_i)$ we introduce a new variable vector

$$\boldsymbol{q}_{i+\tau} - \boldsymbol{q}_i - \tau \boldsymbol{K}(\boldsymbol{q}_i) = \xi \qquad \text{which yields} \tag{9.57}$$

$$F = \int d^N\xi\, \mathcal{N}(\boldsymbol{q}_i)\exp\left(-\xi\frac{G}{\tau}\xi\right) g[\boldsymbol{q}_i + \tau\boldsymbol{K}(\boldsymbol{q}_i) + \xi] \,. \tag{9.58}$$

In order to evaluate the integral up to terms linear in $\tau$ we expand $g$ in (9.58) into a power series of $\xi$ up to second order and into a power series in $\tau$ up to first order. Then, using

$$\int \mathcal{N}(\boldsymbol{q}_i)\exp\left(-\xi\frac{G}{\tau}\xi\right) d^N\xi = 1 \qquad \text{and} \tag{9.59}$$

$$\int \mathcal{N}(\boldsymbol{q}_i)\exp\left(-\xi\frac{G}{\tau}\xi\right) \xi_k\xi_l\, d^N\xi = \frac{\tau}{2}(G^{-1})_{kl} \tag{9.60}$$

and knowing that integrals over odd powers of $\xi$ vanish we readily obtain

$$F(\boldsymbol{q}_i) = g(\boldsymbol{q}_i) + \tau \boldsymbol{K}(\boldsymbol{q}_i) \nabla_{\boldsymbol{q}_i} g(\boldsymbol{q}_i) + \frac{\tau}{4} \sum_{kl} (G^{-1})_{kl} \frac{\partial^2 g}{\partial q_{ik} \partial q_{il}} . \tag{9.61}$$

Inserting (9.61) into (9.56) and performing a partial integration we obtain instead of (9.56) the relation

$$A = \int d^N q_i g(\boldsymbol{q}_i) \left\{ 1 - \tau \nabla_{\boldsymbol{q}_i} K(\boldsymbol{q}_i) + \frac{\tau}{4} \sum_{kl} \frac{\partial^2}{\partial q_{ik} \partial q_{il}} [G^{-1}(\boldsymbol{q}_i)]_{kl} \right\} P(\boldsymbol{q}_i) . \tag{9.62}$$

We now introduce the notation

$$P(\boldsymbol{q}_{i+\tau}) \Rightarrow f(\boldsymbol{q}; t + \tau) , \tag{9.63}$$

$$P(\boldsymbol{q}_i) \Rightarrow f(\boldsymbol{q}; t) . \tag{9.64}$$

Because $g$, which occurs in the middle part of (9.56) and on the r.h.s. of (9.62), is an arbitrary function, the corresponding expressions must be equal even without integration. This leads us to the relation

$$f(\boldsymbol{q}; t + \tau) = f(\boldsymbol{q}; t) + \tau L f(\boldsymbol{q}; t) \tag{9.65}$$

where $L$ is the Fokker-Planck operator

$$Lf = -\nabla_{\boldsymbol{q}}(K(\boldsymbol{q})f) + \frac{1}{4} \sum_{kl} \frac{\partial^2}{\partial q_k \partial q_l} ([G^{-1}(\boldsymbol{q})]_{kl} f) \tag{9.66}$$

which occurs in the Îto calculus. We thus recognize that we have found the short-time propagator for the Îto-Fokker-Planck equation.

## 9.5 Can We Derive Newton's Law from Experimental Data?

In this section we wish to discuss whether our procedure of Sects. 9.1,2,4 can enable us to derive Newton's law

$$\dot{\boldsymbol{p}} = \boldsymbol{K}(\boldsymbol{q}) \tag{9.67}$$

from experimental data concerning a specific motion (e.g. of planets) and a force $\boldsymbol{K}$ giving rise to that motion.

First of all, it is clear that we must consider the limiting case of vanishing noise, for which the conditional probability is a $\delta$-function, so that in the one-dimensional case

$$P(q_{i+\tau} | q_i) = \delta \left( \frac{q_{i+\tau} - q_i}{\tau} - K(q_i) \right) \tag{9.68}$$

or in its multidimensional generalization,

$$P(\boldsymbol{q}_{i+\tau}|\boldsymbol{q}_i) = \delta\left(\frac{\boldsymbol{q}_{i+\tau} - \boldsymbol{q}_i}{\tau} - \boldsymbol{K}(\boldsymbol{q}_i)\right) . \tag{9.69}$$

At first sight it appears that in this way we shall not obtain Newton's law (9.67) but rather

$$\dot{\boldsymbol{q}} = \boldsymbol{K}(\boldsymbol{q}) . \tag{9.70}$$

Such laws had indeed been used in science before the advent of Newtonian mechanics and represented overdamped motion. But we may easily recover Newton's law when we include in the set of observed variables $\boldsymbol{q}$ not only the positions of the particles (or celestial bodies), *but also their momenta (or velocities)*! It is a simple exercise to check what the conditional moments (9.5,6) then look like (when we assume Newtonian mechanics) and to convince oneself that now indeed (9.67) results.

By a translation of our procedure into quantum mechanics one may hope to derive fundamental laws for elementary particles.

**Exercise:** Start from Newton's law $\dot{p} = K(q)$ and evaluate (9.21,22)

$$\langle q_{l,i+\tau}\rangle_{q_{k,i}} , \qquad \langle q_{l,i+\tau}^2\rangle_{q_{k,i}} , \qquad l = 1, 2; k = 1, 2$$

with

$$q_1 \equiv q , \qquad q_2 \equiv p ,$$

for infinitesimal $\tau$. Apply the above formalism to derive (9.69)!

# 10. Application to Some Physical Systems

Today, numerous nonequilibrium phase transitions are known which occur in a wide variety of systems. In Sects. 5.2,3 the single mode and the multimode laser (without phase relations) served as an explicit illustration of our new ideas. In this section we shall first elaborate further on the laser, where we shall treat still more complicated cases. Then we shall turn our attention to the convection instability in fluid dynamics.

## 10.1 Multimode Lasers with Phase Relations

We consider the ring laser in which the electric field strength $E$ can be represented as

$$E(x,t)=\sum_{l} B_l(t)\exp(\mathrm{i}k_l x-\mathrm{i}\omega_l t)+\text{c.c.} \tag{10.1}$$

where $l$ is the index of the cavity modes and $B_l(t)$ their time-dependent amplitudes which have the physical dimension of electric field strength.

We assume that the waves are propagating in one direction only and the values of $k$ and $\omega$ are taken from a certain interval small compared to the central wave number $k_0$ and the central frequency $\omega_0$, respectively

$$k_l \in k_0 \pm \Delta k\ , \qquad \omega_l \in \omega_0 \pm \Delta\omega\ . \tag{10.2}$$

We now consider spatial and temporal averages of $E$ and its powers where the temporal average is defined as in Chap. 5. We readily obtain

$$\bar{E}=0\ , \tag{10.3}$$

$$\overline{E_l(x,t)E_{l'}(x,t)}=2\delta_{ll'}B_l B_{l'}^* \tag{10.4}$$

$$\overline{E_l E_{l'} E_{l''}}=0\ . \tag{10.5}$$

By performing the integration over plane waves and using $k$-selection rules we readily obtain the result

$$\overline{E_l E_{l'} E_{l''} E_{l'''}}=C\sum_{\pm}\delta(\pm k_l \pm k_{l'} \pm k_{l''} \pm k_{l'''})B_l^{\pm}B_{l'}^{\pm}B_{l''}^{\pm}B_{l'''}^{\pm}\ . \tag{10.6}$$

where $B^+=B^*$ and $B^-=B$.

The following identifications can now be made (cf. Sect. 3.3):

$$\text{index } i \text{ of } p_i\text{: } i = (B_1, B_1^*, B_2, B_2^*, \ldots, B_M, B_M^*) \ , \tag{10.7}$$

$$\text{constraints: } f_l = \langle B_l^* B_l \rangle \ , \tag{10.8}$$

$$f_{l,l',l'',l'''} = \langle B_l^* B_{l'}^* B_{l''} B_{l'''} \rangle \delta(k_l + k_{l'} - k_{l''} - k_{l'''}) \ . \tag{10.9}$$

Application of the maximum information entropy principle yields the final result

$$P(B, B^*) = \exp\Bigg[ -\lambda - \sum_l \lambda_l B_l^* B_l - \sum_{ll'l''l'''} \lambda_{l,l',l'',l'''} B_l^* B_{l'}^* B_{l''} B_{l'''} \delta(k_l + k_{l'} - k_{l''} - k_{l'''}) \Bigg] \ . \tag{10.10}$$

This result has previously been derived from a microscopic theory for the case where the principle of detailed balance holds.

## 10.2 The Single-Mode Laser Including Polarization and Inversion

We treat here a single-mode laser for which the polarization and inversion of the atomic system are included as dynamic variables. For simplicity we shall consider travelling waves. We then have the following variables

$$E(x, t) = B(t)\exp(ikx - i\omega t) + \text{c.c.} \qquad \text{field,} \tag{10.11}$$

$$P(x, t) = P(t)\exp(ikx - i\omega t) + \text{c.c.} \qquad \text{polarization,} \tag{10.12}$$

$$D(x, t) \qquad \text{slowly varying inversion,} \tag{10.13}$$

where $B(t)$, $P(t)$ and $D$ are slowly varying functions of time. When we perform spatial averages we readily obtain

$$\bar{E} = \bar{P} = 0 \ , \qquad \bar{D} = 0 \ , \tag{10.14}$$

$$\overline{E^*E} = 2B^*B, \ \overline{E^*E^*EE} = 4B^*B^*BB \ , \tag{10.15}$$

$$\overline{|P(x,t)|^2} = 2P^*(t)P(t), \ \overline{P^*E} = 2P^*(t)B(t) \ . \tag{10.16}$$

In extension of our previous approach, we shall keep all non-vanishing terms up to fourth order. The distribution function will then acquire the form

$$P = \exp(\lambda + \lambda_1 |E|^2 + \cdots + \lambda_{27} D^4) \ . \tag{10.17}$$

Unfortunately there is no microscopic theory in which a distribution function for the present problem can be calculated exactly. Therefore let us again make the adiabatic approximation, where, however, we shall keep the distribution functions for $P$ and $D$ explicitly. We again start from the laser equations (cf. Sect. 7.11)

$$\dot{E} = -\kappa E + gP + F \ , \tag{10.18}$$

$$\dot{P} = -\gamma P + gED + \Gamma \ , \tag{10.19}$$

$$\dot{D} = \gamma_{\parallel}(D_0 - D) - g(EP^* + E^*P) + \Gamma_D \ . \tag{10.20}$$

When we invoke the adiabatic approximation

$$\dot{P} = \dot{D} \simeq 0 \tag{10.21}$$

where we include the fluctuating forces $F$, $\Gamma$ and $\Gamma_D$ we obtain from (10.19) the relation

$$\Gamma = \gamma P - gED \ . \tag{10.22}$$

But because we know that the fluctuating force $\Gamma$ has a Gaussian distribution, we find that the expression

$$\gamma P - gED \tag{10.23}$$

has the same distribution function. Similarly we find from (10.20) a distribution function for the difference of the first two brackets on the r.h.s. of that equation. Finally, eliminating $P$ and $D$ adiabatically close to threshold, we obtain instead of (10.18)

$$\dot{E} = -\kappa E + GE - CE|E|^2 + F_{\text{tot}} \ , \tag{10.24}$$

where $G$ and $C$ are constants.

Equation (10.24) can easily be transformed into a Fokker-Planck equation and, as we have seen before, this equation can readily be solved in the steady state. The total distribution function for $E$, $P$ and $D$ is then the product of the distribution functions belonging to $E$, $P$, $D$, and it reads

$$f = \mathrm{e}^{\lambda} \exp(a|E|^2 - b|E|^4) \times \exp\left\{-\left(\frac{|\gamma P - gED|^2}{Q}\right) - \left(\frac{[g(EP^* + E^*P) + \gamma_{\parallel}(D - D_0)]^2}{Q'}\right)\right\} \ . \tag{10.25}$$

When we compare this distribution function (10.25) with the guessed function (10.17) we find that a number of the Lagrange parameters $\lambda$ vanish identically as is exhibited in Table 10.1. Whether this vanishing of the $\lambda$'s is a consequence of our adiabatic elimination procedure or whether it holds even in the case of the general equations (10.18–20) remains at present an open question.

**Table 10.1.** Vanishing (/) and nonvanishing (×) Lagrange parameters

| | 1 | $\lvert E\rvert^2$ | $\lvert P\rvert^2$ | $EP^*$ | $E^*P$ | $D$ | $D^2$ |
|---|---|---|---|---|---|---|---|
| 1 | × | × | × | × | × | × | × |
| $\lvert E\rvert^2$ | × | × | × | / | / | / | × |
| $\lvert P\rvert^2$ | × | × | / | / | / | / | / |
| $EP^*$ | × | / | / | × | × | × | / |
| $E^*P$ | × | / | / | × | × | × | / |
| $D$ | × | / | / | × | × | × | / |
| $D^2$ | × | × | / | / | / | / | / |

## 10.3 Fluid Dynamics: The Convection Instability

We begin by briefly reminding the reader of the convection instability. Let us consider a fluid in a rectangular vessel which is heated from below. If the temperature difference between the lower and the upper surface of the fluid is small, heat is transported by heat conduction and no macroscopic motion is visible. If however, the temperature difference exceeds a critical value, a macroscopic motion in the form of a roll pattern sets in. It is characterized by a vertical velocity field with a wavelength $\lambda$ which equals twice the diameter of a roll. Because this sinusoidal variation of the vertical velocity field represents the evolving pattern, it is tempting to introduce the amplitude of the corresponding sine-wave (or plane wave) as order parameter. Using its moments up to fourth order and observing that for symmetry reasons the odd moments must vanish, we almost immediately obtain the distribution function

$$f(\xi) = N\exp(\lambda\xi^2 - \beta\xi^4) \tag{10.26}$$

which indeed agrees with the one found from the microscopic theory.

If we assume that the first four spatial correlation functions of the velocity field are measured, we may apply the method of Chap. 6 to determine the joint distribution function $P(\xi_u, \xi_s)$ in the form

$$P = f(\xi_u)\prod_s P(\xi_s|\xi_u) \ . \tag{10.27}$$

But since this approach is rather lengthy, and does not give us any new insights, we shall not pursue this problem any further.

We consider instead another phenomenon in which, for instance in a circular vessel, and for certain fluids, hexagons may be formed. These can be interpreted as superpositions of plane waves $\xi_k \exp[\mathrm{i}\,\boldsymbol{k}\cdot\boldsymbol{x}]$ whose wave vectors form an equilateral triangle. Therefore we shall use the complex amplitudes of these plane waves as order parameters, $\xi$, and distinguish them by the indices 0, $\pi/3$, $2\pi/3$ according to their orientation in the plane parallel to the surface of the fluid. By arguments analogous to the selection rules derived above (Sect. 10.1) for the multimode laser, we may show that a number of moments vanish. As a consequence, only certain moments and their Lagrange multipliers appear. These non-vanishing moments are the following

$$\langle|\xi_0|^2\rangle\,, \qquad \langle|\xi_{\pi/3}|^2\rangle\,, \qquad \langle|\xi_{2\pi/3}|^2\rangle\,, \tag{10.28}$$

$\langle\xi_0^*\xi_{\pi/3}\xi_{2\pi/3}\rangle$ and cyclic permutations, and finally

$$\langle|\xi_j|^2|\xi_k|^2\rangle\,, \qquad \text{with } j, k = \text{any of } 0, \frac{\pi}{3}, \frac{2\pi}{3}\,. \tag{10.29}$$

It is then obvious how to construct the distribution function – which is again in agreement with the microscopic theory.

# 11. Transitions Between Behavioral Patterns in Biology. An Example: Hand Movements

In this chapter we want to show how the methods developed in the previous chapters can be applied to biological systems. Here it will become particularly clear how powerful our approach can be for the modeling of the behavior of complex systems.

## 11.1 Some Experimental Facts

While researching into the voluntary oscillatory motions of the two index fingers, *Kelso* observed an interesting phenomenon. Under instructions to increase the frequency of out-of-phase, antisymmetric motion (involving simultaneous flexor and extensor muscle activities), the subject's finger movements shifted abruptly to an in-phase symmetric mode that involved simultaneous activation of homologous muscle groups (Figs.11.1,2). This finding was not restricted to finger movements. In later work employing similar experimental manipulations, modal transitions in hand motions around the wrist were also observed: the antisymmetric phase relationship between the hands was replaced by symmetric phasing. Moreover, although the phase transition occurred at very different frequencies of hand motion for different subjects, it was nevertheless predictable. When the transition frequency was expressed in units of preferred frequency, i.e., an independent measure of the rate at which each subject was content to cycle the hands "as if he or she were going to do it all day", the resulting dimensionless ratio or "critical value" was constant for all subjects. Introducing a frictional resistance to movement systematically changed both the preferred and the transition frequencies for each subject, but did not change the critical value across all subjects.

The most dramatic aspect of these simple experiments is the sudden and completely involuntary change in the ordering or phasing among muscle groups that occurs at a critical, intrinsically defined frequency. In this respect, the hand movement data share a likeness to gait transitions in locomotion. For example *Shik* et al. showed that a steady increase in electrical stimulation to the midbrain region of the decerebrate cat is sufficient not only to induce an increase in locomotion rate, but, above a certain value of current, gait shifts as well. As in the hand experiments in which "flipping" from one mode to another occasionally occurred at higher movement frequencies, the latter experiments also showed the presence of unstable regions in which the cat shifted from trotting to galloping and back again. Though the hand data as well as these findings on quadruped gait strongly suggest that

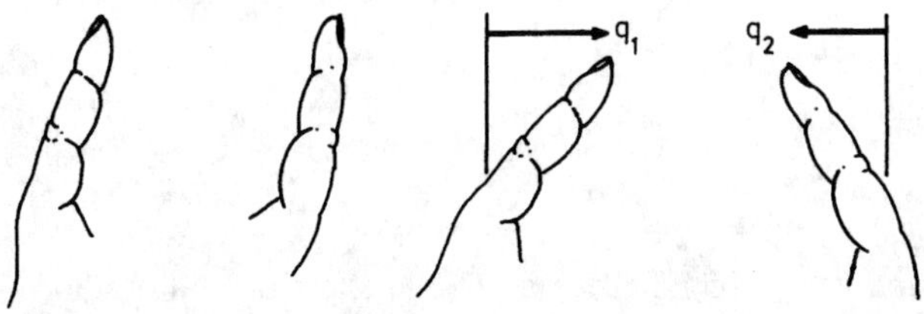


**Fig. 11.1.** The involuntary change of the parallel motion of fingers (l.h.s.) to an antiparallel symmetric motion (r.h.s.)

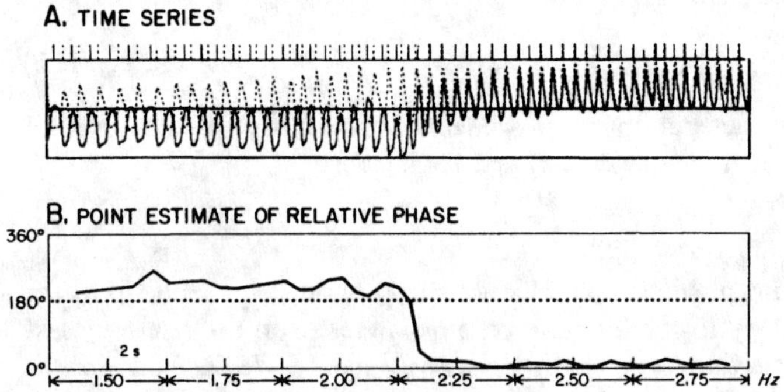


**Fig. 11.2.** Top: Displacements over time of left (*solid line*) and right (*dashed line*) hands. The subject is simply increasing cycling frequency in an antisymmetric mode in response to a verbal cue from the experimenter. Bottom: Phase relationship between the two hands. The peaks of one hand movement act as a "target" file and their phase position is calculated continuously relative to the peak-to-peak period of the other "reference" file. The graphic display repeats the phase curve so that phase lags and leads can be noted

changes in coordination may be ordered by changes in a single parameter, the neural processes underlying such motoric phase transitions are still poorly understood.

To summarize, the main features of the experiments described briefly above are: (i) the presence of only two stable phase (or "attractor") states between the hands (which one is observed is a function of how the system is prepared, i.e., an instruction to move the hands in the out-of-phase or in-phase mode); (ii) the abrupt transition from one attractor state to the other at a critical cycling frequency; (iii) beyond the transition, only one mode (symmetric in-phase) is observed; and (iv) when the cycling frequency is reduced, the system stays in the symmetric mode, i.e., it does not return to its initially prepared state – a result that suggests coexistence of the basins of attraction for the symmetric and antisymmetric modes and the depletion of one of them.

## 11.2 How to Model the Transition

Let us now try to model these findings by means of the methods introduced above. Let us denote the displacement of the two fingertips by $q_1$ and $q_2$ as shown in

Fig.11.1. Then we put

$$q_1 = r_1 \cos(\omega t + \phi_1) \tag{11.1}$$

$$q_2 = r_2 \cos(\omega t + \phi_2) \ , \tag{11.2}$$

where $r_1$ and $r_2$ are the amplitudes which according to the experimental findings can be considered as independent of time but which depend on the frequency $\omega$ prescribed for the finger movement. The phases $\phi_1$ and $\phi_2$ may be slowly varying functions of time.

In the following we shall put

$$r_1 = r_2 = r \ . \tag{11.3}$$

In order to apply our general approach we consider $q_1$ and $q_2$ as the observed quantities whose moments can be determined, for instance

$$\langle q_j \rangle \quad \text{or} \quad \langle q_j q_l \rangle \ . \tag{11.4}$$

In accordance with our previous approach we shall consider moments up to the fourth order. In analogy to the laser we are not interested in the relatively fast oscillation but rather in the slowly varying phases $\phi_1$ and $\phi_2$. We first consider the time average as defined by

$$\langle q_j \rangle_t = \frac{1}{T} \int_0^T q_j \, dt \ , \tag{11.5}$$

where the time $T$ is given by

$$T = \frac{2\pi}{\omega} \ . \tag{11.6}$$

Because we assume in accordance with the experiments that $\phi_1$ and $\phi_2$ are slowly varying functions of time we may use

$$\frac{1}{T} \int_0^T \cos(\omega t + \phi_j) \, dt = 0 \tag{11.7}$$

or, when the cosine function is decomposed into exponential functions, equivalently

$$\frac{1}{T} \int_0^T \exp(\mathrm{i}\omega t + \mathrm{i}\phi_j) \, dt = 0 \ . \tag{11.8}$$

On account of (11.8) we may even require

$$\frac{1}{T} \int_0^T \exp[\mathrm{i}n(\omega t + \phi_j)] \, dt = 0 \ , \tag{11.9}$$

where $n$ is an integer. From (11.7) we obtain

$$\frac{1}{T}\int_0^T q_j\,dt = 0\ . \tag{11.10}$$

Let us now consider the second moments

$$\frac{1}{T}\int_0^T q_j q_k\,dt\ , \tag{11.11}$$

where we shall use the decomposition

$$q_j = \frac{r}{2}\{\exp[\mathrm{i}(\omega t + \phi_j)] + \exp[-\mathrm{i}(\omega t + \phi_j)]\}\ . \tag{11.12}$$

Let us discuss the various terms which we obtain when we insert (11.12) into (11.11). The term

$$\frac{1}{T}\int_0^T \exp(2\mathrm{i}\omega t + \mathrm{i}\phi_j + \mathrm{i}\phi_k)\,dt = 0 \tag{11.13}$$

evidently vanishes because of (11.9). On the other hand in

$$\frac{1}{T}\int_0^T \exp[\mathrm{i}(\phi_j - \phi_k)]\,dt = \exp[\mathrm{i}(\phi_j - \phi_k)] \tag{11.14}$$

the rapidly oscillating parts are no longer present and therefore we may replace the left-hand side by the right-hand side. Taking the results (11.13) and (11.14) together, we obtain

$$\frac{1}{T}\int_0^T q_j q_k\,dt = \frac{r^2}{2}\cos(\phi_j - \phi_k)\ . \tag{11.15}$$

Let us now distinguish two cases.

a) $j = k$. Because of (11.15) we readily obtain a time independent constant.

$$(11.15) = \text{const.}\ . \tag{11.16}$$

Since this term no longer contains any variables, we may ignore it in the following considerations.

b) $j \neq k$. The result (11.15) suggests that we introduce the abbreviation

$$\phi_2 - \phi_1 = \phi \tag{11.17}$$

so that we may write

$$(11.15) = \frac{r^2}{2}\cos\phi\ . \tag{11.18}$$

Let us now consider the moments of third order:

$$\frac{1}{T}\int_0^T q_{j_1} q_{j_2} q_{j_3}\, dt = 0 \ . \tag{11.19}$$

We have immediately written down the result as it turns out that in (11.19) on the left-hand side only expressions of the form

$$\exp(\mathrm{i}\omega t)\ , \qquad \exp(3\mathrm{i}\omega t) \tag{11.20}$$

or their complex conjugates occur which according to (11.9) will give rise to vanishing contributions. Therefore let us finally consider the moments of fourth order

$$\frac{1}{T}\int_0^T q_{j_1} q_{j_2} q_{j_3} q_{j_4}\, dt \ . \tag{11.21}$$

Inserting (11.12) into (11.21) we readily obtain expressions containing

$$\exp(4\mathrm{i}\omega t)\ , \qquad \exp(2\mathrm{i}\omega t) \tag{11.22}$$

or their complex conjugates. When averaged over a time interval $T$ these expressions will vanish. But finally we obtain expressions in which the rapidly oscillating terms no longer appear so that we are only concerned with expressions of the form

$$\exp[\mathrm{i}(\phi_{j_1} + \phi_{j_2} - \phi_{j_3} - \phi_{j_4})] \tag{11.23}$$

or expressions which are obtained by any permutation of the indices $j_1 \ldots j_4$.

**Table 11.1.** Combinations of the $j$'s and the resulting exponents

| $j_1$ | $j_2$ | $j_3$ | $j_4$ | Exponent | |
|---|---|---|---|---|---|
| 1 | 1 | 1 | 1 | 0 | |
| 2 | 2 | 2 | 2 | 0 | |
| 1 | 2 | 1 | 2 | 0 | |
| 1 | 2 | 2 | 1 | 0 | |
| 2 | 1 | 2 | 1 | 0 | |
| 2 | 1 | 1 | 2 | 0 | |
| 1 | 1 | 2 | 1 | $\phi_1 - \phi_2$ | $= -\phi$ |
| 1 | 1 | 1 | 2 | $\phi_1 - \phi_2$ | |
| 2 | 1 | 2 | 2 | $\phi_1 - \phi_2$ | |
| 1 | 2 | 2 | 2 | $\phi_1 - \phi_2$ | |
| 2 | 1 | 1 | 1 | $\phi_2 - \phi_1$ | $= \phi$ |
| 1 | 2 | 1 | 1 | $\phi_2 - \phi_1$ | |
| 2 | 2 | 1 | 2 | $\phi_2 - \phi_1$ | |
| 2 | 2 | 2 | 1 | $\phi_2 - \phi_1$ | |
| 1 | 1 | 2 | 2 | $2(\phi_1 - \phi_2) = -2\phi$ | |
| 2 | 2 | 1 | 1 | $2(\phi_2 - \phi_1) = 2\phi$ | |

Let us now study which kind of expressions we shall obtain when we assign the values 1 or 2 to the indicies $j_1 \dots j_4$. This is shown in Table 11.1.

Using the results of this table we may write

$$\langle q_{j_1} q_{j_2} q_{j_3} q_{j_4} \rangle_t = C_1 + C_2 \cos\phi + C_3 \cos(2\phi) \ . \tag{11.24}$$

Here the coefficients $C_1$, $C_2$ and $C_3$ are proportional to $r^4$. Since $C_1$ no longer contains any variables of the system, this term is uninteresting. The second term is already taken care of by the second-order moments and can be subtracted from (11.24). Therefore the only relevant constraint is contained in the last term of (11.24). Collecting our results we may formulate the following constraints which follow from the second- and fourth-order moments

$$r^2 \langle \cos\phi \rangle = f_1 \tag{11.25}$$

$$r^4 \langle \cos(2\phi) \rangle = f_2 \ . \tag{11.26}$$

It is now a simple task to apply the maximum information entropy principle and to obtain the distribution function $P$ in the form

$$P(\phi) = \exp[-\lambda - \lambda_1 \cos\phi - \lambda_2 \cos(2\phi)] \ . \tag{11.27}$$

Note that the Lagrange multipliers $\lambda$, $\lambda_1$ and $\lambda_2$ will be functions of $\omega$, not only because of the factors $r^2$ and $r^4$ in front of (11.25,6) but also because the average values indicated by the brackets depend on $\omega$ as exhibited by the experiments. The exponent appearing in (11.27) can be represented by a potential function

$$\hat{V}(\phi) = \lambda_1 \cos\phi + \lambda_2 \cos(2\phi) \ . \tag{11.28}$$

This potential function determines the stable and unstable states of the system. It is plotted for various ratios of $\lambda_1$ and $\lambda_2$ in Fig.11.3. Quite evidently in the upper left

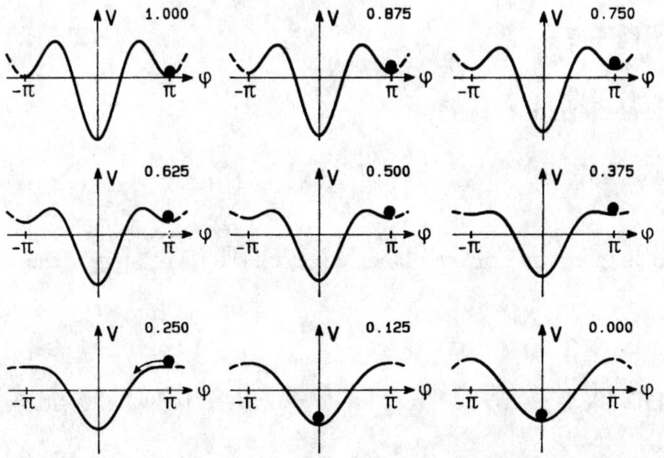


**Fig. 11.3.** The potential $\hat{V}$ for the varying values of $\lambda_2/\lambda_1$. The numbers refer to the ratio $\lambda_2/\lambda_1$

corner there is a local minimum which corresponds to the parallel movement of the fingers. But when the parameter $\lambda_2/\lambda_1$ decreases this minimum becomes flatter and flatter and eventually disappears so that the ball which represents the state of the system will fall down to the lower minimum which now represents the antiparallel (or in other words symmetric) motion of the two fingers. So our result can easily model the transition of the fingers which we described above. The only assumption is that the ratio of $\lambda_2/\lambda_1$ depends on the driving frequency $\omega$. But our model can also do much more. Let us consider a situation where the driving frequency is so high that only the state with the symmetric motion of the fingers is realized. Then when we lower the frequency we easily deduce from Fig.11.3 that the ball, i.e. the system, will now remain in the state with $\phi = 0$. That means the test persons are expected to continue moving their fingers in a symmetric fashion even below the critical frequencies. This prediction was tested by Kelso and found to be realized.

In order to describe time-dependent processes in which $\phi$ varies we now try to guess the Fokker-Planck equation which belongs to the distribution function (11.27). As we have seen in Sect. 6.6 this guessing is not unique because the noise source may depend on the variables. But let us make the simplest assumption, namely that the noise is independent of $\phi$. In such a case we may immediately identify (11.27) with

$$P(\phi) = \mathcal{N} \mathrm{e}^{-\hat{V}} \tag{11.29}$$

where we use the abbreviation

$$\hat{V} = \frac{2}{Q} V \tag{11.30}$$

and where $Q$ is the strength of the noise source. According to the microscopic theory presented in Chap. 2, the Fokker-Planck equation reads

$$\dot{P}(\phi, t) = -\frac{\partial}{\partial \phi}[K(\phi)P] + \frac{Q}{2}\frac{\partial^2}{\partial \phi^2}P \ , \tag{11.31}$$

where we have used the abbreviation

$$K(\phi) = -\frac{\partial V}{\partial \phi} \ . \tag{11.32}$$

As we have seen in that chapter, the Fokker-Planck equation (11.31) belongs to the Langevin equation

$$\dot{\phi} = K(\phi) + F(t) \ , \tag{11.33}$$

where by means of (11.27,28,30,32,33) we may write down the right-hand side explicitly

$$\dot{\phi} = a \sin \phi + b \sin(2\phi) + F(t) \tag{11.34}$$

where $a = Q\lambda_1/2$ and $b = Q\lambda_2$. This equation implies new predictions about the behavior of the finger movement: We may now solve (11.34) and thereby calculate correlation functions for $\phi$ at various times. In other words we may now study transient phenomena from the theoretical point of view and compare them with the experimental findings.

## 11.3 Critical Fluctuations

Let us now start with equation (11.34) as our model equation. It contains three parameters, $a$, $b$, and, implicitly, the strength of the fluctuating force $F(t)$ which we shall write in the form $\sqrt{Q}\,\xi$ where $\xi$ has the following properties:

$$\langle \xi_t \rangle = 0\ , \tag{11.35}$$

$$\langle \xi_t \xi_{t'} \rangle = \delta(t - t')\ . \tag{11.36}$$

Let us first study the fluctuations of the phase in the symmetric mode. Since in this case the system is close to $\phi = 0$ we may linearize the Langevin equation (11.34) or equivalently the Fokker-Planck equation (11.31). The Langevin equation for example, acquires the form

$$\dot{\phi} = -(4b + a)\phi + \sqrt{Q}\,\xi_t\ . \tag{11.37}$$

It is not difficult to solve the linearized Fokker-Planck equation and then to calculate the mean value of the absolute phase

$$\langle |\phi| \rangle_{\text{stat}} = \int_{-\infty}^{\infty} d\phi |\phi| P_{\text{stat}}(\phi) = \frac{1 - \exp(-\pi^2 d^2)}{\sqrt{\pi}\, d\, \text{erf}\{\pi d\}} \tag{11.38}$$

where $P_{\text{stat}}(\phi)$ is the time-independent probability distribution of $\phi$, and the mean square deviation

$$\sigma_{\text{stat}} = \langle \phi^2 \rangle_{\text{stat}} - \langle |\phi| \rangle^2_{\text{stat}} = \frac{1}{2d^2} - \frac{\sqrt{\pi}\exp(-\pi^2 d^2)}{d\, \text{erf}\{\pi d\}} - \langle |\phi| \rangle^2_{\text{stat}}\ . \tag{11.39}$$

Quite evidently the results (11.38) and (11.39) depend on a single parameter $d$ which is given by

$$d = \sqrt{\frac{4b + a}{Q}}\ , \tag{11.40}$$

Our results will allow us to compare the experimental findings on the phase fluctuations with our model. We first refer the reader to Fig.11.4 which represents the experimental results.

In Fig.11.5 we have plotted (11.38) and (11.39) versus $d$ as solid lines, with the experimental values, which may be derived from Fig.11.4, shown as dashed lines.

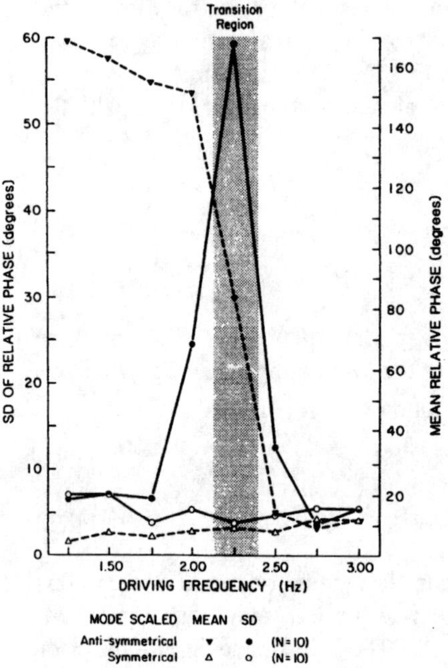


◀ **Fig. 11.4.** The mean phase modulus $\langle|\phi|\rangle$ (▾ AMS, △ SMS) and its SD $= \{\langle\phi^2\rangle - \langle|\phi|\rangle^2\}^{1/2}$ (● AMS, ○ SMS) were determined in the stationary limit on each frequency plateau (i.e. for the last 3 s at each frequency). Each point on the graph represents an average from 10 runs of the experiment. AMS = antisymmetric mode scaled, SMS = symmetric mode scaled (from Kelso and Scholz 1985)

**Fig. 11.5**

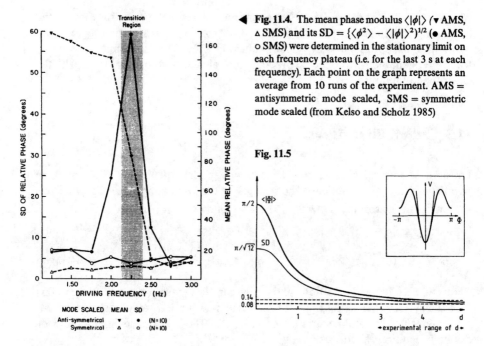

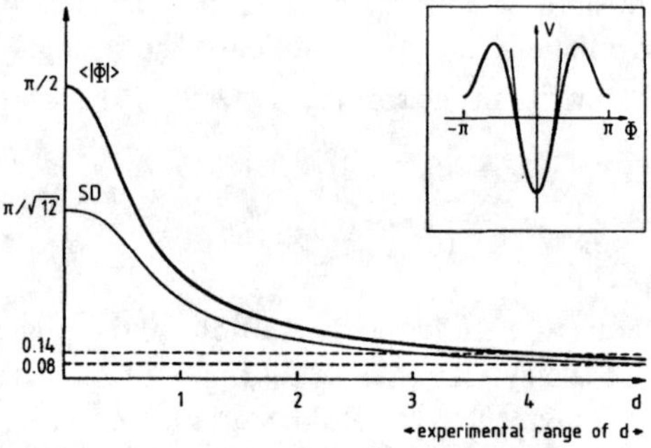


**Fig. 11.5.** The mean phase modulus and its SD as calculated from the local model of the symmetric mode ($\phi_{start} = 0$), as a function of $d = \{(a + 4b)/Q\}^{1/2}$. Two lines indicate the experimental values for mean (= 0.14) and SD (= 0.08), so that experimentally realistic values for $d$ can be read off. The inset shows a sketch of the potential (fat) and the local model (thin) at $a = b = 1$ Hz

In this way we can fix $d$. From the experimental value

$$\langle|\phi|\rangle \approx 8^\circ\ , \qquad \text{we find} \tag{11.41}$$

$$3 \lesssim d \lesssim 5\ . \tag{11.42}$$

From

$$\sigma_{\text{stat}} \approx 4.5^\circ \qquad \text{we obtain} \tag{11.43}$$

$$d \approx 3\ . \tag{11.44}$$

Quite evidently both experimental results can be matched by the assumption that

$$d \approx 4\ . \tag{11.45}$$

The relaxation time of the system around $\phi = 0$ can easily be obtained by solving the linearized Langevin equation in the deterministic case, i.e. without fluctuating forces. We readily obtain

$$\phi(t) = \phi(0)\exp[-(4b + a)t]\ . \tag{11.46}$$

The inverse of the factor multiplying $t$ in the exponent defines the relaxation time so that we obtain by means of (11.40)

$$\tau_{\text{rel}} = \frac{1}{4b+a} = \frac{1}{d^2 Q} \; . \tag{11.47}$$

This relaxation time occurs in a wide variety of different relaxation and reaction phenomena, for instance when we have to lift our foot from the gas-pedal when the traffic-lights turn red. A rough estimate of $\tau_{\text{rel}}$ is 0.25 s. From this value and the value $d \approx 4$ we readily obtain by means of (11.47)

$$Q \simeq 0.25 \text{ Hz} \; . \tag{11.48}$$

The same result can be obtained from a theory which takes into account fluctuations and studies the correlation function of $\phi$ at two different times. Let us now do the same for the antisymmetric mode in which case $\phi$ is centered around $+\pi$ or $-\pi$ so that we may introduce the small quantity $\varepsilon$ which is defined by

$$\varepsilon = \begin{cases} \phi - \pi & \text{for} \quad 0 < \phi \leq \pi \\ \phi + \pi & \text{for} \quad -\pi < \phi \leq 0 \; . \end{cases} \tag{11.49}$$

Using $\varepsilon$ and keeping only linear terms in the Langevin equation (11.37) we immediately obtain

$$\dot{\varepsilon} \equiv -(4b - a)\varepsilon + \sqrt{Q}\,\xi_t \; . \tag{11.50}$$

The corresponding linearized Fokker-Planck equation is easily obtained. Its stationary solution reads

$$P_{\text{stat}}(\varepsilon) = \frac{f}{\sqrt{\pi}\,\text{erf}\{\pi f\}} \mathrm{e}^{-f^2\varepsilon^2} \; , \tag{11.51}$$

where we have introduced the parameter

$$f = \sqrt{\frac{4b-a}{Q}} \; . \tag{11.52}$$

Using the same definition as before for the mean value of the phase and for the standard deviation, and using (11.51) we readily obtain

$$\langle|\phi|\rangle_{\text{stat}} = \pi - \langle|\varepsilon|\rangle_{\text{stat}} = \pi - \frac{1 - \exp(-\pi^2 f^2)}{\sqrt{\pi}\, f\, \text{erf}\{\pi f\}} \tag{11.53}$$

$$\begin{aligned} \sigma_{\text{stat}} &= \langle\phi^2\rangle_{\text{stat}} - \langle|\phi|\rangle^2_{\text{stat}} = \langle\varepsilon^2\rangle_{\text{stat}} - \langle|\varepsilon|\rangle^2_{\text{stat}} \\ &= \frac{1}{2f^2} - \frac{\sqrt{\pi}\exp(-\pi^2 f^2)}{f\,\text{erf}\{\pi f\}} - \langle|\varepsilon|\rangle^2_{\text{stat}} \; . \end{aligned} \tag{11.54}$$

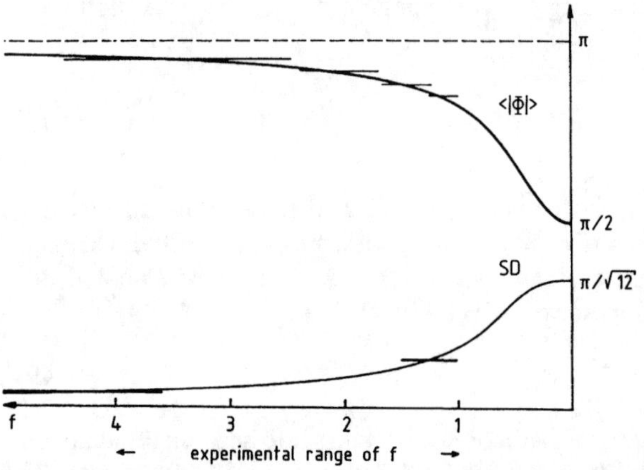


**Fig. 11.6.** The mean absolute phase and its SD for the local model of the antisymmetric mode as a function of $f = [(4b - a)/Q]^{1/2}$. Approach of the transition corresponds to $f \to 0$. The $f$-axis was oriented to the left to illustrate this. The curves have been used to determine from experimental values of mean and SD the corresponding values of $f$. The range of such values is indicated

The values of (11.53) and (11.54) are plotted versus the parameter $f$ in Fig.11.6 According to Fig.11.4 we may read off the mean value of $\phi$ and the standard deviation for each value of $f$. Introducing such pairs of values into Fig.11.6 we may readily find the corresponding $f$. The relaxation time is now given in analogy to (11.47) by

$$\tau_{\mathrm{rel}} = \frac{1}{4b - a} = \frac{1}{f^2 Q} \ . \tag{11.55}$$

Again adopting the value $\tau_{\mathrm{rel}} \approx 0.25$ s and using a value of $f$ in the range of 1–4 according to the experimental data, we obtain $Q$ in the range of 0.25–4 which agrees in order of magnitude with the previous result on the symmetric mode. From (11.40,52) we have the relations

$$4b + a = Qd^2 \ , \qquad 4b - a = Qf^2(\omega) \ . \tag{11.56}$$

Since $f$ is now known as a function of the oscillation frequency $\omega$ of the fingers, we may deduce the frequency dependence of $a$ and $b$ from (11.56):

$$a(f) = \frac{Qd^2}{2} - \frac{Q}{2} f^2(\omega) \tag{11.57}$$

$$b(f) = \frac{Qd^2}{8} + \frac{Q}{8} f^2(\omega) \ . \tag{11.58}$$

The corresponding results are plotted in Fig.11.7. Taking the value of the frequency at which the transition occurs and the corresponding value of $f$ we find the following results: for $d \simeq 4$, $Q = 0.25$ Hz it follows $a_c = 4b_c$ and thus we have the approximate values

$$a_c = 2.0 \ \mathrm{Hz} \tag{11.59}$$

$$b_c = 0.5 \ \mathrm{Hz} \ . \tag{11.60}$$

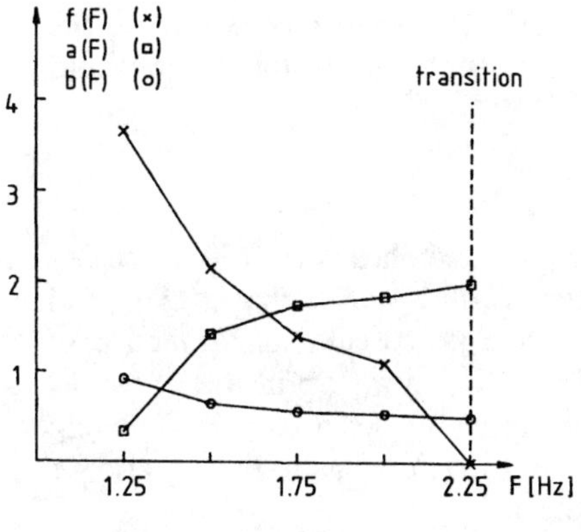


**Fig. 11.7.** The model parameters $f$, $a$ and $b$ as a function of the experimental control parameter frequency $F \equiv \omega$ as determined from the data on mean and SD and an estimate of relaxation time

To summarize we may state that it has been possible to fix the parameters in a reasonable fashion and that we are able in each case to match four experimentally determined values, namely the mean absolute phases and the standard deviations of the symmetric and the antisymmetric modes, by means of three model parameters. Of course the model can still be improved but I believe it shows in a convincing manner that our approach is capable of modeling the behavior of a complex system.

## 11.4 Some Conclusions

### i) Motor-Program Versus Self-Organization

In biology, in order to explain the high coordination between muscles, the idea of a "motor-program" has quite often been proposed. According to this idea the neurons act more or less like a computer in which a specific program is stored and which, after having been invoked once, starts to steer individual motions. But in such a case it is hard to understand why fluctuations should occur at all. After all, a motor-program is a fixed program which does not allow any fluctuations. On the other hand, critical fluctuations are quite typical for nonequilibrium phase transitions which occur when self-organization happens. For this reason we believe that the fluctuations, found in these experiments, strongly support the idea that muscles and neurons form a self-organizing system, perhaps in analogy to the laser which may also show transitions between different kinds of behavior upon the change of a single control parameter. The experiments do not allow us to tell whether self-organisation occurs in the total system of neurons *and* muscles or merely in the neuronal subsystem.

### ii) Information Compression

The high coordination of muscles and neuróns is manifest in the occurrence of specific kinds of macroscopic motion which, in the case dealt with above, can be described by a single order parameter. Connected with the occurrence of this single order parameter is a very small amount of information in comparison to the information necessary to describe the individual states of all the neurons and muscles.

### iii) Morphogenesis of Behavior

As we know from numerous examples in physics and chemistry, self-organizing systems may produce specific temporal patterns. Here we see that a biological system may produce a certain behavioral pattern e.g. a specific kind of motion of fingers. We have seen that the maximum information entropy principle allows us to find the adequate order parameter and, furthermore, even its corresponding equation. It may be hoped that more complex behavioral patterns can also be described by a small number of order parameters.

# 12. Pattern Recognition. Unbiased Guesses of Processes: Explicit Determination of Lagrange Multipliers

In this chapter we first wish to show how my general approach allows us to deal with central problems in pattern recognition. In Sect. 12.1 we will rederive relations allowing feature selection that will be explained below. In Sect. 12.2 we present the basic scheme for the construction of a parallel computer for pattern recognition. The subsequent sections, 12.3 and 12.4, show how such a system can learn patterns to be recognized. Finally, Sects. 12.5–12.11 extend these results to the learning of processes. Section 12.1 is a bit formal so that readers who are only interested in the most important topics of this chapter need to read this section only as far as equation (12.1) and may then proceed to the subsequent sections.

## 12.1 Feature Selection

Since we will not assume that the reader is familiar with the problem of pattern recognition we start with a brief survey. In this section we shall confine our attention to digital pattern recognition. In this approach a pattern is decomposed into so-called features. Consider as an example letters and let us consider a particularly simple case, namely that we have only two letters $X$ and $O$ (Figs.12.1,2). Then we may attribute to the bracket which is open to the left the number 0 and to the bracket which is open to the right the number 1. Using this coding we can attribute the combination $(0,1)$ to the letter $X$ and the combination $(1,0)$ to the letter $O$.

More generally speaking, we may decompose a pattern into features and attribute to each feature a component $q_j$. In this way the total pattern is described by a pattern vector $\boldsymbol{q} = (q_1, q_2, \ldots, q_n)$ (Fig.12.3). Of course, a vector can be represented by its end points in a multi-dimensional space. In this way we attribute a vector or correspondingly a point to each pattern. The components of the vector $\boldsymbol{q}$ need not only acquire the values 0 or 1, but they can also be continuous. For instance when we decompose a pattern into individual cells and attribute an intensity $q_j$ to each point $j$. So at this stage we may say that the recognition of patterns is performed in two steps. Once the features have been selected, we must attribute the vector $\boldsymbol{q}$ to the features by specific measurements. Then we have to look within the feature space to establish to which pattern this vector $\boldsymbol{q}$ belongs.

In practice, there are some difficulties, however. First of all, features can be selected in various manners. Therefore, one is confronted with the problem of selecting features in an adequate fashion. Here, adequate means that the feature

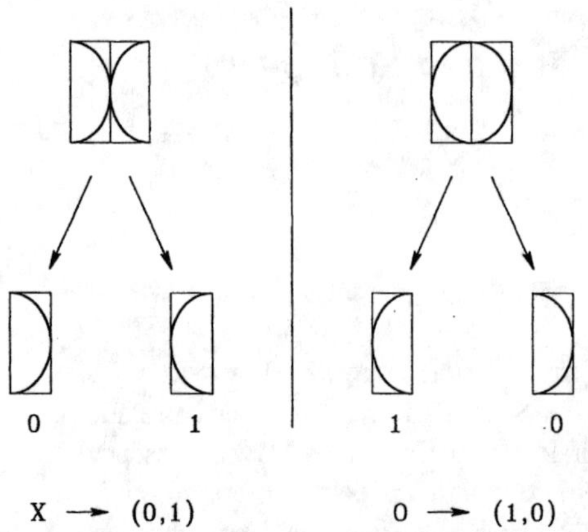


**Fig. 12.1.** Example of a decomposition of the letters $X$ and $O$ into features to which numbers 0 or 1 are attributed

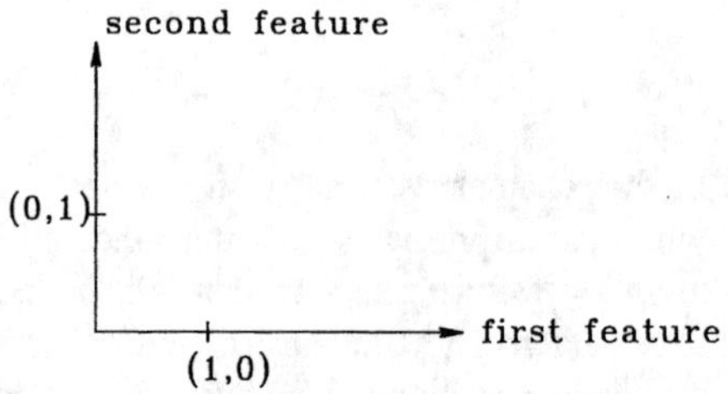


**Fig. 12.2.** The feature space belonging to the symbols of Fig.12.1. Here the point (1,0) corresponds to the letter $O$, whereas the point (0,1) corresponds to the letter $X$

space should not be too large but that the patterns can nonetheless be clearly distinguished. Small dimension means that not so much information must be processed. A further difficulty lies in the measurement itself, or in the patterns, because measurements may introduce random noise or patterns may fluctuate internally. Thus it turns out that we are not dealing with deterministic vectors, $\boldsymbol{q}$, but rather with vectors having components which are random variables with a certain probability distribution function.

In the following we shall be concerned with two problems:

1) How can we determine this probability distribution function from experiments?
2) How can we extract features in such a manner that the information to be processed becomes a minimum?

In order to derive the probability distribution function $P(\boldsymbol{q})$, we invoke the maximum information principle. Namely, we assume that we have made measurements on the correlation functions

$$\langle q_i q_j \rangle = Q_{ij} \tag{12.1}$$

over the ensemble of patterns we wish to classify or to study. Without loss of generality we may assume that the average $\langle \boldsymbol{q} \rangle$ vanishes:

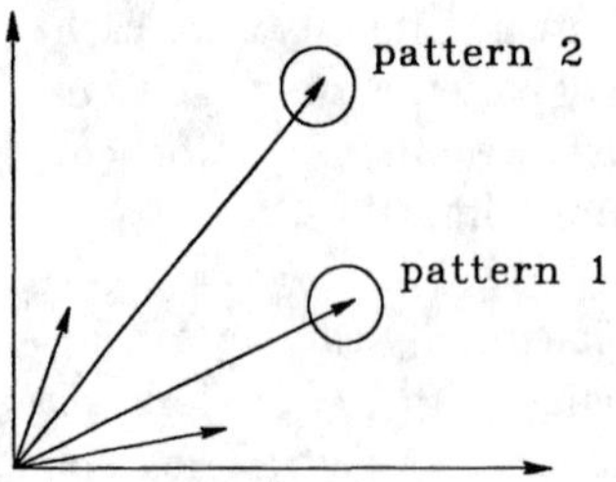


**Fig. 12.3.** Visualization of a general feature space. The end points of the vectors indicate a prototype pattern, whereas the surrounding of them represents patterns which deviate slightly from the prototype pattern and can then be recognized as the specific patterns

$$\langle \boldsymbol{q} \rangle = 0 \ . \tag{12.2}$$

Otherwise we may always introduce new pattern vectors by simply subtracting the non-vanishing average value from these vectors. The application of the maximum information entropy principle immediately yields

$$P(\boldsymbol{q}) = N \exp(-\boldsymbol{q}^T M \boldsymbol{q}) \ , \tag{12.3}$$

where we have used (12.1,2) as constraints. The Lagrange multipliers can be arranged within a matrix $M$ which is related to the correlation matrix $Q$ via

$$M = Q^{-1} \ , \tag{12.4}$$

where $Q$ is defined by (12.1).

We now transform (12.3) in much the same way as we did before in Sect. 6.3 where we wanted to derive the pattern associated with the modes close to a non-equilibrium phase transition. We first seek the eigenvectors $\boldsymbol{u}_j$ and the eigenvalues $\lambda_j$ belonging to the equation

$$Q\boldsymbol{u}_j = \lambda_j \boldsymbol{u}_j \ . \tag{12.5}$$

These eigenvectors diagonalize the matrix $Q$. We wish to show that these eigenvectors also diagonalize the matrix $M$. For this purpose we form the matrix $U$ consisting of the eigenvectors $\boldsymbol{u}_j$ according to

$$(\boldsymbol{u}_1 \boldsymbol{u}_2 \dots \boldsymbol{u}_N) = U \ , \tag{12.6}$$

where the matrix elements of $U$ are now given by

$$u_{kj} = U_{kj} \ . \tag{12.7}$$

The index $k$ distinguishes the individual components of each vector, whereas the index $j$ distinguishes the different eigenvectors. Using $U$ we may write the individual equations (12.5) as a single matrix equation of the form

$$QU = U\Lambda \ , \tag{12.8}$$

where $\Lambda$ has components

$$\Lambda_{kj} = \lambda_j \delta_{jk} \ . \tag{12.9}$$

The matrix equation (12.8) reads in components

$$\sum_k Q_{ik} U_{kj} = \sum_l u_{il} \Lambda_{lj} = \lambda_j u_{ij} \ . \tag{12.10}$$

In the following we shall assume that $Q$ is non-singular as is already implied in (12.4). We now multiply (12.8) from the left by $Q^{-1}$ so that we obtain

$$U = Q^{-1} U \Lambda \ . \tag{12.11}$$

Multiplying this equation by $\Lambda^{-1}$ from the right we obtain

$$U\Lambda^{-1} = Q^{-1}U \ . \tag{12.12}$$

As a result of (12.4) we may rewrite this equation in the form

$$MU = U\tilde{\Lambda} \ , \tag{12.13}$$

where we have used the abbreviation

$$\tilde{\Lambda} = \Lambda^{-1} \ . \tag{12.14}$$

Because $Q$ is a symmetric matrix we know that

$$U^T U = 1 \tag{12.15}$$

where $U^T$ is the transpose of $U$. We now expand the general vector $\boldsymbol{q}$ into the eigenvectors $\boldsymbol{q}_j$. This decomposition may be written in the general form

$$\boldsymbol{q} = U\xi \tag{12.16}$$

or more explicitly in the form

$$\boldsymbol{q} = \sum_j \xi_j \boldsymbol{u}_j \ . \tag{12.17}$$

Note that the vector $\boldsymbol{q}$ is a random variable whereas the eigenvectors $\boldsymbol{u}_j$ are deterministic. Consequently, the vector $\xi$ which appears in (12.16) or (12.17) must have components which are random variables. We now wish to transform the exponent occuring in (12.3), i.e.

$$\boldsymbol{q}^T M \boldsymbol{q} \tag{12.18}$$

to the new variable $\xi$. To this end, we introduce the transposed vector belonging to (12.17), namely

$$\boldsymbol{q}^T = \xi^T U^T \ . \tag{12.19}$$

In this way we obtain

$$\boldsymbol{q}^T M \boldsymbol{q} = \xi^T U^T M U \xi \ . \tag{12.20}$$

Making use of (12.13,14), we transform (12.20) into

$$MU = \xi^T U^T U \Lambda^{-1} \xi \ , \tag{12.21}$$

which because of (12.15,14) can be simplified to yield

$$\boldsymbol{q}^T M \boldsymbol{q} = \xi^T \tilde{\Lambda} \xi \ . \tag{12.22}$$

As is well known from the transformation of probability distribution functions, we have not only to transform the exponent of the probability distribution function but also the volume element. This transformation is given by the Jacobian which in our case turns out to be

$$\mathrm{Det}\, U = 1 \tag{12.23}$$

so that this transformation leaves the volume elements unchanged. In this way we obtain our new probability distribution function: Instead of (12.3) we now have

$$\tilde{P}(\xi) = \mathcal{N} \exp(-\xi^T \tilde{\Lambda} \xi) \ . \tag{12.24}$$

Making use of the fact that $\tilde{\Lambda}$ is a diagonal matrix, we may write (12.24) in the form

$$\tilde{P}(\xi) = \mathcal{N} \exp\left(-\sum_j \tilde{\lambda}_j \xi_j^2\right) \ . \tag{12.25}$$

The right-hand side may now be written as a product

$$\tilde{P}(\xi) = \prod_j P_j(\xi_j) \ , \tag{12.26}$$

where each factor is given by

$$P_j(\xi_j) = \mathcal{N}_j \exp(-\tilde{\lambda}_j \xi_j^2) \tag{12.27}$$

and the normalization constant $\mathcal{N}_j$ is chosen in such a way that $P_j$ is normalized to unity. The specific form (12.26), together with (12.27), yields the following correlation function

$$\langle \xi_j \xi_k \rangle = \lambda_j \delta_{jk} \ , \tag{12.28}$$

i.e. the correlation matrix has now become diagonal. We further note that the smallest $\tilde{\lambda}_j$ corresponds to the biggest $\lambda_j$.

Let us now examine what we have achieved by the transformation from $\boldsymbol{q}$ to $\xi$ in terms of the information. The information is given quite generally by

$$i = -\int d^N q\, P(\boldsymbol{q}) \ln P(\boldsymbol{q}) \ , \tag{12.29}$$

which can be transformed using the above procedure into the form

$$i = -\int d^N \xi\, \tilde{P}(\xi) \ln \tilde{P}(\xi) \ . \tag{12.30}$$

Using the specific form (12.25) we may rewrite (12.30) as a sum over individual contributions

$$i = -\sum_j \int d\xi_j\, P_j(\xi_j) \ln P_j(\xi_j) \tag{12.31}$$

or in an abbreviated form

$$i = \sum_j i_j \ . \tag{12.32}$$

Making use of (12.27), we can immediately calculate the information and obtain

$$i_j = -\tfrac{1}{2}\ln\tilde{\lambda}_j + \tfrac{1}{2}\ln\pi + \tfrac{1}{2} \tag{12.33}$$

in terms of the eigenvalues $\tilde{\lambda}_j$. Since $\tilde{\lambda}_j = 1/\lambda_j$, we may rewrite the information $i_j$ as

$$i_j = \tfrac{1}{2}\ln\lambda_j + \tfrac{1}{2}\ln\pi + \tfrac{1}{2}\ . \tag{12.34}$$

By means of (12.33,34) we recognize that the largest contribution to the information is that belonging to the largest $\lambda_j$ and the total information appears as a sum of uncorrelated information contributions. Let us now discuss what these results mean for pattern recognition.

We started from specifically selected features to which we attributed a vector $\boldsymbol{q}$. By means of specific linear combinations in the form (12.17) we now make a different selection of features. Or in other words we extract now a new type of features. But these new features have the very pleasant property that the information becomes a sum over individual contributions (12.34). At this stage we may make contact with methods used in digital pattern recognition. The expansion (12.17) with the $\boldsymbol{u}$'s chosen as in (12.5) is called the Karhunen-Loeve expansion. In that context the quantity $i$ is called the population entropy. Our results may now be interpreted in the following manner: The Karhunen-Loeve expansion provides us with a method for decomposing the pattern into uncorrelated components (features). In this way a reduction of the number of components or features is achieved, as we shall discuss below. The coordinate system $\boldsymbol{u}_j$ belonging to the Karhunen-Loeve (K-L) expansion is deterministic. The components or more precisely speaking the coefficients contain all the information needed for the reconstruction of the pattern. The coefficients with largest $\lambda_j$ contain most of the information about the pattern vector $\boldsymbol{q}$.

Thus, in order to recognize a pattern we need only consider the coefficients with the largest eigenvalues, while a coefficient with a small variance (identical to the eigenvalue) convey a negligible amount of information. In other words we may say that the transformation of $\boldsymbol{q}$ into the K-L coordinate system yields an information compression. Once we have determined the K-L coordinate system, it is identical for all patterns we wish to put into classes, where patterns belonging to the same class are considered to be equivalent (identical).

In this way all the discriminatory information must be carried by the coefficients $\xi_j$ of the expansion because most of the information is compressed into a small number of these coefficients $\xi$. The problem is less complex computationally when we carry out the subsequent decision making in a lower dimensional feature space. The use of the information has another very nice feature. The greater the dispersion of the eigenvalues $\lambda_j$ the smaller is the corresponding population entropy $i$ and the greater the information compression. Conversely when we have uniformly distributed $\lambda$'s we have a great uncertainty, a large entropy and the observation of one coefficient $\xi$ reduces the uncertainty or the entropy only little.

It is interesting to interpret this procedure from the point of view of non-equilibrium phase transitions. When a system is first in the disordered state and approaches the transition point, the fluctuations of the order parameters grow and become particularly large compared to all the other modes. Recalling this interpretation we may say that the $\xi_j$'s with the largest variance or, correspondingly, with the largest values $\lambda_j$ correspond to the order parameters which determine the pattern. In this region the probability distribution function is well characterized by its moments up to second order in accordance with the above treatment of this section. The next section will be devoted to pattern recognition "above the transition point" (to use the interpretation of pattern recognition as a nonequilibrium phase transition).

## 12.2 An Algorithm for Pattern Recognition

As explained in Sect. 12.1, we describe a pattern by an $N$-dimensional vector, $\boldsymbol{q}$, whose individual components represent features encoded by numbers. We assume that a number $M$ of prototype patterns are given and represented by the state vectors

$$\boldsymbol{v}^{(k)} = \begin{pmatrix} v_1^{(k)} \\ \vdots \\ v_N^{(k)} \end{pmatrix} \tag{12.35}$$

where $k$ runs from 1 to $M$. Depending on their number, these vectors span the feature space fully or in part. In the latter case additional, "idle" vectors must be introduced. The patterns (12.35) are stored in the computer in a way we shall describe below. When a pattern to be recognized by the computer is offered, it may not exactly coincide with one of the vectors (12.35) because of noise, incomplete data (lack of some features), etc., but it may be close to one of the prototype patterns, i.e. close to one of the axes of the feature space. Now our basic idea is this: Let us describe the offered pattern by a vector $\boldsymbol{q}(0)$ where the argument 0 refers to the initial time $t = 0$ of the pattern recognition process. Then we devise equations for a time-dependent vector $\boldsymbol{q}(t)$ so that it develops from its initial state $\boldsymbol{q}(0)$ into a final state $\boldsymbol{q}_f$ that agrees with one of the prototype vectors $\boldsymbol{v}^{(k)}$. The equations will be constructed in such a way that this specific vector $\boldsymbol{v}^{(k)}$ is precisely the one, to which $\boldsymbol{q}(0)$ comes closest, i.e. for which $(\boldsymbol{v}^{(k)+} \cdot \boldsymbol{q}(0))/(|\boldsymbol{v}^{(k)}||\boldsymbol{q}(0)|)$ (12.35) had the smallest value. We assume

$$(\boldsymbol{v}^{(k)+} \cdot \boldsymbol{v}^{(k')}) = \delta_{kk'} \ .$$

This procedure has two pleasant consequences. (1) Even if an initial pattern is incomplete, it will be completed, i.e. our system acts as "associative memory". For instance in a telephone directory, the name (corresponding to $\boldsymbol{q}(0)$) will be complemented by the telephone number, where name + number correspond to a specific $\boldsymbol{v}^{(k)}$. (2) If we store faces + names as prototype patterns $\boldsymbol{v}^{(k)}$ and offer only a face ($\boldsymbol{q}(0)$), the system will drive $\boldsymbol{q}(t)$ into face + name. When we read off the name, the face has been recognized (by the machine).

Our equations by which all this is achieved read

$$\dot{\boldsymbol{q}} = -\mathrm{grad}_{\boldsymbol{q}} V + \boldsymbol{F}(t) \ . \tag{12.36}$$

In it the potential function $V$ is given by

$$V(\boldsymbol{q}) = V_0 + V_1 + V_2 \qquad \text{where} \tag{12.37}$$

$$V_0 = -\frac{1}{2}\boldsymbol{q}^+ \sum_k \lambda \boldsymbol{v}^{(k)} \cdot \boldsymbol{v}^{(k)+} \boldsymbol{q} \tag{12.38}$$

$$V_1 = \sum_{k \neq k'} C_{kk'} (\boldsymbol{v}^{(k)+} \boldsymbol{q})^2 (\boldsymbol{v}^{(k')+} \boldsymbol{q})^2 \ , \tag{12.39}$$

$$C_{kk'} > 0 \qquad \text{and} \tag{12.40}$$

$$V_2 = C\boldsymbol{q}^4 \ . \tag{12.41}$$

$V_0$ serves to pull $\boldsymbol{q}$ into the subspace spanned by the prototype patterns. In the parlance of nonequilibrium phase transitions, this is simply the *order parameter space*. $V_1$ serves for the discrimination of $\boldsymbol{q}$-vectors within that subspace. This can be easily seen from the property that the minimum of $V_1$ is adopted according to

$$V_{1,\min} = 0 \qquad \text{for } \boldsymbol{q} \parallel \boldsymbol{v}^{(k_0)} \tag{12.42}$$

i.e. when the state vector $\boldsymbol{q}$ becomes parallel to one of the prototype vectors and thus the initial state vector $\boldsymbol{q}(0)$ is identified with such a prototype pattern. $V_2$ provides for saturation, i.e. $|\boldsymbol{q}|$ is eventually pulled into a fixed point attractor on the $\boldsymbol{v}^{(k_0)}$ axis. The constant $\lambda$ in (12.38) plays the role of a control parameter. $\lambda < 0$ determines the region below "threshold", whereas $\lambda > 0$ determines the region above "threshold". In this section we assume $\lambda > 0$.

In order that $\boldsymbol{q}$ is more easily pushed into one of the prototype vectors, we have added a fluctuating force $\boldsymbol{F}$ in (12.36). As usual we assume that the fluctuating forces have the properties

$$\langle \boldsymbol{F}(t) \rangle = 0 \ , \tag{12.43}$$

$$\langle F_i(t) F_j(t') \rangle = Q\delta_{ij}\delta(t - t') \ . \tag{12.44}$$

Our approach has been tested by computer calculations for the recognition of faces and excellent results were obtained. Here, however, we shall be concerned with the fundamental aspects. Namely, while in our above equations we had to implement the prototype vectors in the equations, we wish to show how a computer can learn the prototype patterns. As vehicle for this we employ an analogy between pattern recognition as described by the above equations and nonequilibrium phase transitions.

We proceed by transforming the Langevin-type equations (12.36) into the Fokker-Planck equation

$$\dot{f} = \sum_k \frac{\partial}{\partial q_k}\left(\frac{\partial V}{\partial q_k}\cdot f\right) + \frac{Q}{2}\sum_k \frac{\partial^2 f}{\partial q_k^2} \ . \tag{12.45}$$

The stationary solution can be found explicitly and reads [cf. (2.44)]

$$f = N\exp\left(-\frac{2V}{Q}\right) \tag{12.46}$$

which means that the stationary solution is explicitly given by the potential function $V$ (12.37) and the size of the fluctuations $Q$. The maxima of $f(\boldsymbol{q})$ correspond to the minima of $V(\boldsymbol{q})$. A simple analysis shows that the minima of $V(\boldsymbol{q})$ are precisely at the positions $\boldsymbol{q} = \boldsymbol{v}^{(k)}$, i.e. at the prototype patterns. While this holds for $\lambda > 0$, for $\lambda < 0$ the minimum lies at $\boldsymbol{q} = 0$, i.e. even if initially patterns are presented to the system, they will decay.

## 12.3 The Basic Construction Principle of a Synergetic Computer

Let us assume that a system can receive a set of data that are described by a state vector $\boldsymbol{q}$. We further assume that the detecting system is composed of elements $j$, where element $j$ measures the component $q_j$ (feature) of $\boldsymbol{q}$. In the following we shall study what properties these elements *and their connections* must have in order to perform pattern recognition. If there is a stationary process the incoming signals will obey a probability distribution $f(\boldsymbol{q})$. In the spirit of much teaching in schools we shall assume that some specific patterns are offered again and again so that the maxima of

$$f(\boldsymbol{q}) = \max! \tag{12.47}$$

correspond to these patterns.

How can a system determine $f(\boldsymbol{q})$ from measurements? To this end we assume that the system can measure moments

$$\langle q_i\rangle, \langle q_i q_j\rangle, \langle q_i q_j q_k\rangle, \langle q_i q_j q_k q_l\rangle \ . \tag{12.48}$$

In order to guess $f(\boldsymbol{q})$ from given moments we employ the maximum information entropy principle and discuss the minimum order of moments necessary to arrive at a sensitive guess of $f$. If we employ only linear moments, then according to the maximum information entropy principle $f$ must be of the form $f(\boldsymbol{q}) = N\exp(-\boldsymbol{\lambda q})$ where $\boldsymbol{\lambda}$ is real. Quite evidently $f(\boldsymbol{q})$ cannot be normalized in the space of $\boldsymbol{q}$ if the $\boldsymbol{q}$ variables run from minus infinity to plus infinity, so this approach is not possible. When we employ moments up to second order, the general form of the distribution function will be

$$f(\boldsymbol{q}) = N\exp(\bar{\boldsymbol{\alpha}}\boldsymbol{q} + \bar{\boldsymbol{q}}\boldsymbol{\alpha} - \bar{\boldsymbol{q}}\boldsymbol{B}\boldsymbol{q}) \ . \tag{12.49}$$

Introducing the new variable $\boldsymbol{\xi}$ via

$$q = \xi + B^{-1}\alpha \ , \tag{12.50}$$

(12.49) can be cast into the form

$$\tilde{f}(\xi) = N' \exp(-\bar{\xi} B \xi) \ . \tag{12.51}$$

Because $B$ must be a positive definite matrix, the maximum of (12.51) can be acquired for one single value of $\xi$ only, namely $\xi = 0$.

By using (12.50) we realize that there is only one maximum, i.e. only one pattern which is in practically all cases in contradiction to the *set of patterns* actually offered. This leads us to consider moments up to fourth order. For the sake of simplicity we shall assume that the moments of odd order vanish. The distribution function then acquires the form

$$f(q) = N \exp\left( -\sum_{ij} \lambda_{ij} q_i q_j - \sum_{ijmn} \lambda_{ijmn} q_i q_j q_m q_n \right) \tag{12.52}$$

according to the maximum information entropy principle. We now wish to show how we can construct a network which reproduces (12.52). For this we assume that (12.52) is the result of a continuous stationary Markov process that is described by the Fokker-Planck equation (12.45) where, in our present case, $V$ is a still unknown potential function. To make contact between (12.52) and the desired quantities $V$ and $Q$ in (12.45) we abbreviate the bracket in (12.52) by

$$(\dots) = -\tilde{V}(q, \lambda) \tag{12.53}$$

and require that the stochastic forces $F_j$ of the Langevin equation belonging to (12.45) obey the relations (12.43,44). We further set

$$\tilde{V} = \frac{2V}{Q} \ . \tag{12.54}$$

The Langevin equation belonging to the Fokker-Planck equation (12.45) has the same general form as (12.36), namely

$$\dot{q}_j = -\frac{\partial V}{\partial q_j} + F_j(t) \ . \tag{12.55}$$

However, $V$ is now determined by (12.53,54), i.e. by means of measurements. Thus we obtain

$$-\frac{\partial V}{\partial q_i} = Q \left\{ \sum_j \lambda_{ij} q_j + 2 \sum_{jmn} \lambda_{ijmn} q_j q_m q_n \right\} \ . \tag{12.56}$$

We now observe that (12.55) with (12.56) has a form similar to that of the network we introduced in the previous section, or in other words we are capable of constructing a network that reproduces the distribution function $f(q)$, provided that this function is known by its second and fourth order moments. The only function the network must perform is to transfer the value of a quantity $q_j$ of the element $j$, multiplied by a "synaptic strength" $\lambda_{ij}$ to the element $i$ [first term in (12.56)], or to transfer the value $q_j q_m q_n$ from elements $j, m, n$, multiplied by $\lambda_{ijmn}$ to the element $i$ [second term in (12.56)]. Then in the corresponding element $i$ a summation is performed.

There is a further quite general and important conclusion: If the network can measure only a specific set of correlation functions (12.48) (or higher order), we need only include the corresponding terms in (12.56). Because of the attractor states of the potential dynamics, any initial state will be pulled into any of the local minima of $V(\boldsymbol{q})$ that are located at the points of the prototype patterns. If the initial state is close enough to one of these minima, that minimum will be realized and in this way an initially incomplete pattern will be completed, i.e. the whole formalism acts as "associative memory" and thus as a pattern recognizer.

## 12.4 Learning by Means of the Information Gain

In this section we will show how the system can learn the strengths of $\lambda_{ij}$ and $\lambda_{ijmn}$. Let us identify the given distribution function of the incoming patterns by $f(\boldsymbol{q})$ and the distribution function generated by the system by $\tilde{f}(\boldsymbol{q})$. Let us introduce information gain (Kullback information) as a measure of the distance between these two distribution functions

$$K = \int \tilde{f} \ln\left(\frac{\tilde{f}}{f}\right) d^N q \geq 0 \tag{12.57}$$

where we have to observe the constraints

$$\int f \, d^N q = 1 \tag{12.58}$$

$$\int \tilde{f} \, d^N q = 1 \, . \tag{12.59}$$

Because $f$ is a fixed quantity and (12.57) can be written in the form

$$K = \int f \ln f \, d^N q - \int f \ln \tilde{f} \, d^N q \tag{12.60}$$

it will suffice to maximize the expression

$$\int f \ln \tilde{f} \, d^N q = \text{max!} \tag{12.61}$$

To be specific, let us assume $\tilde{f}$ to have the form

$$\tilde{f} = \exp\left[-\tilde{\lambda} - \sum_j \tilde{\lambda}_j V_j(\boldsymbol{q})\right] \tag{12.62}$$

where $V_j$ may contain polynomials in $\boldsymbol{q}$ up to an order to be fixed by us. $\tilde{\lambda}_j$ are parameters which can be varied. The left-hand side of (12.61) (multiplied by $-1$) can be cast into the form

$$W \equiv -\int f \ln \tilde{f} \, d^N q = \tilde{\lambda} + \int f \sum_j \tilde{\lambda}_j V_j(\boldsymbol{q}) \, d^N q \, . \tag{12.63}$$

We now assume that the $\tilde{\lambda}_j$ are subject to an evolutionary strategy, e.g. the network may undergo fluctuations of its connectivities. A particularly elegant evolution strategy is that involving gradients of some potential function. We consider (12.63) as a potential function to be minimized in which we express $\tilde{\lambda}$ by means of

$$\tilde{\lambda} = \ln \int \exp\left[-\sum_j \tilde{\lambda}_j V_j(\boldsymbol{q})\right] d^N q \, . \tag{12.64}$$

The gradient strategy now amounts to subjecting the Lagrange parameters $\tilde{\lambda}_j$ to the equation

$$\dot{\tilde{\lambda}}_j = -\gamma \frac{\partial W}{\partial \tilde{\lambda}_j} \, . \tag{12.65}$$

In order to evaluate the right-hand side we insert (12.64) into (12.63) and form the derivative

$$\frac{\partial W}{\partial \tilde{\lambda}_j} = -\left[\left\{\int \exp\left\{-\sum_j \tilde{\lambda}_j V_j(\boldsymbol{q})\right\} d^N q\right\}^{-1} \int V_j(\boldsymbol{q}) \times \exp\left\{-\sum_j \tilde{\lambda}_j V_j(\boldsymbol{q})\right\} d^N q\right] + \int f(\boldsymbol{q}) V_j(\boldsymbol{q}) \, d^N q \, . \tag{12.66}$$

A simple consideration shows that the expression in the square bracket of (12.66) can be interpreted as the average value:

$$[\;] = \langle V_j(\boldsymbol{q}) \rangle_{\tilde{f}} \tag{12.67}$$

whereas the last expression in (12.66) is the average value

$$\langle V_j(\boldsymbol{q}) \rangle_f \, . \tag{12.68}$$

Equation (12.65) can thus be written in the very concise form

$$\dot{\tilde{\lambda}}_j = \gamma(\langle V_j \rangle_{\tilde{f}} - \langle V_j \rangle_f) \tag{12.69}$$

where the first expression on the right-hand side is $V_j$ averaged over the distribution function $\tilde{f}$, whereas the second part is the same function but averaged over the distribution function prescribed by the outer world. By comparing these two values the system can adjust its connections $\tilde{\lambda}$ correspondingly and thus learn its task. Equation (12.69) is the basis of a machine built by Sejnowski et al. and called the Boltzmann machine. These authors used correlation functions $q_i q_k$ where the $q$'s can acquire only two values, $\pm 1$ in the sense of a "spin-glass" model.

## 12.5 Processes and Associative Action

In the previous sections I have shown how a physical system can learn patterns of signals and reproduce their probability distribution. In the following I wish to show how a system can learn to reproduce or to recognize *processes*. While in the preceding sections we did not prescribe the path along which the system reaches the final attractor states corresponding to the originally offered patterns, we now wish to show that the system can even learn to reproduce a specific path. To this end we make a basic assumption, namely that the processes which the system has to learn are Markovian. We assume that the process is stationary and continuous. A Markovian process is characterized by the conditional probabilities

$$P(\boldsymbol{q}_{i+1}|\boldsymbol{q}_i) \ , \tag{12.70}$$

which represent the probability of finding the system at a position $\boldsymbol{q}_{i+1}$ at time $t_{i+1} = t_i + \tau$ provided the system was at time $t_i$ at the position $\boldsymbol{q}_i$. We denote the components of the state vectors at times $t_{i+1}$ and $t_i$ by

$$q_l(i+\tau) \ , \qquad q_k(i) \ . \tag{12.71}$$

We introduce the following first and second conditional moments

$$f_{1,l} = \langle q_l(i+\tau)\rangle_{\boldsymbol{q}(i)} \tag{12.72}$$

$$f_{2,l,k} = \langle q_l(i+\tau)q_k(i+\tau)\rangle_{\boldsymbol{q}(i)} \ . \tag{12.73}$$

We now use some results of Chap. 9. The application of the maximum information entropy principle to the conditional probability allows us to obtain

$$P(\boldsymbol{q}(i+\tau)|\boldsymbol{q}(i)) = \exp\left[-\lambda - \sum_l \lambda_l q_l(i+\tau) - \sum_{kl} \lambda_{kl} q_k(i+\tau) q_l(i+\tau)\right] \tag{12.74}$$

where the Lagrange multipliers may still be functions of $\boldsymbol{q}(i)$. [Note the change of signs as compared with (9.23)]. We introduce the column vector

$$\boldsymbol{\lambda} = \begin{pmatrix} \lambda_1 \\ \lambda_2 \\ \vdots \\ \lambda_N \end{pmatrix} \tag{12.75}$$

and the matrix

$$\Lambda = (\lambda_{kl}) \tag{12.76}$$

and define a new vector $\boldsymbol{h}$ by

$$\boldsymbol{h} = -\tfrac{1}{2}\Lambda^{-1}\boldsymbol{\lambda} \ . \tag{12.77}$$

We further define the quantities $G_{kl}$ and $K_k$ by

$$\lambda_{kl} = \frac{1}{\tau} G_{kl}(\boldsymbol{q}(i)) \ , \tag{12.78}$$

and for $\tau \to 0$

$$h_k = q_k(i) + \tau K_k(\boldsymbol{q}(i)) \ , \tag{12.79}$$

respectively. Then according to Sect. 9.2 the Fokker-Planck equation reads

$$\dot{f} = Lf \tag{12.80}$$

where the linear operator $L$ is defined by

$$Lf = -\nabla_{\boldsymbol{q}}(K(\boldsymbol{q})f) + \frac{1}{4}\sum_{kl} \frac{\partial^2}{\partial q_k \partial q_l}\{(G^{-1}(\boldsymbol{q}))_{kl} f\} \ . \tag{12.81}$$

This Fokker-Planck equation refers to the Îto-calculus from which we can easily construct the Îto-Langevin equation (see below). In contrast to the results of Sects. 12.3 and 12.4, the fluctuating forces will become at least in general $\boldsymbol{q}$ dependent. But otherwise the analogy with the results of Sect. 12.3 is entirely retained and thus we may construct our physical system. If we have a physical system in which only a finite number of links are established, we may proceed in analogy to Sect. 9.3. Since in the case of several variables, $q_i$, this procedure is not trivial we shall describe it. In a number of practical applications we may not know $\boldsymbol{q}_i$ but rather we may measure correlation functions of a type to be discussed next. We assume that the measurements are made under steady state conditions. In this case we can express the joint probability by a product of the conditional probability and the steady-state probability distribution

$$P(\boldsymbol{q}_{i+\tau}, \boldsymbol{q}_i) = P(\boldsymbol{q}_{i+\tau} | \boldsymbol{q}_i) P_{\mathrm{st}}(\boldsymbol{q}_i) \ . \tag{12.82}$$

In order to define moments or other correlation functions we shall introduce the functions $U_j(\boldsymbol{q}_i)$ which may be assumed e.g. in the form

$$U_j(\boldsymbol{q}_i) \equiv U_{j,i} = q_{1,i}^{\mu_1} q_{2,i}^{\mu_2} \dots q_{N,i}^{\mu_N}; \ \mu_1 + \mu_2 + \dots + \mu_N = R \ . \tag{12.83}$$

Note that the index $i$ refers to the time index whereas the other indices 1, 2, ... refer to components of a vector. We then introduce the following constraints

$$\langle U_{j,i}^{(1)} \rangle \tag{12.84}$$

$$\langle q_{k,i+\tau} U_{m,i}^{(2)} \rangle \ , \quad k = 1, \dots, N \ , \quad m = 1, 2, \dots \quad \text{and} \tag{12.85}$$

$$\langle q_{k,i+\tau} q_{l,i+\tau} U_{n,i}^{(3)} \rangle \quad k = 1, \dots, N, \quad l = 1, \dots, N, \quad n = 1, 2, \dots \ . \tag{12.86}$$

For a more detailed discussion of these constraints we refer the reader to Sects. 9.7–9.11.

Using our previous results we may immediately determine the steady state distribution in the form

$$P_{st}(\boldsymbol{q}_i) = \exp\left(-\lambda_{st} - \sum_j \lambda_{j,st} U_{j,i}^{(1)}\right) . \tag{12.87}$$

We now make use of a generalization of the maximum information entropy principle to the space-time domain, i.e. we may now subject a multi-time joint probability to this principle. For our present approach it is sufficient to consider the two-time joint probability, which, from the maximum information entropy principle, acquires the form

$$P(\boldsymbol{q}_{i+\tau}, \boldsymbol{q}_i) = \exp[-\lambda_0 + A(\boldsymbol{q}_i) + \boldsymbol{B}(\boldsymbol{q}_i)\boldsymbol{q}_{i+\tau} + \bar{\boldsymbol{q}}_{i+\tau} C(\boldsymbol{q}_i)\boldsymbol{q}_{i+\tau}] , \tag{12.88}$$

where we have used the following abbreviations

$$A(\boldsymbol{q}_i) = -\sum_j \lambda_j^{(1)} U_j^{(1)}(\boldsymbol{q}_i) , \tag{12.89}$$

$$B_k(\boldsymbol{q}_i) = -\sum_m \lambda_{km}^{(2)} U_m^{(2)}(\boldsymbol{q}_i) , \tag{12.90}$$

$$C_{kl} = -\sum_n \lambda_{kln} U_n^{(3)}(\boldsymbol{q}_i) , \tag{12.91}$$

where the $\lambda$'s are the Lagrange multipliers.

Using (12.82,87,88) we readily obtain (12.70) in the form

$$P(\boldsymbol{q}_{i+\tau}|\boldsymbol{q}_i) = \exp[-\lambda + \tilde{A}(\boldsymbol{q}_i) + \boldsymbol{B}(\boldsymbol{q}_i)\boldsymbol{q}_{i+\tau} + \bar{\boldsymbol{q}}_{i+\tau} C(\boldsymbol{q}_i)\boldsymbol{q}_{i+\tau}] \tag{12.92}$$

where we have used the abbreviations

$$\lambda = \lambda_0 - \lambda_{st} \quad \text{and} \tag{12.93}$$

$$\tilde{A} = A + \sum_j \lambda_{j,st} U_j^{(1)}(\boldsymbol{q}_i) . \tag{12.94}$$

A comparison of the result (12.92) with (12.74) reveals that we may now make the following identifications between the previous Lagrange parameters and those which appear now, namely

$$\lambda(\boldsymbol{q}_i) \leftrightarrow \lambda - A(\boldsymbol{q}_i) \tag{12.95}$$

$$\lambda_l(\boldsymbol{q}_i) \leftrightarrow -B_l(\boldsymbol{q}_i) \quad \text{and} \tag{12.96}$$

$$\lambda_{kl}(\boldsymbol{q}_i) \leftrightarrow -C_{kl}(\boldsymbol{q}_i) . \tag{12.97}$$

In this way we may again derive an Îto-Fokker-Planck equation.

If we use the constraints in the form of the functions (12.83) we have polynomials in the drift and diffusion coefficients. Because of the formal analogy between two-time probability distribution functions and the stationary distribution function we may transcribe the learning procedure of Sect. 12.4 to our present case (cf. Sects. 12.6–12.11).

We now derive the Îto-equation [cf. (2.12)] belonging to the Fokker-Planck equation. The Îto-equation reads

$$dq_l(t) = K_l(\boldsymbol{q}(t))\,dt + \sum_m g_{lm}(\boldsymbol{q}(t))\,dw_m(t) \tag{12.98}$$

where the stochastic process is defined by [cf. (2.13, 14)]

$$\langle dw_m \rangle = 0 \quad \text{and} \tag{12.99}$$

$$\langle dw_m(t)\,dw_l(t) \rangle = \delta_{lm}\,dt \ . \tag{12.100}$$

The diffusion coefficients of the Fokker-Planck equation (12.80,81), $(G^{-1})_{kl}$, are connected with the functions $g_{lm}$ by the following formula [cf. (2.27)]:

$$\tfrac{1}{2}(G^{-1})_{kl} = \sum_m g_{km} g_{lm} \ . \tag{12.101}$$

We introduce the matrices $G$ and $g$ by means of

$$(G_{kl}) = G; \qquad (g_{km}) = g \tag{12.102}$$

so that (12.101) can be written in the form

$$G^{-1} = g\bar{g} \ , \tag{12.103}$$

where $\bar{g}$ means the transposed matrix. For the solution of (12.101) it will be sufficient to assume that $g$ is a square matrix and also that $g$ and $G^{-1}$ are symmetric. Equation (12.103) then acquires the form

$$G^{-1} = g^2 \ . \tag{12.104}$$

The solution matrix $g$ can be easily determined by assuming that $G$, which is positive definite, can be diagonalized by means of the orthogonal matrix $U$. We introduce the diagonal matrix $D$ with matrix elements $D_l$ and have

$$D = UG^{-1}\bar{U} = (Ug\bar{U})^2 \qquad \text{where} \tag{12.105}$$

$$U\bar{U} = 1 \tag{12.106}$$

holds. We then immediately find

$$(Ug\bar{U})_{kl} = \delta_{kl}\sqrt{D_l} \tag{12.107}$$

from which we may calculate $g$ immediately.

We have shown how a physical system that reproduces a Markov process can be constructed. It is to be expected that such a system cannot only reproduce the Markov process under initial conditions which are completely determined, but in the sense of associative memory may restore incomplete or partly incorrect data to complete sets. In this way we may expect that the system runs through a self-correcting trajectory. This behavior may be called associative action.

## 12.6 Explicit Determination of the Lagrange Multipliers of the Conditional Probability. General Approach for Discrete and Continuous Processes

A major task in the application of the maximum information (entropy) principle consists in the explicit calculation of the Lagrange multipliers. At first sight it might seem that in the preceding sections we have accomplished that task by deriving, for instance, the Îto-Fokker-Planck equation. But when we scrutinize our approach more closely, we shall observe the following: We employed only a few general properties the Lagrange multipliers must have, namely their functional dependence on the time-interval $\tau$ in the limit $\tau \to 0$. But so far the question remained open as to how we can determine the functional dependences of these multipliers or, equivalently, of the drift coefficients $\boldsymbol{K}_\ell$ and of the diffusion coefficients $\left(G_{k\ell}^{-1}\right)$ on $\boldsymbol{q}(i)$ explicitly, based on the experimentally available data.

In this and the following sections I want to demonstrate how these multipliers that occur in the conditional probability can be calculated. As it will turn out, this can be done if the variables are measured at a discrete time-interval $\tau$ ("stroboscopic" measurements) or at practically continuous time-intervals, where we may assume $\tau \to 0$. In the first case, the conditional probability (12.74) can then be used to calculate the probability distribution $P(t)$ for arbitrary later times $t = n\tau$ and arbitrary initial conditions by means of path integrals (9.35), provided we may assume that the underlying stochastic process is stationary. If the measurements are such that we can extrapolate to $\tau \to 0$, we may even derive the Fokker-Planck or Îto-Langevin equation with *explicitly* determined drift and diffusion coefficients. If not otherwise stated, we assume stationarity also in this case.

We assume that the (unknown) conditional probability $\tilde{P}(\boldsymbol{q}(i+\tau) \mid \boldsymbol{q}(i))$ is experimentally, but implicitly given by (12.72), (12.73). We wish to approximate $\tilde{P}$ by (12.74), where the Lagrange multipliers $\lambda_\ell$ and $\lambda_{k\ell}$ are unknown, and are, in general, functions of $\boldsymbol{q}(i)$, i.e. $\lambda_\ell = \lambda_\ell(\boldsymbol{q}(i)), \lambda_{k\ell} = \lambda_{k\ell}(\boldsymbol{q}(i))$. We note that $\lambda$ in (12.74) is fixed by the normalization condition

$$\int \mathrm{d}^N q(i+\tau) P(\boldsymbol{q}(i+\tau) \mid \boldsymbol{q}(i)) = 1, \tag{12.108}$$

so that

$$\begin{aligned} N(\boldsymbol{q}(i))^{-1} &\equiv \exp[\lambda(\boldsymbol{q}(i))] \qquad (12.109)\\ &= \int \mathrm{d}^N q(i+\tau) \exp\left[-\sum_\ell \lambda_\ell q_\ell(i+\tau) - \sum_{k\ell} \lambda_{k\ell} q_k(i+\tau) q_\ell(i+\tau)\right]. \end{aligned}$$

We define the Kullback information by

$$K = \int \tilde{P} \ln\left(\frac{\tilde{P}}{P}\right) \mathrm{d}^N q(i+\tau) \geq 0, \tag{12.110}$$

where we integrate only over $\boldsymbol{q}(i+\tau)$, but not over $\boldsymbol{q}(i)$ so that $K$ is still a function of $\boldsymbol{q}(i)$. In addition to (12.108) we assume

$$\int \tilde{P} \mathrm{d}^N q(i+\tau) = 1. \tag{12.111}$$

Instead of minimizing (12.110), it is sufficient to minimize

$$W = -\int \tilde{P} \ln P \mathrm{d}^N q(i+\tau), \tag{12.112}$$

which due to (12.74) and (12.111) acquires the form

$$W = \lambda + \int \tilde{P}\left(\sum_\ell \lambda_\ell q_\ell(i+\tau) + \sum_{k\ell} \lambda_{k\ell} q_k(i+\tau) q_\ell(i+\tau)\right) \mathrm{d}^N q(i+\tau). \tag{12.113}$$

We invoke a gradient dynamics (cf. (12.65)) and obtain, in analogy to (12.69),

$$\dot{\lambda}_\ell = \gamma_\ell\Big(\langle q_\ell(i+\tau)\rangle_{P,\boldsymbol{q}(i)} - \langle q_\ell(i+\tau)\rangle_{\tilde{P},\boldsymbol{q}(i)}\Big), \tag{12.114}$$

$$\dot{\lambda}_{k\ell} = \gamma_{k\ell}\Big(\langle q_k(i+\tau) q_\ell(i+\tau)\rangle_{P,\boldsymbol{q}(i)} - \langle q_k(i+\tau) q_\ell(i+\tau)\rangle_{\tilde{P},\boldsymbol{q}(i)}\Big). \tag{12.115}$$

Because of the use of conditional probabilities $\tilde{P}, P$, the brackets on the r.h.s. of (12.114), (12.115) are functions of $\boldsymbol{q}(i)$ and so are $\lambda_\ell, \lambda_{k\ell}$ in (12.114), (12.115). While the first term on the r.h.s. of each equation is considered as experimentally given, the second term is a function of the Lagrange parameters. We shall show how we can express these brackets by means of $\lambda_\ell, \lambda_{k\ell}$. To this end we recall the definitions of (12.72), (12.73), or equivalently (9.21), (9.22) by means of a generalization of (9.5)–(9.7) to the multi-dimensional case. Using the explicit form of $P$ (12.74), we thus have

$$\langle q_\ell(i+\tau)\rangle_{P,\boldsymbol{q}(i)} = \int \mathrm{d}^N q(i+\tau) q_\ell(i+\tau) \exp[...],$$

$$\text{where } [...] = \left[-\lambda - \sum_\ell \lambda_\ell q_\ell(i+\tau) - \sum_{k\ell} \lambda_{k\ell} q_k(i+\tau) q_\ell(i+\tau)\right], \tag{12.116}$$

which in analogy to the transition from (9.23) to (9.25) can be cast into the form

$$\langle q_\ell(i+\tau)\rangle_{P,\boldsymbol{q}(i)} = \int \mathrm{d}^N q(i+\tau) q_\ell(i+\tau) \exp[..], \tag{12.117}$$

where now

$$\exp[..] = \exp\left[-\tilde{\lambda} - \sum_{k\ell}(q_k(i+\tau) - h_k)\lambda_{k\ell}(q_\ell(i+\tau) - h_\ell)\right]. \tag{12.118}$$

(Note the change of sign of the $\lambda$'s). While $\tilde{\lambda}$ is again determined by the normalization condition (12.108), the vector

$$\boldsymbol{h} = \begin{pmatrix} h_1 \\ \vdots \\ h_N \end{pmatrix} \tag{12.119}$$

is connected with the vector

$$\boldsymbol{\lambda} = \begin{pmatrix} \lambda_1 \\ \vdots \\ \lambda_N \end{pmatrix} \tag{12.120}$$

by means of

$$\boldsymbol{h} = -\frac{1}{2}\boldsymbol{\Lambda}^{-1}\boldsymbol{\lambda}, \tag{12.121}$$

where $\boldsymbol{\Lambda}$ is a matrix defined by

$$\boldsymbol{\Lambda} = (\lambda_{k\ell}). \tag{12.122}$$

These simple algebraic relations can be directly checked by inserting (12.121) into (12.118) and absorbing a resulting constant term (i.e. independent of $\boldsymbol{q}(i+\tau)$) in $\tilde{\lambda}$.

In order to evaluate (12.117) with (12.118), we make the substitution

$$\boldsymbol{q}(i+\tau) = \boldsymbol{\xi} + \boldsymbol{h}, \tag{12.123}$$

which yields

$$\begin{aligned} \langle q_\ell(i+\tau)\rangle_{P,\boldsymbol{q}(i)} &= h_\ell \int \mathrm{d}^N\xi \exp\left[-\tilde{\lambda} - \xi\Lambda\xi\right] \\ &+ \int \mathrm{d}^N\xi\xi \exp\left[-\tilde{\lambda} - \xi\Lambda\xi\right], \end{aligned} \tag{12.124}$$

where the first integral is equal to one because of normalization and the second vanishes for symmetry reasons so that

$$\langle q_\ell(i+\tau)\rangle_{P,\boldsymbol{q}(i)} = h_\ell(\boldsymbol{q}(i)). \tag{12.125}$$

In a similar fashion, we obtain (cf. (9.60))

$$\langle q_k(i+\tau)q_\ell(i+\tau)\rangle_{P,q(i)} = \frac{1}{2}\left(\Delta^{-1}\right)_{k\ell} + h_k h_\ell, \tag{12.126}$$

or because of (12.125)

$$\begin{aligned}\langle q_k(i+\tau)q_\ell(i+\tau)\rangle_{P,q(i)} - \langle q_k(i+\tau)\rangle_{P,q(i)}\langle q_\ell(i+\tau)\rangle_{P,q(i)}\\ = \frac{1}{2}\left(\Delta^{-1}\right)_{k\ell}.\end{aligned} \tag{12.127}$$

Now we are able to write down our first important result: The equations for the Lagrange multipliers read explicitly

$$\dot{\lambda}_\ell = -\gamma_\ell\Big(\langle q_\ell(i+\tau)\rangle_{\tilde{P},q(i)} - h_\ell\Big), \tag{12.128}$$

where $h_\ell$ can be expressed by the $\lambda$'s via (12.121), (12.122), and

$$\dot{\lambda}_{k\ell} = -\gamma_{k\ell}\left(\langle q_k(i+\tau)q_\ell(i+\tau)\rangle_{\tilde{P},q(i)} - h_k h_\ell - \frac{1}{2}\left(\Delta^{-1}\right)_{k\ell}\right). \tag{12.129}$$

Since we are interested in the values of $\lambda_\ell, \lambda_{k\ell}$ at the minimum of their potential function, we put $\dot{\lambda}_\ell = \dot{\lambda}_{k\ell} = 0$. Using a vector notation, we obtain

$$\langle \boldsymbol{q}(i+\tau)\rangle_{\tilde{P},\boldsymbol{q}(i)} - \boldsymbol{h} = 0, \tag{12.130}$$

and, replacing the $h$'s in (12.129) by means of the components of (12.130),

$$\begin{aligned}\langle q_k(i+\tau)q_\ell(i+\tau)\rangle_{\tilde{P},q(i)} - \langle q_k(i+\tau)\rangle_{\tilde{P},q(i)}\langle q_\ell(i+\tau)\rangle_{\tilde{P},q(i)}\\ - \frac{1}{2}\left(\Delta^{-1}\right)_{k\ell} = 0.\end{aligned} \tag{12.131}$$

Using an obvious matrix notation with $Q = (Q_{k\ell})$, we can write (12.131) in the form

$$Q - \frac{1}{2}\Delta^{-1} = 0. \tag{12.132}$$

Since the brackets are experimentally given quantities, we can easily first solve equation (12.131), (12.132) for $\Delta$ by matrix inversion and then the equation (12.130) by means of (12.121). Thus we are able of determining the Lagrange multipliers. It is illuminating to check what the equations (12.130) and (12.131) mean for the drift and diffusion coefficients of, e.g., the Fokker-Planck equation. To this end, we recall the relationships (9.31) between the drift coefficients $K_k[\boldsymbol{q}(i)]$ and $h_k$, i.e.

$$h_k = q_k(i) + \tau K_k[\boldsymbol{q}(i)], \tag{12.133}$$

and the diffusion matrix $D = (D_{k\ell})$ (cf. also (9.30), (9.66))

$$D(\boldsymbol{q}(i)) = \frac{1}{2}G^{-1} = \frac{1}{2\tau}\Delta^{-1}. \tag{12.134}$$

Using (12.130), we obtain, in the limit $\tau \to 0$,

$$\begin{aligned} K_k[(\boldsymbol{q}(i))] &= \frac{1}{\tau}\Big(\langle q_k(i+\tau)\rangle_{\tilde{P},\boldsymbol{q}(i)} - q_k(i)\Big) \\ &\equiv \frac{1}{\tau}\langle (q_k(i+\tau) - q_k(i))\rangle_{\tilde{P},\boldsymbol{q}(i)}, \end{aligned} \tag{12.135}$$

which coincides with the conventional definition of the drift coefficients of the Fokker-Planck equation. Let us consider

$$\begin{aligned} D_{k\ell}(\boldsymbol{q}(i)) =& \frac{1}{\tau}\{\langle q_k(i+\tau)q_\ell(i+\tau)\rangle_{\tilde{P},\boldsymbol{q}(i)} \\ &- \langle q_k(i+\tau)\rangle_{\tilde{P},\boldsymbol{q}(i)}\langle q_\ell(i+\tau)\rangle_{\tilde{P},\boldsymbol{q}(i)}\}. \end{aligned} \tag{12.136}$$

This expression is formally different from

$$\frac{1}{\tau}\langle (q_k(i+\tau) - q_k(i))(q_\ell(i+\tau) - q_\ell(i))\rangle_{\tilde{P},\boldsymbol{q}(i)}, \tag{12.137}$$

which in the limit $\tau \to 0$ conventionally serves as the definition of the diffusion coefficients

$$D_{k\ell} = \lim_{\tau\to 0}(12.137). \tag{12.138}$$

Therefore we shall denote (12.136) as generalized diffusion coefficients. We note, however, that in the limit $\tau \to 0$ the expression (12.136) reduces to (12.138). To this end, we again recall (12.130) and (12.133). Taking the difference between (12.136) and (12.137) and inserting (12.130), (12.133), we readily obtain $\tau K_\ell K_k$, which in the limit $\tau \to 0$ vanishes. (The interested reader can easily verify this by using paper and pencil.) Thus our optimization approach leads us to the conventional definitions of the drift and diffusion coefficients via the quantities appearing on the r.h.s. of (12.135) and (12.136) (or (12.137)).

On the other hand, we must note that there is still a gap between theory and practical applications. Theory requires that the quantities in (12.135) and (12.137) are known for the continuum of $\boldsymbol{q}$-values, and precisely. In practice, neither requirement is fulfilled: the experimental data are finite and they are of limited precision. In a number of cases it may also happen that the measured range of $\boldsymbol{q}$-values is limited and that we must extrapolate to a wider range. Furthermore quite often we do not want just a numerical list of the functions $\boldsymbol{K}$ and $D_{k\ell}$, but explicit functions in the sense of data-compression by means of algorithms, or to get a deeper insight into the nature of the process. For all these purposes we must make an unbiased guess on the stochastic process underlying the *available* observed data. We shall do this in the next sections (Sects. 12.7–12.9).

A comment regarding the limit $\tau \to 0$ is in order at this moment. In the case of a continuous Markov process, i.e. with $\tau \to 0$, the process is fully determined by the drift and diffusion coefficients, i.e. in the frame of our present approach by the constraints (12.72) and (12.73) (see above). Thus in the limit $\tau \to 0$ and if all required data in $\boldsymbol{q}$-space are available, our procedure is exact. In practice, however, $\tau$ is always a finite quantity. In this case we can consider our procedure as an unbiased guess under the knowledge of the constraints (12.72), (12.73). In this way, we give approaches that directly start from the definition of the drift and diffusion coefficients a deeper foundation. In addition, in the case of finite $\tau$, we may improve our guess, or check its validity by including constraints containing higher moments (or cumulants) of $\boldsymbol{q}(i+\tau)$. For example, in the case of a single variable, such constraints read $\langle q(i+\tau)^{\nu}\rangle_{q(i)}$. To give the reader a feeling of what the conditional probability then looks like, at least formally in the frame of the present approach, we quote the result of the one-dimensional case

$$P(q(i+\tau) \mid q(i)) = \exp\left[\sum_{\nu=0}^{M} \lambda_{\nu}(q(i))q(i+\tau)^{\nu}\right].$$

The Lagrange multipliers $\lambda_{\nu}(q(i))$ can be determined in a way analogous to Sects. 12.6–12.9. A further discussion of this approach, however, is beyond the scope of this book.

## 12.7 Approximation and Smoothing Schemes. Additive Noise

For reasons that will become obvious later, we start from the Kullback information for *joint* probabilities. To distinguish between conditional probabilities and joint probabilities, we add an index $j$ to the latter.

$$\tilde{P}_j(\boldsymbol{q}(i+\tau), \boldsymbol{q}(i)) \tag{12.139}$$

refers to the experimentally observed process, whereas

$$P_j(\boldsymbol{q}(i+\tau), \boldsymbol{q}(i)) \tag{12.140}$$

refers to the joint probability by which we want to model the observed process. The integration in the Kullback information now refers both to $\boldsymbol{q}(i+\tau)$ *and* $\boldsymbol{q}(i)$. As in Sect. 12.6 it suffices to minimize

$$W = -\int\int \tilde{P}_j(\boldsymbol{q}(i+\tau), \boldsymbol{q}(i)) \ln P_j(\boldsymbol{q}(i+\tau), \boldsymbol{q}(i)) \mathrm{d}^N q(i+\tau) \mathrm{d}^N q(i). \tag{12.141}$$

We shall use the relationship between joint, conditional, and possibly stationary probability distributions, namely

$$P_j(\boldsymbol{q}(i+\tau),\boldsymbol{q}(i)) = P(\boldsymbol{q}(i+\tau) \mid \boldsymbol{q}(i))P_{st}(\boldsymbol{q}(i)), \tag{12.142}$$

$$\tilde{P}_j(\boldsymbol{q}(i+\tau),\boldsymbol{q}(i)) = \tilde{P}(\boldsymbol{q}(i+\tau) \mid \boldsymbol{q}(i))\tilde{P}_{st}(\boldsymbol{q}(i)). \tag{12.143}$$

As we shall see below, the explicit form of $P_{st}$ is not needed, whereas our results depend on $\tilde{P}_{st}$. This distribution function depends on the past history of the system including its initial condition and on the kind of measurements, for instance on a single system or on an ensemble. In the ideal case, if the measurements are made after a sufficiently long transit time, $\tilde{P}_{st}$ may be identified with the steady state distribution, but this requirement is not needed for the following analysis. We shall discuss the practical determination of $\tilde{P}_{st}$ after (12.178). Inserting (12.142) into (12.141) yields

$$W = W_c + W_{st}, \tag{12.144}$$

where

$$W_c = -\int\int \tilde{P}_j \ln P(\boldsymbol{q}(i+\tau) \mid \boldsymbol{q}(i+\tau))\mathrm{d}^N q(i+\tau)\mathrm{d}^N q(i), \tag{12.145}$$

$$W_{st} = -\int\int \tilde{P}_j \ln P_{st}(\boldsymbol{q}(i))\mathrm{d}^N q(i+\tau)\mathrm{d}^N q(i). \tag{12.146}$$

Since, for instance, for the derivation of drift and diffusion coefficients we need only the conditional probability, we focus our attention on (12.145). We make again the hypothesis (12.74) for $P(\boldsymbol{q}(i+\tau) \mid \boldsymbol{q}(i))$, but make a special hypothesis for $\lambda_\ell(\boldsymbol{q}(i))$ in the form

$$\lambda_\ell(\boldsymbol{q}(i)) = \sum_m \lambda_{\ell m}^{(2)} U_m^{(2)}(\boldsymbol{q}(i)), \tag{12.147}$$

where $U_m^{(2)}(\boldsymbol{q}(i))$ are appropriately chosen functions, e.g. in the form (12.83). Because of (12.74), we may decompose $W_c$ (12.145)

$$W_c = \overline{\lambda} + \hat{W}_c, \tag{12.148}$$

where

$$\overline{\lambda} = \int\int \tilde{P}_j(\boldsymbol{q}(i+\tau),\boldsymbol{q}(i))\lambda(\boldsymbol{q}(i))\mathrm{d}^N q(i+\tau)\mathrm{d}^N q(i) \tag{12.149}$$

and

$$\begin{aligned}\hat{W}_c = \int\int \tilde{P}_j(\boldsymbol{q}(i+\tau),\boldsymbol{q}(i))\Big\{&\sum_{\ell m} \lambda_{\ell m}^{(2)} U_m^{(2)}(\boldsymbol{q}(i))q_\ell(i+\tau)\\ &+\sum_{k\ell} \lambda_{k\ell} q_k(i+\tau)q_\ell(i+\tau)\Big\}\mathrm{d}^N q(i+\tau)\mathrm{d}^N q(i).\end{aligned} \tag{12.150}$$

Because of (12.109) and (12.147), $\lambda(\boldsymbol{q}(i))$ in (12.149) reads

$$\lambda(\boldsymbol{q}(i)) = \ln\Big\{\int \mathrm{d}^N q(i+\tau)\exp\Big[-\sum_{\ell m}\lambda_{\ell m}^{(2)}U_m^{(2)}(\boldsymbol{q}(i))q_\ell(i+\tau) - \sum_{k\ell}\lambda_{k\ell}q_k(i+\tau)q_\ell(i+\tau)\Big]\Big\}. \tag{12.151}$$

Equation (12.149) can be further simplified. By means of (12.143) we transform it into

$$\overline{\lambda} = \int\int \tilde{P}(\boldsymbol{q}(i+\tau)\mid \boldsymbol{q}(i))\tilde{P}_{st}(\boldsymbol{q}_i)\lambda(\boldsymbol{q}_i)\mathrm{d}^N q(i+\tau)\mathrm{d}^N q(i), \tag{12.152}$$

which because of the normalization of $\tilde{P}(\boldsymbol{q}(i+\tau)\mid \boldsymbol{q}(i))$ reduces to

$$\overline{\lambda} = \int \tilde{P}_{st}(\boldsymbol{q}(i))\lambda(\boldsymbol{q}(i))\mathrm{d}^N q(i), \tag{12.153}$$

where (12.151) must be inserted.

We now apply a gradient strategy to determine the parameters $\lambda_{\ell m}^{(2)}$ that actually play the role of Lagrange multipliers belonging to the constraints (12.85). The sole constraints we employ with respect to $\lambda_{k\ell}$ are (12.86) with $U^{(3)} = 1$.

We form (putting the usual constants $\gamma = 1$)

$$\dot{\lambda}_{\ell m}^{(2)} = -\frac{\partial W_c}{\partial \lambda_{\ell m}^{(2)}}. \tag{12.154}$$

According to the decomposition (12.148), we consider the individual terms, whereby we use for (12.147) the abbreviation

$$\lambda_\ell = f_\ell\left(q(i), \lambda_{\ell m}^{(2)}\right). \tag{12.155}$$

We first obtain from (12.150)

$$-\frac{\partial \hat{W}_c}{\partial \lambda_{\ell m}^{(2)}} = -\int\int \tilde{P}_j(\boldsymbol{q}(i+\tau), \boldsymbol{q}(i))\Big\{\left(\partial f_\ell/\partial\lambda_{\ell m}^{(2)}\right)q_\ell(i+\tau)\Big\}\mathrm{d}^N q(i+\tau)\mathrm{d}^N q(i), \tag{12.156}$$

or, in short,

$$-\frac{\partial \hat{W}_c}{\partial \lambda_{\ell m}^{(2)}} = -\langle q_\ell(i+\tau)\partial f_\ell/\partial\lambda_{\ell m}^{(2)}\rangle_{\tilde{P}_j}, \tag{12.157}$$

where $\langle\ldots\rangle_{\tilde{P}_j}$ refers to the joint probability (12.139). The negative derivatives of $\overline{\lambda}$, (12.153) with (12.151), can be evaluated in an obvious way, which yields consecutively

$$-\frac{\partial \bar{\lambda}}{\partial \lambda_{\ell m}^{(2)}} = -\int \tilde{P}_{st}(\boldsymbol{q}(i)) \frac{\partial}{\partial \lambda_{\ell m}^{(2)}} \ln \left\{ \int \mathrm{d}^N q(i+\tau) \right.$$

$$\left. \times \exp\left[ -\sum_{\ell} f_\ell q_\ell(i+\tau) - \sum_{k\ell} \lambda_{k\ell} q_k(i+\tau) q_\ell(i+\tau) \right] \right\} \mathrm{d}^N q(i) \tag{12.158}$$

$$= \int \tilde{P}_{st}(\boldsymbol{q}(i)) \frac{1}{\{..\}} \int \mathrm{d}^N q(i+\tau) \left( \partial f_\ell / \partial \lambda_{\ell m}^{(2)} \right) q_\ell(i+\tau) \exp[..] \mathrm{d}^N q(i), \tag{12.159}$$

where the curly bracket coincides with that of (12.158). The integral over $\boldsymbol{q}(i+\tau)$ divided by the curly bracket can be abbreviated by

$$\langle \partial f_\ell / \partial \lambda_{\ell m}^{(2)} q_\ell(i+\tau) \rangle_{P,\boldsymbol{q}(i)} = \partial f_\ell / \partial \lambda_{\ell m}^{(2)} \langle q_\ell(i+\tau) \rangle_{P,\boldsymbol{q}(i)} \tag{12.160}$$

so that we finally obtain

$$-\frac{\partial \bar{\lambda}}{\partial \lambda_{\ell m}^{(2)}} = \int \tilde{P}_{st}(\boldsymbol{q}(i)) \partial f_\ell / \partial \lambda_{\ell m}^{(2)} \langle q_\ell(i+\tau) \rangle_{P,\boldsymbol{q}(i)} \mathrm{d}^N q(i). \tag{12.161}$$

We now add the contributions (12.161) and (12.157). In order to obtain a result as symmetric as possible with respect to $P$ and $\tilde{P}$, we rewrite the r.h.s. of (12.157) using (12.143). Our final result then can be written in the rather concise form

$$\dot{\lambda}_{\ell m}^{(2)} = \int \mathrm{d}^N q(i) \tilde{P}_{st}(\boldsymbol{q}(i)) \left( \partial f_\ell / \partial \lambda_{\ell m}^{(2)} \right)$$

$$\times \left\{ \langle q_\ell(i+\tau) \rangle_{P,\boldsymbol{q}(i)} - \langle q_\ell(i+\tau) \rangle_{\tilde{P},\boldsymbol{q}(i)} \right\}. \tag{12.162}$$

The derivation of equations for $\dot{\lambda}_{k\ell}$ is even simpler and yields

$$\dot{\lambda}_{k\ell} = \int \tilde{P}_{st}(\boldsymbol{q}(i)) \left\{ \langle q_k(i+\tau) q_\ell(i+\tau) \rangle_{P,\boldsymbol{q}(i)} \right.$$

$$\left. - \langle q_k(i+\tau) q_\ell(i+\tau) \rangle_{\tilde{P},\boldsymbol{q}(i)} \right\}. \tag{12.163}$$

Now we have to evaluate the brackets $\langle ... \rangle_{P,\boldsymbol{q}(i)}$ that occur in (12.162) and (12.163), where we may use the results of Sect. 12.6, (12.125), (12.126). We thus obtain

$$\langle q_\ell(i+\tau) \rangle_{P,\boldsymbol{q}(i)} = h_\ell, \tag{12.164}$$

and because of (12.121)

$$(12.164) = -\frac{1}{2} (\Delta^{-1} \lambda)_\ell \equiv -\frac{1}{2} \left( \sum_{\ell'} (\Delta^{-1})_{\ell\ell'} \lambda_{\ell'} \right). \tag{12.165}$$

In the following, we shall express $\lambda_{\ell'} \equiv f_{\ell'}$ by means of (12.147). Because of (12.133), we can express (12.164) by means of the drift coefficients $K_\ell$

$$h_\ell - q_\ell(i) = \tau K_\ell(\boldsymbol{q}(i)). \tag{12.166}$$

Having this relation in mind, we subtract and add $q_\ell(i)$ in the curly bracket of (12.162). Furthermore, at the minimum of $W_c$ we may assume $\dot{\lambda}^{(2)}_{\ell m} = 0$. Then, after performing the steps described after (12.164), we derive from (12.162) the following equation

$$\int \mathrm{d}^N q(i) \tilde{P}_{st}(\boldsymbol{q}(i)) U^{(2)}_{\ell m}(\boldsymbol{q}(i)) \left\{ -\frac{1}{2} \sum_{\ell'} (\Delta^{-1})_{\ell\ell'} \sum_{m'} \lambda^{(2)}_{\ell' m'} U^{(2)}_{\ell' m'}(\boldsymbol{q}(i)) - q_\ell(i) \right\}$$
$$= \int \mathrm{d}^N q(i) \tilde{P}_{st}(\boldsymbol{q}(i)) U^{(2)}_{\ell m}(\boldsymbol{q}(i)) \left\{ \langle q_\ell(i+\tau) \rangle_{\tilde{P},\boldsymbol{q}(i)} - q_\ell(i) \right\}. \tag{12.167}$$

In order to simplify this equation, we introduce the abbreviation

$$\int \mathrm{d}^N q(i) \tilde{P}_{st}(\boldsymbol{q}(i)) U(\boldsymbol{q}(i)) = \langle U \rangle_{\tilde{P}_{st}} \tag{12.168}$$

and use the relation

$$\int \mathrm{d}^N q(i) \tilde{P}_{st}(\boldsymbol{q}(i)) U(\boldsymbol{q}(i)) \langle q_\ell(i+\tau) \rangle_{\tilde{P},\boldsymbol{q}(i)} = \langle q_\ell(i+\tau) U(\boldsymbol{q}(i)) \rangle_{\tilde{P}_j,\boldsymbol{q}(i)}, \tag{12.169}$$

which results from (12.146) and the definition of the brackets. When we further assume that $\Delta^{-1}$ is independent of $\boldsymbol{q}(i)$ (see below the discussion following (12.175)), we can cast (12.167) into the concise form

$$-\frac{1}{2} \sum_{\ell'} (\Delta^{-1})_{\ell\ell'} \sum_{m'} \lambda^{(2)}_{\ell' m'} \langle U^{(2)}_{\ell m} U^{(2)}_{\ell' m'} \rangle_{\tilde{P}_{st}} - \langle U^{(2)}_{\ell m} q_\ell \rangle_{\tilde{P}_{st}}$$
$$= \langle (q_\ell(i+\tau) - q_\ell(i)) U^{(2)}_{\ell m}(\boldsymbol{q}(i)) \rangle_{\tilde{P},\boldsymbol{q}(i)}. \tag{12.170}$$

If $\Delta^{-1}$ is known, (12.170) provides us with as many linear equations for $\lambda^{(2)}_{\ell' m'}$ as there are these unknowns. In view of (12.166) these equations must be solved in the limit $\tau \to 0$ after dividing both sides by $\tau$. Equation (12.170) contains averages over $P_{st}$ according to the abbreviation (12.168). Thus we may concentrate the evaluation of the drift coefficients (or equivalently $\lambda_\ell$'s (12.147)) on those $\boldsymbol{q}$-regions that are most important for the process under study. In order to achieve such a result, we used the joint probabilities in the Kullback information at the beginning of this section.

We are left with the further treatment of (12.163). We first recapitulate the result (12.127) and insert it on the r.h.s. of (12.163). Again we put $\dot{\lambda}_{k\ell} = 0$. We thus obtain:

$$0 = \int \mathrm{d}^N q(i) \tilde{P}_{st}(q(i)) \Big\{ \langle q_k(i+\tau) q_\ell(i+\tau) \rangle_{\tilde{P},q(i)} - \langle q_k(i+\tau) \rangle_{P,q(i)} \langle q_\ell(i+\tau) \rangle_{P,q(i)} - \frac{1}{2} \left( \varDelta^{-1} \right)_{k\ell} \Big\}. \tag{12.171}$$

In this equation, there are two sets of unknowns, namely $\varDelta^{-1}$ and $\langle q_k(i+\tau) \rangle_{P,q(i)}$. We may assume that by means of the solution of (12.170) we have already determined $\langle q_k(i+\tau) \rangle_{P,q(i)}$, which means we may put

$$\langle q_k(i+\tau) \rangle_{P,q(i)} = \langle q_k(i+\tau) \rangle_{\tilde{P},q(i)}, \tag{12.172}$$

where the r.h.s. is experimentally given. When we further assume that $\varDelta^{-1}$ is independent of $q(i)$, we may transform (12.171) into

$$\frac{1}{\tau} \frac{1}{2} \left( \varDelta^{-1} \right)_{k\ell} = \frac{1}{\tau} \Big\{ \int \mathrm{d}^N q(i) \tilde{P}_{st}(q(i)) \Big[ \langle q_k(i+\tau) q_\ell(i+\tau) \rangle_{\tilde{P},q(i)} \Big] - \langle q_k(i+\tau) \rangle_{\tilde{P},q(i)} \langle q_\ell(i+\tau) \rangle_{\tilde{P},q(i)} \Big\}, \tag{12.173}$$

where we have divided both sides by $\frac{1}{\tau}$ to secure that both sides do not vanish (at least in general) in the limit $\tau \to 0$.

Let us write (12.173) as

$$-\frac{1}{2\tau} \left( \varDelta^{-1} \right)_{k\ell} = Q_{k\ell}, \tag{12.174}$$

where $Q_{k\ell}$ is an abbreviation for the r.h.s. of (12.173). Obviously the determination of $\varDelta$ amounts to an inversion of the matrix $Q = (Q_{k\ell})$ which is, indeed,independent of $q(i)$. Our procedure assumes that the experimentally given generalized diffusion matrix

$$\frac{1}{\tau} \Big\{ \langle q_k(i+\tau) q_\ell(i+\tau) \rangle_{\tilde{P},q(i)} - \langle q_k(i+\tau) \rangle_{\tilde{P},q(i)} \langle q_\ell(i+\tau) \rangle_{\tilde{P},q(i)} \Big\} \tag{12.175}$$

is independent of $q(i)$ ("additive noise"). We are now in a position to discuss the meaning of (12.173). Because of (12.81) with (12.76) and (12.78), the l.h.s. of (12.173) is the diffusion matrix that occurs in the Îto-Fokker-Planck equation, whereas the r.h.s. of that equation represents the experimentally given (generalized) diffusion matrix that in the limit $\tau \to 0$ coincides with the conventional diffusion matrix. Thus in the case of additive noise, we are only concerned with an approximation of the drift coefficients. Whether or not the process under study is governed by additive noise may be checked by looking at the dependence of (12.175) on $q(i)$ for $\tau \to 0$, though a certain scatter of values of (12.175) must be admitted in practice.

Before we illustrate our method by means of an explicit example, we discuss how to evaluate the expressions that occur in our above equations. These are of the form

$$\int \tilde{P}_{st}(q) \tilde{F}(q) \mathrm{d}^N q, \tag{12.176}$$

$$\langle q_\ell(i+\tau)\rangle_{\tilde{P},q(i)}, \tag{12.177}$$

$$\langle q_k(i+\tau)q_\ell(i+\tau)\rangle_{\tilde{P},q(i)}. \tag{12.178}$$

In all cases we decompose, according to the experimental resolution, the $q$ space into cubes with indices $j$ and edge lengths $\epsilon$ so that $q_j$ belongs to cube $j$. Then

$$(12.176) = \frac{1}{Z}\epsilon^N \sum_j n(q_j)\tilde{F}(q_j) \tag{12.179}$$

with

$$Z = \epsilon^N \sum_j n(q_j), \tag{12.180}$$

where $n(q_j)$ is the number of $q$'s found in cube $j$. The plot $n(q_j)$ versus $q_j$ is called a *histogram*. The practical determination of $n(q_j)$ depends on the experimental conditions, where we mention the following cases:

1. Measurements over a time interval $\Delta T$ on a single system that undergoes a stationary process and has reached steady state conditions. In this case we sample all $q_j$ that are observed over $\Delta T$.
2. If the process observed on a single system is nonstationary or transient, or both, but has a slow time-dependence of its drift and diffusion coefficients, we may scan the process by a shifting time-window $\Delta T$ of suitable length.
3. If we are in a position to make measurements on an ensemble of systems under the same conditions for each member, we can cover both stationary and instationary processes, where we sample the $q_j$'s at the same time $t \leftrightarrow i$ (or small time-interval $\tau$). Then $n(q_j)$ is the number of $q$'s found in cube $j$ in all the members of the ensemble.

To determine (12.177), we start from a specific cube $j_0$ in which $q(i)$ lies. Then

$$(12.177) = \frac{1}{Z_c}\epsilon^N \sum_j m(q_j)q_{\ell,j} \tag{12.181}$$

with

$$Z_c = \epsilon^N \sum_j m(q_j). \tag{12.182}$$

In (12.181) $q_{\ell j}$ is the component $\ell$ of the vector $q$ that lies in cube $j$ and $m(q_j)$ the number of $q$'s found in cube $j$ after time $\tau$, when in each case the system had been found before in cube $j_0$. Thus instead of $m(q_j)$ we could write $m(q_j \mid q_{j_0})$. The plot of $m(q_j \mid q_{j_0})$ against $q_j$ and $q_{j_0}$ is again a histogram. In the same fashion we can determine $\langle q_\ell(i+\tau) - q_\ell(i)\rangle_{\tilde{P},q(i)}$ by replacing $q_{\ell,j}$ in (12.181) by $q_{\ell,j} - q_{\ell,j_0}$. The evaluation of (12.178) is analogous.

## 12.8 An Explicit Example: Brownian Motion

We now discuss a simple example, namely that of Brownian motion. This example may also be used as a check, because in this case we know the stationary solution $P_{st}$ and the conditional probability $P$ explicitly. We treat the one-dimensional case. The Langevin equation reads (in the sense of (12.98))

$$dq(t) = -aq\mathrm{d}t + g\mathrm{d}w(t), \tag{12.183}$$

where the stochastic process is defined by (12.99) and (12.100). The Fokker-Planck equation belonging to (12.183) reads (cf. 12.80, 81)

$$\dot{f} = -\frac{\partial}{\partial q}(-aqf) + \frac{1}{4}G^{-1}\frac{\partial^2 f}{\partial q^2}, \tag{12.184}$$

where according to (12.101)

$$\frac{1}{2}G^{-1} = g^2. \tag{12.185}$$

The stationary probability distribution can be determined from (12.184) with $\dot{f} = 0$ and reads

$$\tilde{P}_{st} = \sqrt{\frac{a}{\pi g^2}}\exp\left(-\frac{a}{g^2}q^2\right), \tag{12.186}$$

while the conditional probability density $\tilde{P}$ is given by

$$\tilde{P}(q(i+\tau) \mid q(i)) = \frac{1}{\sqrt{\pi a}}\exp\left(-\frac{(q(i+\tau)-b)^2}{a}\right), \tag{12.187}$$

where

$$a = \frac{g^2}{a}(1 - \exp(-2a\tau)) \tag{12.188}$$

and

$$b = q(i)\exp(-a\tau). \tag{12.189}$$

In (12.186) and (12.187) we denoted the $P$'s by means of a tilde to indicate that in our model we consider these quantities as experimentally given but, so to speak, unknown to us. Thus our formalism should recover these results.

In principle, to provide the "experimental" data, we should run the solutions of the Langevin equation (12.183) on a computer with randomly generated realizations of $dw(t)$. For our present purposes may it suffice to calculate $\langle q(i+\tau)\rangle_{\tilde{P},q(i)}$ and

$\langle q(i+\tau)^2 \rangle_{\tilde{P},q(i)}$ directly by means of (12.187). From Sect. 12.6 we know how to do this and we obtain

$$\langle q(i+\tau) \rangle_{\tilde{P},q(i)} = b \equiv q(i) \exp(-a\tau), \tag{12.190}$$

or, for small $\tau$,

$$\langle q(i+\tau) \rangle_{\tilde{P},q(i)} - q(i) = -a\tau q(i). \tag{12.191}$$

We further obtain

$$\langle q(i+\tau)^2 \rangle_{\tilde{P},q(i)} - \langle q(i+\tau) \rangle^2_{\tilde{P},q(i)} = \frac{1}{2}a, \tag{12.192}$$

and, for small $\tau$,

$$(12.192) \approx g^2\tau. \tag{12.193}$$

The expressions (12.186), (12.191), (12.192) with (12.193) are now our "experimental" data. Our task will be to guess $K(q)$ and $g^2 = \frac{1}{2}G^{-1}$ from these data. To this end we choose "appropriate" constraints (12.85), or correspondingly (12.147). Since the forces $K$ are quite often of a polynomial character with a small power, we choose (with $\ell = 1$),

$$U^{(2)}_{1,m} = \sum_{m=0}^{3} \lambda^{(2)} q^m. \tag{12.194}$$

We first consider (12.173) and obtain, due to $\left(\Delta^{-1}\right) = \lambda_{11}^{-1}$,

$$-\frac{1}{\tau}\frac{1}{2}\lambda_{11}^{-1} = \frac{1}{\tau}g^2\tau. \tag{12.195}$$

Because of (9.30), (12.104), we obtain, quite correctly, the "experimental" diffusion constant $g^2$. Let us now treat (12.170). We first calculate

$$\langle U^{(2)}_{1m} U^{(2)}_{1m'} \rangle_{\tilde{P}_{st}} = \langle q^\mu \rangle_{\tilde{P}_{st}} \equiv c_\mu, \quad \mu = m + m'. \tag{12.196}$$

We readily obtain (using (12.186) and putting $\gamma = a/g^2$)

$$(12.196) = 0 \quad \text{for} \quad \mu \quad \text{odd} \tag{12.197}$$

and

$$\begin{aligned} (12.196) &= 1 && \text{for} \quad \mu = 0 \\ &= \tfrac{1}{2}\gamma^{-1} && \text{for} \quad \mu = 2 \\ &= \tfrac{3}{4}\gamma^{-1} && \text{for} \quad \mu = 4 \\ &= \tfrac{15}{8}\gamma^{-3} && \text{for} \quad \mu = 6. \end{aligned} \tag{12.198}$$

Finally we calculate

$$\begin{aligned}&\langle (q(i+\tau)-q(i))U^{(2)}_{1m}q(i)\rangle_{\tilde{P}_j}\\&\quad\equiv \int \mathrm{d}q(i)\tilde{P}_{st}(q(i))\langle (q(i+\tau)-q(i))\rangle_{\tilde{P},q(i)}U^{(2)}_{1m}(q(i)).\end{aligned} \tag{12.199}$$

Using (12.191), we obtain

$$(12.199)=\langle -a\tau q(i)q(i)^{m}\rangle_{\tilde{P}_{st}}, \tag{12.200}$$

$$\equiv -a\tau\langle q(i)^{m+1}\rangle_{\tilde{P}_{st}}, \tag{12.201}$$

which can be evaluated by means of (12.196)–(12.198). Using the results (12.195)–(12.201), we formulate the equations (12.170) explicitly

$$g^2\sum_{m'=0}^{3}\lambda^{(2)}_{1m'}c_{m+m'}-c_{m+1}=-a\tau c_{m+1},\quad m=0,1,2,3. \tag{12.202}$$

The solution of (12.202) is a simple matter (and perhaps a nice exercise) and reads

$$\lambda^{(2)}_0=\lambda^{(2)}_2=\lambda^{(2)}_3=0;\quad \lambda^{(2)}_1=(1-a\tau)/g^2. \tag{12.203}$$

Thus according to (12.147)

$$\lambda_1=(1-a\tau)/g^2\cdot q \tag{12.204}$$

so that finally with aid of the definition (12.135) of the drift coefficient, we obtain

$$K(q)=-aq. \tag{12.205}$$

In this way we have fully recovered the "experimental" process. This example illustrates the applicability of our approach. The general case of (12.170) seems to be clumsy. It is, however, a simple matter to establish a corresponding computer program to automate the data-analysis, and thus to model stochastic processes.

We now turn to processes in which the diffusion coefficients depend on the variables $\boldsymbol{q}$. In such a case it is tempting to employ, besides the constraints (12.85), also the constraints (12.86). In some special cases this is, indeed, possible, provided we allow for negative powers in (12.85), (12.86), as was shown by Borland (cf. references). In order to cover the general case, we proceed in the next section in a different way, where we approximate the drift and diffusion coefficients more directly.

## 12.9 Approximation and Smoothing Schemes, Multiplicative (and Additive) Noise

We first explain our procedure by means of an example of an approximation scheme well-known in mathematics. In order to approximate a given function $a(q)$ by a polynomial $\sum_{\nu=0}^{n} c_\nu q^\nu$, we form

$$\hat{V}(c_0, ..., c_n) = \int dq \left( \sum_{\nu=0}^{n} c_\nu q^\nu - a(q) \right)^2 = \min! \tag{12.206}$$

and seek the minimum of $\hat{V}$ by an appropriate choice of the $c_\nu$'s by means of

$$\frac{\partial \hat{V}}{\partial c_\mu} = 2 \int dq q^\mu \left( \sum_{\nu=0}^{n} c_\nu q^\nu - a(q) \right) = 0. \tag{12.207}$$

This gives rise to a set of $n$ linear algebraic equations for the $n$ unknowns $c_\nu$

$$\sum_{\nu=0}^{n} c_\nu \int q^{\nu+\mu} dq = \int q^\mu a(q) dq. \tag{12.208}$$

The generalization to the multidimensional case and a general set of "test" functions is obvious and yields

$$\tilde{V}(c_0, ..., c_n) = \int d^N q \left( \sum_{\nu=0}^{n} c_\nu U_\nu(\boldsymbol{q}) - a(\boldsymbol{q}) \right)^2 = \min!, \tag{12.209}$$

$$\sum_{\nu=0}^{n} c_\nu \int U_\mu(\boldsymbol{q}) U_\nu(\boldsymbol{q}) d^N \boldsymbol{q} = \int U_\mu(\boldsymbol{q}) a(\boldsymbol{q}) d^N \boldsymbol{q}. \tag{12.210}$$

The approach becomes especially elegant if

$$\int U_\mu(\boldsymbol{q}) U_\nu(\boldsymbol{q}) d^N q = \delta_{\mu\nu}, \tag{12.211}$$

i.e. if the $U_\mu$'s are orthonormal.

In the following we wish to approximate the experimentally observed drift and diffusion coefficients in a suitable way. The results of Sect. 12.7 suggest the inclusion in the approximation scheme of a weight function, namely $\tilde{P}_{st}(\boldsymbol{q}(i))$. We, therefore, minimize the expression

$$\hat{V}_\ell = \int d^N q(i) \tilde{P}_{st}(\boldsymbol{q}(i)) \left\{ K_\ell(\boldsymbol{q}(i)) - \frac{1}{\tau} \left( \langle q_\ell(i+\tau) \rangle_{\tilde{P},\boldsymbol{q}(i)} - q_\ell(i) \right) \right\}^2 \tag{12.212}$$

with

$$K_\ell(\boldsymbol{q}(i)) = \sum_\nu \lambda^{(2)}_{\ell\nu} U^{(2)}_\nu(\boldsymbol{q}(i)) \tag{12.213}$$

by a proper choice of the coefficients $\lambda^{(2)}_{\ell\nu}$. This leads to

$$\sum_\nu \lambda^{(2)}_{\ell\nu} \int \tilde{P}_{st} U^{(2)}_\mu U^{(2)}_\nu \mathrm{d}^N q(i) = \frac{1}{\tau} \int \tilde{P}_{st} U^{(2)}_\mu \Big( \langle q_\ell(i+\tau)\rangle_{\tilde{P},\boldsymbol{q}} - q_\ell(i) \Big) \mathrm{d}^N q(i). \tag{12.214}$$

The solution of this set of linear equations allows us to calculate the drift coefficients (12.213). To obtain the diffusion coefficients, we proceed in an analogous fashion. With the definition

$$\begin{aligned}\hat{V}_{k\ell} = \int \mathrm{d}^N q(i) \tilde{P}_{st}(\boldsymbol{q}(i)) \Big\{ G_{k\ell}(\boldsymbol{q}(i)) - \frac{1}{\tau} \Big( \langle q_k(i+\tau) q_\ell(i+\tau)\rangle_{\tilde{P},\boldsymbol{q}(i)} \\ - \langle q_k(i+\tau)\rangle_{\tilde{P},\boldsymbol{q}(i)} \langle q_\ell(i+\tau)\rangle_{\tilde{P},\boldsymbol{q}(i)} \Big) \Big\}^2\end{aligned} \tag{12.215}$$

and the hypothesis

$$G_{k\ell} = \sum_n \lambda^{(3)}_{k\ell n} U^{(3)}_{k\ell n}(\boldsymbol{q}), \tag{12.216}$$

we obtain

$$\begin{aligned}\sum_n \lambda^{(3)}_{k\ell n} \int \tilde{P}_{st} U^{(3)}_{k\ell m} U^{(3)}_{k\ell n} \mathrm{d}^N q(i) = \frac{1}{\tau} \int \tilde{P}_{st} U^{(3)}_{k\ell m} \Big( \langle q_k(i+\tau) q_\ell(i+\tau)\rangle_{\tilde{P},\boldsymbol{q}}(i) \\ - \langle q_k(i+\tau)_{\tilde{P},\boldsymbol{q}} \cdot \langle q_\ell(i+\tau)\rangle_{\tilde{P},\boldsymbol{q}}(i) \Big) \mathrm{d}^N q(i),\end{aligned} \tag{12.217}$$

whose solution fixes the diffusion coefficients (12.216).

Both in this section as well as in the preceding ones the $\tau$-dependence of the experimental data is not obvious. Rather one has to extrapolate to $\tau \to 0$ either with respect to the experimental data which (occasionally) will be difficult, or extrapolate the $\lambda$'s.

## 12.10 Explicit Calculation of Drift and Diffusion Coefficients. Examples

In Sects. 12.6–12.9 I have demonstrated how, by means of an optimization procedure the conditional probability of an unknown process, that is hypothesized to be Markovian, can be calculated. If the process can be assumed to be continuous Markovian, this optimization procedure automatically leads to the conventional definitions of drift and diffusion coefficients. In this section I present some numerical calculations due to Siegert, Friedrich and Peinke (cf. references), who calculate the drift and diffusion coefficients according to the conventional definitions (e.g.

(12.135), (12.137)) for several models. Their results shed light on the question of whether smoothing according to Sects. 12.7, 12.8 is needed or not. In their first example they solve the Langevin equation (cf. also (2.1), (2.3)–(2.5))

$$\dot{q} = \epsilon q - q^3 + \boldsymbol{\Gamma}(t) \tag{12.218}$$

with $\epsilon = 0, 1$ and fluctuation strength $Q = (0.05)^2$ numerically. They determine the conditional probability distribution (i.e. the histogram) of the noisy time series and by means of them (cf. (12.181)) the drift and diffusion coefficients. As it turns out, the scatter of data around the "true" coefficients is very small, and we need not reproduce their results here. Their next example, however, is more illuminating. They solve the Langevin equation (cf. also the related equations (11.33, 34))

$$\dot{\phi} = \omega + \sin\phi + \boldsymbol{\Gamma}(t) \tag{12.219}$$

with $\omega = 0, 2, Q = (0.6)^2$ and $\omega = 1, 0, Q = (0.005)^2$. Here we represent their results for the latter case. The upper part of Fig. 12.4 shows the time series of the phase $\phi$, whereas in its lower part the numerically determined drift coefficient $D^{(1)}$ is plotted together with the "true" function (solid curve). Two features are remarkable. Even a rough study of the data indicates that the drift coefficient is a periodic function of $\phi$, which suggests using periodic test functions in the smoothing proce-

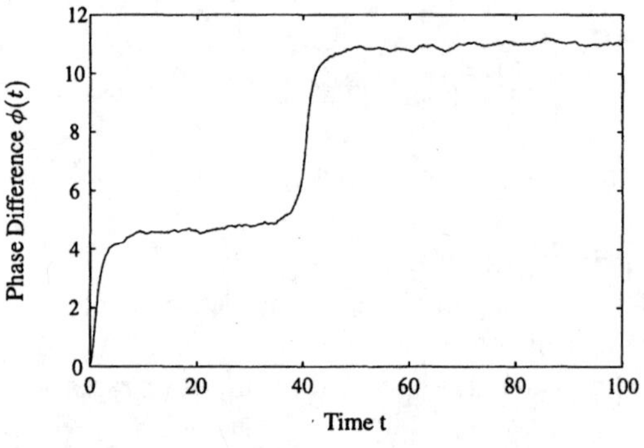


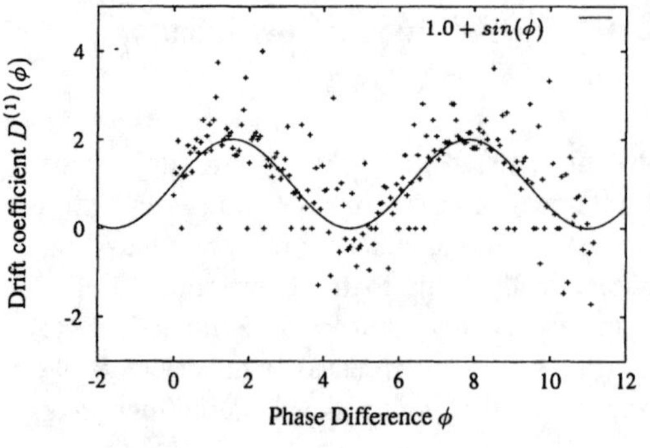


**Fig. 12.4.** Phase difference $\phi$ versus time $t$ and drift coefficient $D^{(1)}$ versus $\phi$. The noisy time series belongs to the Langevin equation $\dot{\phi}(t) = 1.0 + \sin(\phi(t)) + 0.05\,F(t)$, where $F(t)$ is a Gaussian distributed fluctuating force. In the second part of the figure the corresponding numerically determined drift coefficient $D^{(1)}$ is plotted together with the theoretical function (solid curve) (after S. Siegert, R. Friedrich, J. Peinke (1998))

dure of Sects. 12.7 and 12.9. Furthermore, in spite of the relatively weak fluctuating force, the scatter of data is remarkable. Thus it seems at least advisable to invoke a smoothing procedure. Such a procedure is, of course, still more necessary if the experimentally available data are scarcer.

I conclude this section with their third example which refers to two variables $q_1, q_2$ that obey the equations

$$\dot{q}_1 = \epsilon q_1 - \gamma q_2 + (q_1^2 + q_2^2)(\mu q_1 - \omega q_2) + \boldsymbol{\Gamma}_1(t), \tag{12.220}$$

$$\dot{q}_2 = \gamma q_1 + \epsilon q_2 + (q_1^2 + q_2^2)(\omega q_1 + \mu q_2) + \boldsymbol{\Gamma}_2(t), \tag{12.221}$$

where the fluctuating forces $\boldsymbol{\Gamma}_1, \boldsymbol{\Gamma}_2$ are assumed uncorrelated and of equal strengths, $Q = 0.04$. The parameters are chosen as $\epsilon = 0.05, \gamma = 1, \mu = -5, \omega = 7.5$ and thus lead to a Hopf bifurcation, i.e. to an oscillatory solution of (12.220), (12.221). As the authors find, the numerically calculated drift coefficients agree well with the exact ones in the central region, but at the borders differences occur that are caused by too few variable values in these regions. It is likely that smoothing (or, in other words, extrapolation in the sense of Sect. 12.7) may improve these results.

## 12.11 Process Modelling, Prediction and Control, Robotics

Since we are coming close to the end of this chapter and, with the exception of the quantum theoretical chapter (Chap. 13), to the end of this book, let us recall the purpose of our enterprise. We want to gain insight into the mechanisms underlying processes in complex systems. But we have only limited knowledge about the processes for several reasons:

1. Because of the complexity, we cannot (at least in general) derive the properties of the whole system from those of its individual parts. Thus we must rely on directly observed data.
2. The processes can be studied only over a limited time-interval $\Delta T$.
3. They cannot be repeated indefinitely or may even occur once only.
4. The data can be collected only over a limited domain $\Omega$ of variables. For example, the rotation of a robot arm may be studied only over a limited range of angles $\phi_1 < \phi < \phi_2$.
5. The data are noisy.
6. The data may be scarce (in the observed domain $\Delta T, \Omega$).

In this book, and especially in this chapter, we have shown how to make an unbiased guess of the unknown process including a best fit. The formalism is based on the assumption that the process is discrete or continuous Markovian. This assumption can be checked in a self-consistent way as we will discuss in the next section. When we recollect the approach outlined in Sects. 12.6–12.10, we recognize that a crucial task consists in the choice and adaptation of the constraints (12.85) or, more generally, in the test function $U(\boldsymbol{q})$. In this author's opinion, the use of test-functions is indispensable. Even if in a continuous process the data are dense and allow the determination of drift and diffusion coefficients with small scatter (cf. the process of (12.218)), the drift and diffusion coefficients are known only numerically. To gain insight into the nature of the process and to be able to extrapolate outside the known $\Omega$-domain, we need these coefficients in an analytical form. In practice, we often have to deal with scarcer data that possess considerable scatter. Then the use of test-functions is indispensable and we have to discuss this important issue. Quite often a glance at the scattered data of (generalized) drift and diffusion coefficients will provide us with the type of test-functions, e.g. periodic (superpositions of sine- and cos-functions) or low-order polynomials with adaptable coefficients. In a number of cases, we may possess some prior knowledge, for instance about general properties of the process studied. When training robots, we may equip them with a class of test-functions that can be improved by learning through feedback. Once the test-functions and their coefficients are fixed, we may calculate

1. the transition probability $P$ for finite $\tau$ (discrete process) or
2. the drift ($D_\ell$) and diffusion ($D_{k\ell}$) coefficients of the Îto-Fokker-Planck equation and the Îto-Langevin equation (continuous process).

The dependence of $P$ or $D_\ell$ and $D_{k\ell}$ on $q(i)$ via explicit test-functions allows extensions of the guessed process into larger domains. The extrapolation into the future is possible provided the process is stationary, i.e. there is no explicit time-dependenceof the transition probability or of the drift and diffusion coefficients.

In case 1, we form the path integral according to (2.52), but without the limit $\tau \to 0$. In it, the exponent $G$ is of the form

$$\mathcal{L} = \sum_i L(q(i+\tau), q(i)), \quad t = \tau i.$$

The variations $\delta q(i)$ of $\mathcal{L}$ with $\delta\mathcal{L} = 0$ define an extremal path and if $\mathcal{L}$ maximum, the most probable path $q^{(0)}(i)$. The expansion of $\mathcal{L}$ around $q^{(0)}(i)$ yields the probability of the corresponding fluctuations. It may also be possible that several extremal paths with $\mathcal{L}$ (local) maximum occur. In this case we then can study the relative probabilities of various paths and fluctuations around them. We are familiar with such phenomena in the case of nonequilibrium phase transitions.

In case 2, we may resort to direct solutions of the Fokker-Planck equation. Thus we may obtain $f(t)$ for $t > T$ for arbitrary initial conditions and because of the explicit knowledge of the test-functions, we can extrapolate from the region $\Omega$ and interpolate within it. This leads us to one interpretation of the concept of *associative action*. Even if a process had been studied over a domain $\Omega$ (which need not be dense in variable space), we can define it on a dense space, including new initial conditions. Thus the system's behavior is modelled and predicted with respect to processes that haven't been studied or were executed before. I believe that our procedure has considerable potential going beyond the examples of Sect. 12.10. Robots may learn movements, such as walking and grasping, and the coordination of arms, hands, and legs. Since noise is admitted, the robot can learn walking in rough terrain and under uncertainty. Actually it is known from studies of movements, such as walking or running, that movements are not at all exactly periodic; in fact deviations occur from cycle to cycle. The same is true for grasping; each time hand and arm follow a somewhat different trajectory.

In conclusion, one fundamental problem in all inter/extrapolation schemes should be mentioned. Let us consider the movement of a robot arm around an axis and assume we model it around $0 \leq \phi \leq \phi_1$. Quite evidently, we can extrapolate this movement until it hits an obstacle which represents a "singularity". Thus, in the end, the main task of a robot will be to learn singularities and how to get around them. This can be done, e.g., by trial and error, but a more detailed discussion is beyond the scope of this book.

One last word, namely on the control of systems, may be in order. Once we know the test-functions $U$ and their properties, we can change them by adjustment of suitable parameters that via the drift coefficients may lead to desired types of behavior. This implies, of course, that we find a suitable mapping between the model parameters and the real, accessible parameters of the system. A simple, though rather general example is provided by the case in which the drift coefficients can be derived from a potential function, and the diffusion matrix is diagonal and a constant. In this case the fixed points of the system are known. Such an idea lies at the root of the pattern recognition algorithm of the synergetic computer (Sects. 12.2, 12.3).

## 12.12 Non-Markovian Processes. Connection with Chaos Theory

### 12.12.1 Checking the Markov Property

In this book I have made the assumption that the process under study is Markovian, i.e. that the joint probability distribution function (9.1) can be split into the product (9.3). To check whether the experimental data can be attributed to a Markov process, the following procedure, which is equivalent to (9.3), is useful.

We first define the *conditional* probability distribution function by

$$P(q_n, t_n \mid q_{n-1}, t_{n-1}; \ldots; q_0, t_0) = \frac{P_j(q_n, t_n; \ldots; q_0, t_0)}{P_j(q_{n-1}, t_{n-1}; \ldots; q_0, t_0)}, \tag{12.222}$$

where the $P_j$'s are the *joint* probability distribution functions. Then the necessary and sufficient condition for a Markov process is given by

$$P(q_n, t_n \mid q_{n-1}, t_{n-1}; \ldots; q_0, t_0) = P(q_n, t_n \mid q_{n-1}, t_{n-1}) \tag{12.223}$$

for $t_n \geq t_{n-1} \geq \ldots \geq t_0$.

By means of histograms (cf. Sect. 12.7) (12.222) can be checked numerically, provided $n$ remains a small number.

We now discuss the case in which the process is not Markovian. It has long been known that, by the introduction of suitable additional variables, a non-Markovian process can be made Markovian. But, as it seems, it was left to chaos theory to devise practical conceptual and numerical methods that allow us to introduce these variables. The procedure is known as reconstruction of an attractor or, more generally, as time-series analysis. In order to illustrate the basic ideas, we ignore the effect of noise and start with a very simple example.

### 12.12.2 Time-Series Analysis

Let us consider a pendulum. Its motion can be visualized in the phase-plane, where we plot a trajectory as a function of the coordinate $x$ and the velocity $v$ (Fig. 12.5). In view of later generalisations, we shall denote the $x$-axis by the coordinate $q_1$ and the $v$-axis by the coordinate $q_2$. The velocity is, of course, connected with the position $x$ by

$$v = \dot{x}, \tag{12.224}$$

where the dot means as usual the time-derivative. In the new notation we may also write (12.224) in the form

$$q_2 = \dot{q}_1. \tag{12.225}$$

Furthermore we note that the coordinate $x$, or equivalently $q_1$, obeys the oscillator equation

$$\ddot{q}_1 = -kq_1, \tag{12.226}$$

where $k$ is a positive constant.

Let us consider the case where we measure only the $q_1$ coordinate, which we plot as a function of time according to Fig. 12.6. In it $q_1$ is given by

$$q_1 = A \sin \omega t. \tag{12.227}$$

Then we ask ourselves: Can we reconstruct from this time series the trajectory or the attractor in the phase-plane of Fig. 12.5. In the present case this is quite simple, because of the relation (12.225), i.e.

$$q_2 = \dot{q}_1. \tag{12.228}$$

From this we can immediately deduce

$$q_2 = \omega A \cos \omega t. \tag{12.229}$$

By plotting (12.227) and (12.226) in the phase-plane, as time $t$ proceeds, we obtain, of course, the trajectory of that figure.

These relations can be cast into a more general and abstract form. Let us assume that a system described by the variable $q_1$ obeys an equation of the form

$$\ddot{q}_1 = f(q_1, \dot{q}_1), \tag{12.230}$$

where $f$ is a given function of $q_1$ and $\dot{q}_1$. We may deduce the new variable $q_2$ by

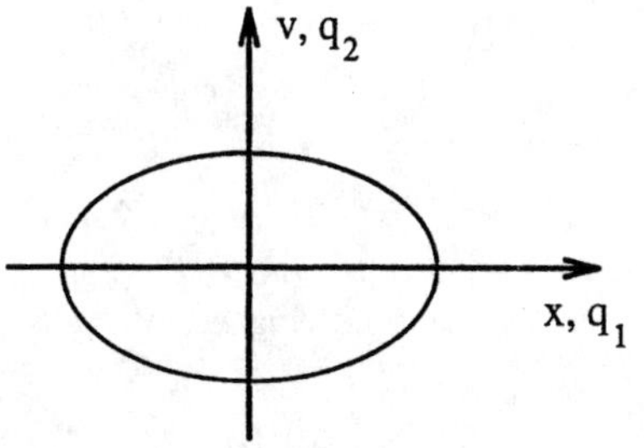


**Fig. 12.5.** A trajectory described by $q_1 = A \sin \omega t$, $q_2 = \dot{q}_1 = \omega A \cos \omega t$

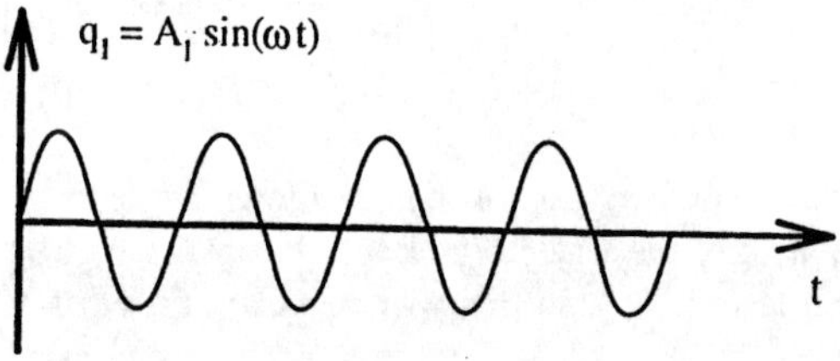


**Fig. 12.6.** Example of a time series $q_1 = A \sin \omega t$

$$q_2 = \dot{q}_1 \tag{12.231}$$

and may now replace the system of equations (12.230) and (12.231) by

$$\dot{q}_1 = q_2 \tag{12.232}$$

and

$$\dot{q}_2 = f(q_1, q_2). \tag{12.233}$$

This means that we can represent the trajectory in the phase-plane by these two equations. Quite evidently, to deduce a trajectory in the phase-plane from a time series is a trivial task.

These considerations can easily be generalized to the case of $n$-dimensions. The basic assumption, however, is that we are dealing with a dynamical system and that we know the dimension of the phase space, i.e. the number of independent variables. We first introduce an abbreviation for the $j$th derivative

$$\frac{d^j q_1}{dt^j} \equiv q_1^{(j)}. \tag{12.234}$$

We assume that the dynamics is described by an equation of the form

$$q_1^{(n)} = f\left(q_1, q_1^{(1)}, \ldots, q_1^{(n-1)}\right). \tag{12.235}$$

We may introduce the new coordinates

$$q_1^{(0)} \equiv q_1 \tag{12.236}$$

and

$$q_j = q_1^{(j-1)}. \tag{12.237}$$

Equation (12.235) can then be immediately replaced by the following set of equations:

$$\dot{q}_1 = q_2, \tag{12.238}$$

$$\ddot{q}_1 \equiv \dot{q}_2 = q_3, \tag{12.239}$$

$$\dot{q}_n = f(q_1, q_2, ..., q_n). \tag{12.240}$$

These define trajectories in a phase space with $n$ dimensions. Following the sequence of time, we may immediately construct the trajectory simply by taking higher and higher derivatives of $q_1$ according to (12.237). In this way one can easily calculate the attractor.

There is, however, a basic difficulty when one wishes to apply this procedure to actually measured time series. Because any time series can consist only of discrete points and is, in reality, neither continuous nor even differentiable, the evaluation of derivatives introduces considerable numerical errors. Therefore, another method for constructing an attractor was introduced by *Takens* and others and has become very useful. Again our goal is to reconstruct the trajectory of Fig. 12.5 from a time series as given in Fig. 12.6. The basic idea is to introduce a time-shift $T$ so that the coordinate $q_1$

$$q_1 = A \sin \omega t \tag{12.241}$$

is transformed into

$$\begin{aligned} q_2(t) &= Cq_1(t+T) \\ &= CA \sin[\omega(t+T)], \end{aligned} \tag{12.242}$$

which is actually the time-sequence of the variable $q_2$, provided $\omega T = \pi/2$. The quantity $C$ in (12.242) is a scaling constant. This example shows that we can reconstruct the trajectory of Fig. 12.5 by means of a suitable time-shift $T$. However, a difficulty arises in practical applications, because we do not know a priori from the experimental data what time-shift $T$ must be used. Thus we consider the case where $T$ is different from that used in (12.242). From the second line in (12.242) we readily obtain

$$q_2 = CA(\sin \omega t \cos \omega T + \cos \omega t \sin \omega T). \tag{12.243}$$

With help of (12.243), we may show that $q_1$ and $q_2$ now obey an equation of the form

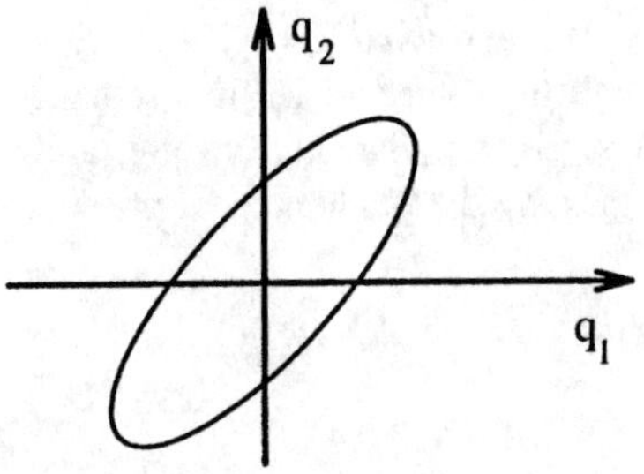


**Fig. 12.7.** Reconstructed trajectory (see text)

$$(q_2 - aq_1)^2 + \beta q_1^2 = (CA)^2 \tag{12.244}$$

with

$$\begin{aligned} a &= C\cos\omega T \\ \beta &= C\sin\omega T. \end{aligned} \tag{12.245}$$

The resulting new trajectory is shown in Fig. 12.7. From a comparison between Figs. 12.7 and 12.5 we may conclude that by using an arbitrary time-shift or time-delay $T$, we obtain essentially the same *attractor* but now rotated and deformed in phase-space. A pathological case occurs if $\beta = 0$ which happens for $T = 2\pi/\omega$. In this case, (12.244) can be replaced by

$$q_2 - Cq_1 = \pm CA \tag{12.246}$$

and the attractor consists of two individual lines only, i.e., the attractor is no longer resolved. This example shows clearly that the reconstruction of an attractor can, at least in some cases, depend sensitively on the choice of $T$. Below we shall show how to reconstruct attractors for more complicated systems. But before we do so, we discuss how to determine the dimensions of an attractor, in particular of a chaotic attractor.

Let us consider the case in which the time-series $q_1(t)$ of a single variable is given, but where we have to introduce additional variables to obtain the Markov property. Since we do not know a priori the dimension of the underlying phase space, we have to try different dimensions in phase space. The procedure of choosing a dimension $n$ of phase space is called embedding, i.e. we embed the trajectory in an $n$-dimensional phase space by constructing an adequate number of $n$ variables. The procedure is as follows: We choose a time delay $T$ and form the vector

$$\boldsymbol{q}(t) = (q_1(t), q_1(t-T), ..., q_n(t-nT)). \tag{12.247}$$

When time $t$ runs, this vector forms a trajectory in that $n$-dimensional space. $n$ must be chosen big enough so that the trajectory does not hit itself. As it turns out, such a trajectory does not fill the phase space so that the concept of fractal dimension must be used. These and related questions are beyond the scope of our book, and it must be mentioned that a number of problems have not yet found a definite solution. Such problems include how to cope with noisy data and what to do if the measured time-series are not stationary over a long enough time interval. A final remark is in order: In many cases the attractors are found to be irregular, i.e. chaotic, and thus chaos theory has been developed to study and classify the mathematical properties of these attractors.

# 13. Information Compression in Cognition: The Interplay between Shannon and Semantic Information

In this chapter[1] we return to our discussion of different aspects of "information" where we wish to shed new light on the relation between Shannon information and what we called semantic information. (Actually, in the literature the latter is occasionally called "pragmatic information".) As we have seen, Shannon information does not carry any meaning. On the other hand, we were able to define semantic information by invoking some basic concepts of dynamic systems theory, in particular that of attractors. These two concepts of information are not independent of each other, as one might think at first sight, but they are closely interlinked. This link is established by cognition. First, recall that in order to calculate Shannon information, we have to *label* objects which implies that we have to *distinguish* them. The objects may be quite simple, such as letters or numbers, but they may be also complicated, such as buildings in a city. At any rate, distinction relies on pattern recognition where we may invoke the attractor concept. Thus, Shannon information depends on semantic information. In our opinion, one of the main tasks of our cognitive system (in other words: our brain) consists of *compressing* information, i.e. of reducing the amount of Shannon information. This can be done in a variety of ways. Below, as an example we shall demonstrate how symmetries can be used.

## 13.1 Information Compression: A General Formula

In order to elucidate our approach to information compression we consider a series of experiments with outcomes each composed of two joint events which we distinguish by indices $k$ and $j$, respectively. An example is provided by throwing time and again two dice where each throw ("event") defines an outcome with $k$ eyes of one die and $j$ eyes of the other die. For sake of generality we assume that the faces carry different weights. Thus we consider the probabilities

$$p_{kj} \tag{13.1}$$

which are assumed to be normalized

$$\sum_{jk} p_{kj} = 1 \,. \tag{13.2}$$

[1] The results of this chapter are based on part of a joint study with Juval Portugali.

The corresponding Shannon information is given by

$$i = -\sum_{jk} p_{kj} \ln p_{kj} \,. \tag{13.3}$$

(As everywhere in this book we use ln instead of $\log_2$ for the definition of information.)

Now let us assume that $p_{kj}$ is independent of $j$. (In the case of the dice this means, that the outcomes of the throws of the second die have the same probabilities.) How is Shannon information then transformed? We introduce the probability distribution $\tilde{p}_k$ and put

$$p_{kj} = M\tilde{p}_k \,. \tag{13.4}$$

We assume the normalization condition

$$\sum_k \tilde{p}_k = 1 \tag{13.5}$$

and we want to determine the still unknown multiplying factor $M$. We insert (13.4) into (13.2) and observe (13.5). This leads us to the equation

$$\sum_{kj} M\tilde{p}_k = NM \sum_k \tilde{p}_k = 1 \tag{13.6}$$

(where $N$ is the number of indices $j$) and

$$M = 1/N \,. \tag{13.7}$$

Substituting (13.7) in (13.4) and this latter quantity in (13.3) we arrive at

$$i = -\sum_{kj} (\tilde{p}_k/N) \ln(\tilde{p}_k/N) \tag{13.8}$$

which can be transformed into

$$= -\sum_j 1/N \cdot \sum_k \tilde{p}_k \ln(\tilde{p}_k/N) \tag{13.9}$$

or

$$= -\sum_k \tilde{p}_k \ln \tilde{p}_k + \sum_k p_k \ln N \tag{13.10}$$

so that we obtain the final result

$$i = i_r + \ln N \tag{13.11}$$

where $i_r$ (reduced information) is defined by the first sum in (13.10). Rearranging (13.11) we obtain

$$i_r = i - \ln N \tag{13.12}$$

which is an explicit formula for the reduced information as function of the complete information $i$ and the logarithm $N$. Note that in all these cases finally a correction factor $K$ must be added that makes up for the replacement of $\log_2$ by ln.

## 13.2 Pattern Recognition as Information Compression: Use of Symmetries

After this simple but general consideration let us turn to the main issue of this chapter, namely to pattern recognition by the human brain or a computer. To be quite concrete we consider the following case: We decompose a two-dimensional picture into its pixels that are distinguished by an index vector.

$$\boldsymbol{j} = (j_x, j_y) \tag{13.13}$$

with integer components $j_x = 1, 2, ..., j_y = 1, 2, ...$ We attribute a grey value $q_j$ to each pixel. Then a pattern is described by the pattern vector (see also Sect. 12.2)

$$k,j \to \boldsymbol{q} = (q_{11}, q_{12}, ..., q_{nn}) \equiv q(\boldsymbol{j}) = \{q_{\boldsymbol{j}}\} \tag{13.14}$$

whereby the former indices of Sect. 13.1, $k$ and $j$, are replaced according to (13.14). In order not to deal with boundary conditions we assume that the boundary is periodic as it is actually the case when we include a 360° turn of our heads. We also change the notation (cf. Sect. 12.2).

$$p_{kj} \to f(\boldsymbol{q}) \tag{13.15}$$

where $f$ is the corresponding probability distribution of patterns described by the pattern vector $\boldsymbol{q}$.

Now let us assume that a human or a computer is exposed to all sorts of patterns. Then the Shannon information is given by

$$i = -\sum_{\boldsymbol{q}} f(\boldsymbol{q}) \ln(\boldsymbol{q}) \tag{13.16}$$

where we assume as usual the normalization condition

$$\sum_{\boldsymbol{q}} f(\boldsymbol{q}) = 1\,. \tag{13.17}$$

In accordance with the usual assumptions made on learning of patterns we assume that patterns are learned according to their frequency of occurrence.

Now consider the learning of a face by a baby who sees the face of his or her mother again and again but at quite different positions. In accordance with the concept of semantic information we may assume that these faces have the same impression on the baby irrespective of their position in space. Thus the baby will not learn all these faces as different but as a single face that is relevant to the baby. Thus the distinction of the face at different places as different is superfluous. This allows us to reduce Shannon information in the following sense: Let us assume that a face at a specific position in space is described by a prototype vector $\boldsymbol{q}'$. Then we can find the prototype vector of the same face at a place in space by a transformation $T$ via

$$\boldsymbol{q} = T\boldsymbol{q}' . \tag{13.18}$$

Using the transformation (13.18), we split the total sum into

$$\sum_{q} = \sum_{T} \sum_{q'} . \tag{13.19}$$

We note that the transformation $T$ has the effect

$$T\boldsymbol{q}' \equiv Tq'(j) = q(j + \boldsymbol{a}) , \tag{13.20}$$

$\boldsymbol{a}$: displacement vector.

We observe, however, that we may consider also other transformations, e.g. rotations. In complete analogy to what we described above, we assume that the distribution function $f$ is invariant against the transformation $T$ up to a constant factor $M_T$.

$$f(T\boldsymbol{q}') = M_T f(\boldsymbol{q}') \tag{13.21}$$

We determine $M_T$ by the normalization condition

$$\sum_{q'} f(\boldsymbol{q}') = 1 \tag{13.22}$$

so that

$$\sum_{q} f(\boldsymbol{q}') = \sum_{T} \sum_{q'} M_T f(\boldsymbol{q}') = 1 \tag{13.23}$$

follows. Thus we obtain

$$\sum_{T} M_T = 1 \tag{13.24}$$

where the sum over $T$ yields the number of transformations $N_T$. Because of the homogeneity of space we may assume that $M_T$ is independent of $T$ which leads us to the relationship

$$N_T M_T = 1\,. \tag{13.25}$$

These intermediate steps allow us to cast (13.21) into the form

$$f(T\boldsymbol{q}') = f(\boldsymbol{q}')/N_T\,. \tag{13.26}$$

Using (13.26) in (13.16) and (13.19) we obtain our final result

$$i = i_T + \ln N_T \tag{13.27}$$

where we introduced the abbreviation

$$i_T = -\sum_{\boldsymbol{q}'} f(\boldsymbol{q}') \ln f(\boldsymbol{q}')\,. \tag{13.28}$$

To interpret (13.27), we write this relation as

$$i_T = i - \ln N_T\,. \tag{13.29}$$

Thus using the semantic information that the patterns (e.g. faces) are just the same but merely shifted in space or rotated, the decrease of Shannon information is explicitly described by (13.29). To be quite precise, the case of rotation is somewhat tricky, because – in contrast to computer vision – a human observer is not capable of fully distinguishing an upside down face from one in its normal position.

## 13.3 Deformations

We now turn to objects that are deformed with respect to some kind of normal form, e.g. we may observe the emotional expressions on faces which, of course, lead to deformations. We start from the pattern vector

$$\boldsymbol{q}' = v(\boldsymbol{j}) \tag{13.30}$$

where we conceive $v$ as a (practically) continuous function of the two-dimensional vector $\boldsymbol{j}$. In a number of cases these deformations can be visualized by means of patterns drawn on a rubber sheet which is then deformed. An example of this kind is provided by D'Arcy Wentworth Thomson's model of the occurrence of different kinds of fish. We introduce an abstract deformation operator $D$ that transforms the "normal" face into a face with an emotion. In order to go from the deformed pattern to the undeformed one, we have to apply the inverse $D^{-1}$. Now by means of semantic information we identify the deformed pattern with the undeformed one so that we introduce a relationship

$$f(D^{-1}q') = \tilde{f}(\boldsymbol{q}') \cdot M_D \tag{13.31}$$

where as just said $D^{-1}\boldsymbol{q}'$ corresponds to the normal face. Because facial expressions appear not so often as normal faces and this frequency may depend on deformations (laughing, being depressed, etc.) we assume that $M_D$ depends on the kind of deformation $D$. Generally speaking, in spite of the fact that laughing occurs less frequently than a normal face, the effective frequency of $\tilde{f}(\boldsymbol{q}')$ may be enhanced by emotions of the observer, i.e. by an increase of attention. On the other hand in this case more information is required for the storage of associations so that in the end the Shannon information of a face without laughing is smaller than that with laughing. We now replace the sum over $\boldsymbol{q}$ by a double sum over the various displacements $D$ and the normal set $\boldsymbol{q}'$. In this way we introduce the normalization condition

$$\sum_{D,\boldsymbol{q}'} f(D^{-1}\boldsymbol{q}') = 1 \tag{13.32}$$

which because of (13.31) can be cast into the form

$$\sum_D M_D \sum_{\boldsymbol{q}'} \tilde{f}(\boldsymbol{q}') = 1 \tag{13.33}$$

with the result

$$\sum_D M_D = 1 \tag{13.34}$$

because of the normalization requirement for the sum over $\boldsymbol{q}'$. Generally speaking we may say that the dependence of $M$ on $D$ is given by the relative frequency of appearance compared to the normal case up to a common normalization factor that is determined by (13.34). Inserting (13.31) into the expression for the total information $i$ yields

$$i = -\sum_{D\boldsymbol{q}'} M_D \tilde{f}(\boldsymbol{q}') \ln(M_D \tilde{f}(\boldsymbol{q}')) \tag{13.35}$$

from which we deduce

$$= -\sum_{\boldsymbol{q}'} \tilde{f}(\boldsymbol{q}') \ln \tilde{f}(\boldsymbol{q}') - \sum_{\boldsymbol{q}'} M_D \ln M_D \,. \tag{13.36}$$

Thus we obtain the important result

$$i = i_D + \Delta_D \tag{13.37}$$

where we used the abbreviations

$$i_D = -\sum_{\boldsymbol{q}'} \tilde{f}(\boldsymbol{q}') \ln \tilde{f}(\boldsymbol{q}') \tag{13.38}$$

and

$$\Delta_D = -\sum_{\boldsymbol{q}'} M_D \ln M_D \tag{13.39}$$

Resolving (13.37) for $i_D$ we obtain the final result that the Shannon information $i$ is reduced by the amount $\Delta_D$ because of the use of semantic information.

## 13.4 Reinterpretation of the Results of Sects. 13.1–13.3

The relationships we derived above can be interpreted by means of considerations which are closely related to those of Sect. 2.6.3. This will allow us to reduce Shannon information still more. Again, as in 2.6.3, but using a somewhat different notation, we start from the notion of joint probability $p(j, k)$ where $j$ and $k$ both can stand for a whole set of indices. We assume the normalization condition

$$\sum_{jk} p(j, k) = 1 . \tag{13.40}$$

We further use the notion of conditional probability, i.e. the probability for the occurrence of an event characterized by $j$ provided that the event $k$ has occurred. We denote this probability as usual by $p(j|k)$. We assume the normalization condition

$$\sum_{j} p(j|k) = 1 . \tag{13.41}$$

Furthermore we introduce a suitably chosen probability distribution $p(k)$ with the normalization condition

$$\sum_{k} p(k) = 1 . \tag{13.42}$$

Then according to probability theory we may use the decomposition

$$p(j, k) = p(j|k)p(k) . \tag{13.43}$$

The idea is now to express the Shannon information

$$i = -\sum_{jk} p(j, k) \ln p(j, k) \tag{13.44}$$

by means of the decomposition (13.43). The intermediate steps are just generalizations of what we have done before and can easily be performed via

$$= -\sum_{jk} p(j|k)p(k)\ln(p(j|k)p(k)), \tag{13.45}$$

$$= -\sum_{jk} p(j|k)p(k)\ln(p(j|k)) - \sum_{jk} p(j|k)p(k)\ln p(k), \tag{13.46}$$

and because of (13.41)

$$= \sum_{k} p(k)\left(-\sum_{j} p(j|k)\ln p(j|k)\right) - \sum_{k} p(k)\ln p(k). \tag{13.47}$$

This allows us to write our final result in the following specific form

$$i = \sum_{k} p(k)i(\{j\}|k) + i(\{k\}) \tag{13.48}$$

so that the Shannon information is decomposed into the Shannon information for the distribution function $p(k)$ and an average over the conditional Shannon information $i(\{j\})|k)$ where the weights are just given by the distribution function $p(k)$.

In the spirit of our book we now introduce the following interpretation. $k$ refers to the states of one or several order parameters and the index $j$ describes the individual enslaved parts of the system. In other words, once an order parameter $k$ is given this implies a specific probability distribution of $j$, i.e. an order parameter $k$ implies a specific enslaved state. The important step now is the following. By semantic information we "know" that the state of a complex system microscopically described by the indices $j$ is fixed or known once the order parameter or a subset of order parameters $k$ is fixed. This insight allows us to drop the indices $j$ entirely so that the total Shannon information is reduced to the far smaller Shannon information of the order parameters

$$i_{\text{total}} \to i(\{k\}). \tag{13.49}$$

# 14. Quantum Systems

## 14.1 Why Quantum Theory of Information?

So far we have been dealing with classical systems whose dimensions are macroscopic. In this chapter we wish to consider systems that must be described by means of quantum theory. Let us look at a few examples of systems where quantum effects play a role. The first example is provided by the laser. Its individual atoms must be treated by quantum mechanics and in it the field is produced by quantum mechanical events, namely those of spontaneous and stimulated emission. Therefore, a rigorous theory of the laser must be based on quantum mechanics. In the previous chapters we based our laser theory on the so-called semiclassical treatment which can be derived from the fully quantum mechanical treatment by a method called quantum classical correspondence which will not be pursued here.

A further important example is provided by computer elements. These are being made smaller and smaller so that eventually quantum effects will become important. Thus we shall have to consider one or few electrons trapped in so-called quantum wells. We may have oscillations of electron currents in the Gunn oscillator, or we may have formation of filaments by electrons in semiconductors. Electrons or holes in different trapped states may store information and electrons may form temporal or spatial patterns. Biological molecules may store information and we may expect the construction of devices which combine molecules with semiconductors as a means to store and process information. Therefore it would be worthwhile to attempt an extension of our previous results to the world of quantum mechanical processes in these microscopic devices.

In the following we shall see that this goal is, to a large extent, realizable. We shall assume that the reader is familiar with basic notions of quantum theory, but we begin, nonetheless, with a reminder of some of its most important features. To proceed from classical physics to quantum mechanics, we have to replace observables $q_i$ such as the position and momentum of a particle by the corresponding operators, which we shall denote in the following also by $q_i$ (consult Table 14.1). In this way not only can classical mechanics be translated into quantum mechanics, but so can the theory of electromagnetic fields. Here, the amplitude $E(x)$ of the electric field at position $x$ must be replaced by an operator $E(x)$. When we consider the electron wave function $\psi(x)$ which is a solution of the Schrödinger equation, as a classical field, we may replace it by an operator $\psi(x)$. This is the process of so-called second quantization. Accordingly the electron density becomes an operator (cf. Table 14.1).

**Table 14.1.** Analogies between the classical and the quantum-mechanical formulations

| Classical | Quantum mechanical |
|---|---|
| observable $q_i$ | operator $q_i$ |
| field amplitude $E(x)$ at position $x$ | $E(x)$ |
| electron wave function $\psi(x)$ | $\psi(x)$ |
| electron density $\tilde{\rho}(x) = \psi^*(x)\psi(x)$ | $\tilde{\rho}(x) = \psi^+(x)\psi(x)$ |
| distribution function $f(q)$ | density matrix $\rho = (\rho_{ij})$ |
| moments: | |
| $f_i^{(1)} = \langle q_i \rangle = \int f(\boldsymbol{q}) q_i d^N q$ | $\langle q_i \rangle = \mathrm{Tr}\{q_i \rho\}$ |
| $f_{ij}^{(2)} = \langle q_i q_j \rangle = \int f(q) q_i q_j d^N q$ | $\mathrm{Tr}\{q_i q_j \rho\}$ |
| $i = -\int f \ln f d^N q$ | $i = -\mathrm{Tr}\{\rho \ln \rho\}$ |

Let us now make an important step further, namely the inclusion of statistical mechanics. In Chap. 2 we dealt with the distribution function, $f(\boldsymbol{q})$, where $\boldsymbol{q}$ is a vector of dynamic variables, e.g. positions of particles. In quantum theory this distribution function must be replaced by the density matrix, $\rho$.While the distribution function obeys for instance the Fokker-Planck-equation, the density matrix obeys for instance the so-called master equation. In the following we are aiming at a macroscopic derivation of the explicit form of the density matrix $\rho$ in complete analogy to our previous "macroscopic" derivation of $f(\boldsymbol{q})$. Therefore we shall not dwell here on the form of the density matrix equation or the master equation, but will instead elaborate on another analogy, namely that with respect to moments. Examples of moments of the classical theory are presented in the third last and the second last row on the left hand side of Table 14.1. The corresponding expressions on the right hand side tell us how the analogous moments are defined in quantum theory. The abbreviation Tr means trace. For a matrix $A$ with elements $a_{ik}$, the trace is defined by

$$\mathrm{Tr}\{A\} = \sum_j a_{jj} \,. \tag{14.1}$$

A difficulty which arises when we try to translate the results of classical physics into quantum mechanics stems from the fact that, at least in general, the operators $q_i$ do not commute; i.e., in general we find

$$q_i q_j - q_j q_i \neq 0 \,, \quad i \neq j \,. \tag{14.2}$$

As well known from quantum theory, the operators $q_i$ must eventually be applied to a wave function. But depending on the sequence in which we apply these operators to a wave function, different final functions will result. For this reason it will be very important to carefully consider the sequence in which operators occur. In most of the applications we shall consider, we shall use a symmetrized product of $q_i, q_j$ so that the sequence of operators is no longer important. The formal resemblance of the classical and quantum mechanical expression (Table 14.1) immediately suggests the definition of information in quantum theory as given in the last row of Table 14.1.

After these preparatory remarks we may now turn to the quantum mechanical formulation of the maximum information principle.

## 14.2 The Maximum Information Principle

In this section we shall utilize the correspondence scheme presented in Table 14.1. We shall denote the information or entropy by the letter $S$. The individual steps in the further translation of the maximum information principle are summarized in Table 14.2. We now require that not only the expression for $S$ in the classical case becomes a maximum, but also its corresponding quantum-mechanical expression given in Table 14.1, each time under given constraints. The constraints are summarized in the second row of Table 14.2. The normalization is given in the third row. Again we shall make use of Lagrange multipliers in order to perform the maximization. We multiply the expressions in the second row by $\lambda_k$ and in the last row by $\lambda - 1$. We then subtract the resulting expressions from $S$ and require that the variation of the resulting expression vanishes.

$$\delta \left[ \mathrm{Tr}\{\rho \ln \rho\} - (\lambda - 1)\mathrm{Tr}\{\rho\} - \sum_k \lambda_k \mathrm{Tr}\{g^{(k)}\rho\}\right] = 0 \,. \tag{14.3}$$

When we perform the variation, some care must be exercised, because in general the variation $\delta\rho$ will not commute with $\rho$.

$$[\rho, \delta\rho] \equiv \rho\delta\rho - \delta\rho\rho \neq 0 \,. \tag{14.4}$$

On the other hand, we can make use of an important property of the trace, namely its cyclic property:

$$\mathrm{Tr}\{AB\} = \mathrm{Tr}\{BA\} \,. \tag{14.5}$$

In order to perform the variation in (14.3), let us consider the individual terms beginning with

**Table 14.2.** The maximum information entropy principle

| Classical | Quantum mechanical |
|---|---|
| $S = -\int f \ln f \, d^N q = \max!$ | $S = -\mathrm{Tr}\{\rho \ln \rho\} = \max!$ |
| | constraints |
| $f^{(k)} = \int f g^{(k)}(q) dV$ | $f^{(k)} = \mathrm{Tr}\{g^{(k)}(q)\rho\}$ |
| | normalization |
| $\int f d^N q = 1$ | $\mathrm{Tr}\{\rho\} = 1$ |

$$\delta \mathrm{Tr}\,\{\rho\}\,. \tag{14.6}$$

Taking the variation means that we compare the value of $\mathrm{Tr}\,\rho$ taken for a value of the density matrix, $\rho$, with that of another one, $\rho + \delta\rho$, where $\delta\rho$ is small. Thus the variation of (14.6) is defined by

$$\delta \mathrm{Tr}\,\{\rho\} = \mathrm{Tr}\,\{\rho + \delta\rho\} - \mathrm{Tr}\,\{\rho\} = \mathrm{Tr}\,\{\delta\rho\} \tag{14.7}$$

where to obtain the last equality we have made use of the linearity of a trace operation. In a similar fashion we immediately find

$$\delta \mathrm{Tr}\,\{g^{(k)}\rho\} = \mathrm{Tr}\,\{g^{(k)}\delta\rho\}\,. \tag{14.8}$$

The variation of the first term in (14.3) requires some care because of (14.4). Using the definition that the variation of the trace is the difference of the two traces taken for $\rho + \delta\rho$ and $\rho$ we obtain

$$\delta \mathrm{Tr}\,\{\rho \ln \rho\} = \mathrm{Tr}\,\{(\rho + \delta\rho)\ln(\rho + \delta\rho) - \rho \ln \rho\}\,. \tag{14.9}$$

By subtracting and adding a term we transform (14.9) into

$$\begin{aligned} &= \mathrm{Tr}\,\{(\rho + \delta\rho)\ln(\rho + \delta\rho) - \rho \ln(\rho + \delta\rho)\} \\ &\quad + \mathrm{Tr}\,\{(\rho \ln(\rho + \delta\rho) - \rho \ln \rho\}\,. \end{aligned} \tag{14.10}$$

The first difference can immediately be evaluated to give

$$\mathrm{Tr}\,\{\delta\rho \ln(\rho + \delta\rho)\} = \mathrm{Tr}\,\{\delta\rho \ln \rho\}\,, \tag{14.11}$$

where we have kept only terms linear in $\delta\rho$. In order to evaluate the second difference we use a trick, namely we write $\ln \rho$ in a form that allows us to use the usual expansion of the logarithm into a Taylor series:

$$\ln \rho = \ln(1 + \rho - 1) = \sum_{\nu=0}^{\infty} c_\nu (\rho - 1)^\nu\,. \tag{14.12}$$

In quite the same fashion we obtain

$$\ln(\rho + \delta\rho) = \sum_{\nu=0}^{\infty} c_\nu (\rho - 1 + \delta\rho)^\nu\,. \tag{14.13}$$

Now let us consider an individual term of the sum and let us start with $\nu = 1$. Here we have

$$\rho - 1 + \delta\rho\,. \tag{14.14}$$

For $\nu = 2$ we readily obtain

$$(\rho - 1)^2 + (\rho - 1)\delta\rho + \delta\rho(\rho - 1) + (\delta\rho)^2 \tag{14.15}$$

where we have taken care of the precise sequence of the operators $\rho - 1$ and $\delta\rho$. Keeping the leading terms we obtain for the general term $\nu$

$$(\rho-1)^{\nu}+(\rho-1)^{\nu-1}\delta\rho+(\rho-1)^{\nu-2}\delta\rho(\rho-1)+(\rho-1)^{\nu-3}\delta\rho(\rho-1)^{2}+\ldots$$
$$+\delta\rho(\rho-1)^{\nu-1}+(\rho-1)^{\nu-2}(\delta\rho)^{2}+\ldots. \qquad (14.16)$$

In the following we shall retain only terms independent of or linear in $\delta\rho$. In this approximation we now multiply the term (14.16) from the left by $\rho$ and take the trace. To exhibit the essentials we pick out a general term of (14.16) which yields

$$\mathrm{Tr}\,\{\rho(\rho-1)^{\nu-\mu-1}\delta\rho(\rho-1)^{\mu}\}\,. \qquad (14.17)$$

We now may use the cyclic property (14.5) and obtain

$$(14.17)=\mathrm{Tr}\,\{\delta\rho(\rho-1)^{\mu}\rho(\rho-1)^{\nu-\mu-1}\}\,. \qquad (14.18)$$

Because $\rho$ commutes with $\rho - 1$, we can finally write down the result of (14.17) in the form

$$\mathrm{Tr}\,\{\delta\rho\rho(\rho-1)^{\nu-1}\}\,. \qquad (14.19)$$

What we learn from the above transformation is the following: When we take the trace operation we obtain the same result as if $\rho$ and $\delta\rho$ commute. So from now on we can skip all the algebraic details in evaluating

$$\mathrm{Tr}\,\{(\rho\ln(\rho+\delta\rho)-\rho\ln\rho\}\,. \qquad (14.20)$$

We merely may assume that $\rho$ and $\delta\rho$ commute formerly, or in other words, we may treat $\rho$ and $\delta\rho$ as $c$-numbers (i.e. classical numbers, in contrast to operators) for the evaluation of (14.20). Using the usual property of the logarithm we obtain instead of (14.20)

$$\mathrm{Tr}\,\{\rho\ln(1+\delta\rho\rho^{-1})\} \qquad (14.21)$$

or if we retain the leading term

$$(14.20)=\mathrm{Tr}\,\{\delta\rho\}\,. \qquad (14.22)$$

Using the results (14.7, 8, 20, 22) we may evaluate (14.3) and obtain

$$-\mathrm{Tr}\{\delta\rho\ln\rho\}-\lambda\mathrm{Tr}\,\{\delta\rho\}-\sum_{k}\lambda_{k}\mathrm{Tr}\,\{\delta\rho g^{(k)}\}=0\,. \qquad (14.23)$$

In order to perform the variations $\delta\rho$ explicitly we may use any representation of $\rho$ with respect to a set of eigenstates possessing a discrete or a continuous spectrum. Using such a representation we may write $\rho$, $\delta\rho$, and $g^{(k)}$ as matrices.

Let us express the various terms in (14.23) using the definition of the trace (14.1):

$$\sum_{j}(\delta\rho\ln\rho)_{jj}=\sum_{jl}(\delta\rho)_{jl}(\ln\rho)_{lj} \qquad (14.24)$$

$$\sum_{j} (\delta\rho)_{jj} \tag{14.25}$$

$$\sum_{jl} (\delta\rho)_{jl} (g^{(k)})_{lj} \,. \tag{14.26}$$

Inserting the corresponding expressions in (14.23) we obtain

$$\sum_{j} (\delta\rho)_{jl} \{ (\ln\rho)_{lj} + \lambda\delta_{lj} + \sum_{k} \lambda_k (g^{(k)})_{lj} \} = 0 \tag{14.27}$$

where we have made use of the Kronecker symbol $\delta_{lj} = 1$ for $l = j$ and $=0$ for $l \neq j$. Because of the use of the Lagrange multipliers we may assume that the variations $(\delta\rho)_{jl}$ are independent of each other. As a consequence, (14.27) can be fulfilled only if the curly bracket vanishes for each pair $l, j$. But the whole bracket can be considered as the element of a matrix with indices $l, j$. Therefore the vanishing of the curly brackets in (14.27) is equivalent to the following matrix equation

$$\ln\rho + \lambda I + \sum_{k} \lambda_k g^{(k)} = 0 \tag{14.28}$$

where $I$ is the unit matrix. This matrix or operator equation possesses the solution

$$\rho = \exp\left( -\lambda - \sum_{k} \lambda_k g^{(k)} \right) . \tag{14.29}$$

The Lagrange multipliers $\lambda$ and $\lambda_k$ can now be determined, at least in principle, by the normalization condition

$$\mathrm{Tr}\,\{\rho\} = 1 \tag{14.30}$$

and by the constraints

$$\mathrm{Tr}\,\{g^{(k)}\rho\} = f^{(k)} . \tag{14.31}$$

Inserting (14.29) into (14.30) yields

$$\mathrm{Tr}\,\{\rho\} = \mathrm{e}^{-\lambda} \mathrm{Tr} \left\{ \exp\left( -\sum_{k} \lambda_k g^{(k)} \right) \right\} = 1 \,. \tag{14.32}$$

This immediately leads us to the relation

$$\mathrm{e}^{\lambda} = Z = \mathrm{Tr} \left\{ \exp\left( -\sum_{k} \lambda_k g^{(k)} \right) \right\} , \tag{14.33}$$

which also contains the definition of the quantum mechanical partition function, $Z$. A number of the relations which we derived in Sect. 3.3 can be translated into the quantum mechanical case. But as the reader will notice, some care is needed and some tricks must be used. We shall not present the full translation, but will merely indicate some of the main steps and then write down an important result. In order to derive the quantum mechanical analogues of the relations (3.44, 45), we have to learn how to differentiate an exponential function containing operators which, in general, do not commute. Since the rest of this section is of a rather technical nature, the reader not interested in such technical details may just take note of the final formula (14.43) and then proceed to the next section.

Let us define our goal. We wish to evaluate

$$\frac{\partial}{\partial \lambda_{k_0}} \mathrm{Tr}\,\{\rho\}\,, \tag{14.34}$$

i.e., we wish to learn how to differentiate the density matrix $\rho$ under the trace. To this end we consider the exponential function occurring on the left-hand side of

$$\exp(A+\lambda' B) = T\exp\left[\int_0^1 dt(A_t+\lambda' B_t)\right]\,, \tag{14.35}$$

which contains a parameter $\lambda'$ with respect to which we wish to differentiate the exponential function. But since $A$ and $B$ are assumed not to commute with each other, this differentiation cannot be done in the usual way. Instead we have to apply a trick in which we introduce so-called time-ordered operators. We label $A$ and $B$ with an index $t$ and we replace the left-hand side of (14.35) by the expression on the right-hand side of that equation. $T$ means time ordering. When we evaluate the exponential function, operators with a smaller index $t$ must operate prior to operators with a higher value of $t$. Under this provision we may formally treat $A_t$ and $B_t$ as if these operators commute with each other. Therefore we may write (14.35) in the form

$$T\exp\left(\int_0^1 dtA_t\right)\exp\left(\int_0^t \lambda' dtB_t\right)\,. \tag{14.36}$$

We now differentiate (14.35) or equivalently (14.36) with respect to $\lambda'$

$$\frac{d}{d\lambda'} T\exp\left[\int_0^1 dt(A_t+\lambda' B_t)\right]\,, \tag{14.37}$$

and obtain

$$T\int_0^1 B_{t'}dt' \exp\left[\int_0^1 dt(A+\lambda' B)_t\right]\,. \tag{14.38}$$

According to the time ordering convention some of the operators appearing in the exponential function, i.e. those for which $t < t'$, must be applied prior to $B_{t'}$. On the other hand the operators in the exponential function with $t' < t$ must be applied after the operator $B_{t'}$. Thus we may explicitly perform the time ordering by splitting the exponential function accordingly and obtain

$$\int_0^1 dt' \exp\left[\int_{t'}^1 dt(A+\lambda' B)_t\right] B_{t'} \exp\left[\int_0^{t'} dt(A+\lambda' B)_t\right]\,. \tag{14.39}$$

Now that the individual time ordering has been performed, we may replace the operators $A_t$ and $B_t$ by operators without the index $t$ and perform the integrals. This yields

$$\int_0^1 dt' \exp[(1-t')(A+\lambda' B)] B \exp[t'(A+\lambda' B)] \,. \tag{14.40}$$

We now take the trace of (14.40) and utilize the cyclic property (14.5), which allows us to bring the first exponential function in (14.40) to the right-hand side. The two exponential functions can be amalgamated into a single one, which no longer depends on $t'$. We may now perform the integral over $t'$ immediately, and this yields unity. We further observe that the order of the differentiation and trace operations can be exchanged, giving the desired result:

$$\frac{d}{d\lambda'} \mathrm{Tr}\,\{\exp(A+\lambda' B)\} = \mathrm{Tr}\,\{B \exp(A+\lambda' B)\} \,. \tag{14.41}$$

Let us now recall our initial goal. We wanted to differentiate the trace of $\rho$ with respect to $\lambda_k$. To this end we identify $B$ with $g^{(k)}$ and the rest of the exponential function of $\rho$ with $A$. We then immediately obtain our desired result

$$\frac{\partial}{\partial \lambda_k} \mathrm{Tr}\,\{\rho\} = \mathrm{Tr}\,\{g^{(k)} \rho\} \tag{14.42}$$

Using again the explicit form of $\rho$ and the definition of $Z$, we readily obtain our final result

$$f_k = \mathrm{Tr}\,\{g^{(k)} \rho\} = -\frac{\partial}{\partial \lambda_k} \ln Z \,, \tag{14.43}$$

which is the complete analogue of (3.45). When we wish to take second derivatives of $Z$ with respect to $\lambda_k$, some of the above tricks can be used again, but the final result looks somewhat more complicated than the corresponding results of Sect. 3.3.

To conclude this section we mention that the results of Chap. 4 on thermodynamics can also be translated into the quantum mechanical case. For instance when we use the energy of the system as constraint we obtain

$$\rho = Z^{-1} \exp\left(-\frac{H}{kT}\right) \tag{14.44}$$

where $H$ is the Hamiltonian operator, $k$ Boltzmann's constant and $T$ the absolute temperature. The partition function, $Z$, is defined by

$$Z = \mathrm{Tr}\left\{\exp\left(-\frac{H}{kT}\right)\right\} . \tag{14.45}$$

## 14.3 Order Parameters, Enslaved Modes and Patterns

In this section we wish to study how we can translate the results of Chap. 6 into quantum mechanics. For this purpose we make again a specific choice of the constraints

$$f_k = \mathrm{Tr}\,\{g^{(k)}\rho\}\,, \tag{14.46}$$

namely, we shall replace $f_k$ by symmetrized moments, i.e. by

$$\langle q_j \rangle \tag{14.47}$$

$$\langle q_j q_k + q_k q_j \rangle \tag{14.48}$$

etc. In the following we shall include moments up to the fourth order. The density matrix which maximizes the information is then given by

$$\rho = \exp\left(-\lambda - \sum_j \lambda_j q_j - \sum_{jj'} \lambda_{jj'} q_j q_{j'} - \ldots\right) \tag{14.49}$$

where the exponent contains expressions up to fourth order in $q_j$. In the following we shall abbreviate the exponent by $V$, where

$$\rho = \mathrm{e}^V\,. \tag{14.50}$$

We shall follow as closely as possible the procedure of Chap. 6. We thus wish to eliminate the linear terms in (14.49, 50). Therefore we make the hypothesis

$$q_j = c_j + \bar{q}_j \tag{14.51}$$

where $c_j$ is a $c$-number. When we insert (14.51) into (14.49), we obtain in particular, terms which are linear in $q_j$ and the coefficients of these are now required to vanish.

$$\left[\lambda_j + \sum_{j'} (\lambda_{jj'} c_{j'} + \lambda_{j'j} c_{j'}) + \ldots\right] = 0\,. \tag{14.52}$$

One may easily convince oneself that the expression (14.52) can be arrived at if we require

$$\frac{\partial V}{\partial q_i} = 0 \quad \text{for } q_j^0 \tag{14.53}$$

whereby we treat $q_j$ as $c$-numbers.

We make the identification

$$q_j^0 \leftrightarrow \bar{q}_j = 0 \tag{14.54}$$

so that $c_j$ is given by

$$c_j = q_j^0\,. \tag{14.55}$$

In this way we have transformed (14.49) into

$$\exp[-\bar{V}(\bar{\boldsymbol{q}})] \tag{14.56}$$

where $V$ is replaced by

$$\bar{V}(\bar{\boldsymbol{q}}) = \bar{\lambda} + \sum_{jj'} \bar{\lambda}_{jj'} \bar{q}_j \bar{q}_{j'} + \ldots \tag{14.57}$$

Because we have chosen symmetrized moments we can be sure that

$$\bar{\lambda}_{jj'} = \bar{\lambda}_{j'j} \, . \tag{14.58}$$

We now make the hypothesis

$$\bar{q}_j = \sum_k \xi_k v_{kj} \tag{14.59}$$

where $v_{kj}$ are $c$-numbers whereas $\xi_k$ are operators. Inserting the hypothesis (14.59) into (14.57) we obtain for the quadratic term

$$\sum_{kk'} \xi_k \xi_{k'} \left\{ \sum_{jj'} \bar{\lambda}_{jj'} v_{kj} v_{k'j'} \right\} . \tag{14.60}$$

We now choose $v_{kj}$ in such a way that

$$\sum_{jj'} \bar{\lambda}_{jj'} v_{kj} v_{k'j'} = \hat{\lambda}_k \delta_{kk'} \, . \tag{14.61}$$

Because $\bar{\lambda}_{jj'}$ is a symmetric matrix we may choose $v_{kj}$ to be a real matrix, and the values $\hat{\lambda}_k$ are real. Under these transformations (14.49) is replaced by

$$\rho(\xi_u, \xi_s) = \exp\left[ -\bar{\lambda} - \sum_u \hat{\lambda}_u \xi_u^2 - \sum_s \hat{\lambda}_s \xi_s^2 + V_u(\xi_u) + V_s(\xi_u, \xi_s) \right] \tag{14.62}$$

where again we have distinguished between positive $\hat{\lambda}_u$ and negative $\hat{\lambda}_s$ so that we can clearly distinguish between order parameters and enslaved modes.

We now introduce a density matrix which refers to the order parameters $\xi_u$ alone by means of

$$\rho_u(\xi_u) = \mathrm{Tr}_s\{\rho(\xi_u, \xi_s)\} \, . \tag{14.63}$$

Because the total trace is normalized, it follows that

$$\mathrm{Tr}\{\rho_u\} = 1 \, . \tag{14.64}$$

In analogy to the procedure in Chap. 6 we now with to split the density matrix $\rho$, which depends on the operators of the order parameters and enslaved mode, into a product of a density matrix (14.63) and a kind of conditional density matrix according to

$$\rho(\xi_u, \xi_s) = \rho_u(\xi_u) \rho_s(\xi_s | \xi_u) \, . \tag{14.65}$$

When we take the trace on both sides over the variables which refer to the operators $\xi_s$ we obtain on the left-hand side (14.63) by definition. On the right-hand side we obtain

$$\rho_u(\xi_u)\mathrm{Tr}_s\{\rho_s(\xi_s|\xi_u)\}\,. \tag{14.66}$$

Comparing the left- and right-hand sides we immediately obtain the relation

$$\mathrm{Tr}_s\{\rho_s(\xi_s|\xi_u)\} = 1\,. \tag{14.67}$$

We now wish to transform (14.62) in such a way that it can be written in the form (14.65). This is by no means a trivial task because the operators $\xi$ do not commute. In order to solve this problem we write (14.62) in the form

$$\rho(\xi_u, \xi_s) = \exp[\hat{V}_u(\xi_u) + \hat{V}_s(\xi_u, \xi_s)] \tag{14.68}$$

where the notation is obvious and it is left open whether $\bar{\lambda}$ is put into $V_u$ or $V_s$.

We now add and subtract an operator function $h$, so that we obtain

$$(14.68) = \exp[\hat{V}_u + h(\xi_u) - h(\xi_u) + \hat{V}_s(\xi_u, \xi_s)]\,. \tag{14.69}$$

It will be our goal to cast (14.69) into the form (14.65). To achieve this we need an auxiliary theorem. We shall consider the expression

$$T\exp\left[\int_0^1 dt(A_t + B_t)\right] \tag{14.70}$$

which results from (14.35) by putting $\bar{\lambda} = 1$. $T$ is again the time ordering operator and $A_t$ and $B_t$ are operators which must be time ordered. We now utilize Feynman's disentangling theorem. According to this theorem (14.70) can be written in the form

$$= \mathrm{e}^{A}T\mathrm{e}^{\tilde{B}} \quad \text{where} \tag{14.71}$$

$$\tilde{B} = \int_0^1 dt(\mathrm{e}^{-At}B\mathrm{e}^{At})_t\,. \tag{14.72}$$

We now make the identifications

$$\hat{V}_u + h = A\,; \quad -h + \hat{V}_s = B\,. \tag{14.73}$$

This immediately allows us to derive the relation

$$\rho_u(\xi_u) = \exp[\hat{V}_u + h] \tag{14.74}$$

whereas the conditional density matrix is defined by

$$\rho_s(\xi_s|\xi_u) = T\mathrm{e}^{\tilde{B}}\,. \tag{14.75}$$

The still unknown operator function, $h$, must be determined in such a way that

$$\mathrm{Tr}_s\{\rho_s(\xi_s|\xi_u)\} = 1\,. \tag{14.76}$$

Let us summarize what we have achieved in this section. Starting from operator moments up to the fourth order we have constructed the density matrix by means of the maximum information entropy principle. Then by diagonalizing the bilinear part we have identified the operators that belong to the order parameters and to the enslaved modes. Finally, we were able to write the joint density matrix for $\xi_u$, $\xi_s$ as a product, where the factor $\rho_u$ is the density matrix referring to the order parameters alone, whereas $\rho_s$ can be interpreted as a conditional density matrix. In complete analogy to the classical case, we may interpret $v_{kj}$ as the pattern at the space points $j$ belonging to the mode $k$, provided $q_j$ is, for example, an intensity attached to a space point $j$. It is quite remarkable that $v$ is a classical quantity, which means that the emerging patterns are described by classical quantities and can thus be interpreted without knowledge of any quantum mechanical wave function.

To end this chapter we wish to demonstrate that the information can be split into two parts, one containing only the order parameters and the other containing the conditional density matrix, i.e. the enslaved modes.

## 14.4 Information of Order Parameters and Enslaved Modes

We start from the expression for the information of the total system

$$i = \mathrm{Tr}\,\{\rho \ln \rho\} \tag{14.77}$$

and insert into this the form (14.65), where the two factors are defined by (14.74) and (14.75), respectively. Using (14.69) in addition we obtain

$$(14.77) = \mathrm{Tr}\,\{\rho_u \rho_s[(\hat{V}_u + h) + (-h + \hat{V}_s)]\} \tag{14.78}$$

which can be split into the sum of

$$\mathrm{Tr}_u\{\rho_u(\mathrm{Tr}_s\{\rho_s\})(\hat{V}_u + h)\} \quad \text{and} \tag{14.79}$$

$$\mathrm{Tr}_u\{\rho_u \mathrm{Tr}_s\{\rho_s(-h + \hat{V}_s)\}\}\,. \tag{14.80}$$

Because of (14.67) the expression (14.79) simplifies to

$$\mathrm{Tr}_u\{\rho_u(\hat{V}_u + h)\} \tag{14.81}$$

while the trace over $s$ in (14.80) can be interpreted as the information belonging to the enslaved mode $s$. Because it depends on the operators $\xi_u$ it is an operator expression for the information and we write

$$i_{s,\mathrm{op}}(\xi_u) = \mathrm{Tr}_s\{\rho_s(-h + \hat{V}_s)\}\,. \tag{14.82}$$

Abbreviating (14.81), which refers to the order parameters alone, by $i_u$, we obtain the final result for $i$ (14.77):

$$i = i_u + \mathrm{Tr}_u\{\rho_u i_{s,\mathrm{op}}(\xi_u)\}\,. \tag{14.83}$$

Thus in complete analogy to the classical case, it is possible to decompose the information of the total system into that of the order parameters and that which refers to the enslaved modes which is then to be averaged over the distribution or density matrix of the order parameters.

In conclusion we can claim that it is possible to carry over a good many of the results that were obtained in the framework of the classical theory to the case of quantum mechanical operators. In particular our procedure allows us to identify patterns evolving in quantum systems by means of the matrix $v_{kj}$ so that classical patterns can be recognized within quantum systems.

# 15. Quantum Information

In the preceding chapter we have shown in which way we may extend the previous results of our book which refers to classical systems in physics, to the quantum domain. In particular we have seen how we can replace statistical averages by quantum statistical averages where we replaced distribution functions by the density matrix. This allowed us in particular to extend the maximum entropy principle into quantum physics. The use of these averages and corresponding ensembles implies implicitly that the systems under consideration undergo phase-destroying processes, be it because of internal interactions in large systems or be it because of the interaction of the system under consideration with its environment. In the field of quantum information, which we wish to briefly present now, the phase relations, which we shall discuss below, must be strictly conserved and not disturbed by any external influences.

## 15.1 Basic Concepts of Quantum Information. Q-bits

As we have seen in this book, the basic unit of (classical) information theory is one bit. This is a set of two numbers 0 and 1, or the set "no", "yes", or "false", "true". The one bit unit can be realized by a variety of physical systems, e.g. "current off", "current on", or "atom in lower state", "atom in upper state", etc. In the context of classical physics it is important to note that in all cases we know with certainty that the system is in the one or in the other state.

Let us now proceed to the concept of quantum information. Its basic unit is the Q-bit. It can best be visualized by means of a spin of size $\frac{\hbar}{2}$ where $\hbar$ is Planck's constant, divided by $2\pi$. In the context of quantum information we may ignore the factor $\hbar/2$. With respect to a preferential direction, say the z-axis, the spin may show upwards or downwards. Each state is described by a wave function $\phi$ which can best be represented in Dirac's bra- and ket-notation, namely as

$$\text{“spin up”}: \ \phi = |\uparrow\rangle \tag{15.1}$$

$$\text{“spin down”}: \ \phi = |\downarrow\rangle \tag{15.2}$$

To have a more explicit representation of these spin states we may use a vector notation with two components so that

$$|\uparrow\rangle \Leftrightarrow \begin{pmatrix} 1 \\ 0 \end{pmatrix}, \; |\downarrow\rangle \Leftrightarrow \begin{pmatrix} 0 \\ 1 \end{pmatrix}. \tag{15.3}$$

So far nothing has changed as compared to the classical bit unit. However, the crucial difference of a Q-bit compared to the usual bit consists in the fact that we may form coherent superpositions between the spin states,

$$\psi = a|\downarrow\rangle + \beta|\uparrow\rangle\,, \tag{15.4}$$

where $a$ and $\beta$ are complex coefficients which must be normalized so that

$$|a|^2 + |\beta|^2 = 1\,. \tag{15.5}$$

All the states which can be written in the form (15.4) under the condition (15.5) form a Q-bit. The vectors (15.3) with (15.4), (15.5) span a 2-dimensional Hilbert space. Being realized by physical systems such as spins a Q-bit can be manipulated from the outside. E.g. a spin which is connected with a magnetic moment can be changed in its direction by means of an applied oscillating electromagnetic field (for details see Sect. 16.6). Choosing its frequency $\omega = \Delta E/\hbar$, where $\Delta E$ is the energy difference between the two spin-directions "up" and "down" in a constant magnetic field, a specific field strength and duration, a spin which shows initially in the minus z-direction can be flipped to any other chosen direction ranging from the minus z-direction to the plus z-direction. Because the electromagnetic field is applied for only a specific time, one speaks of pulses and in correspondence to the final direction of the spin one speaks e.g. of $\frac{\pi}{2}$- and $\pi$-pulses. Here a $\frac{\pi}{2}$-pulse transfers a spin which initially shows in the +z or –z-direction into the equatorial plane whereas a $\pi$-pulse transfers it to the opposite direction. From the mathematical point of view these flips can be described by operators in form of

$$U_\vartheta = \begin{pmatrix} \cos(\vartheta/2) & \sin(\vartheta/2) \\ -\sin(\vartheta/2) & \cos(\vartheta/2) \end{pmatrix}. \tag{15.6}$$

acting on the states (15.3) or their superpositions (15.4). Readers who want to learn more about the spin formalism are referred to Sections 16.5, 16.6.

These matrices are unitary. We remind the reader of the property of a unitary transformation. Let the matrix $U$ have the matrix elements

$$U = (U_{kj}) \tag{15.7}$$

and the matrix $U^+$

$$U^+ = (U^+_{kj}) \tag{15.8}$$

with the property

$$U^+_{kj} = U^*_{jk}\,, \tag{15.9}$$

(where * denotes the complex conjugate)

then the matrix elements must obey relations so that the matrix product obeys

$$UU^+ = 1 \tag{15.10}$$

where 1 is the unity matrix $\begin{pmatrix} 1 & 0 \\ 0 & 1 \end{pmatrix}$. Other physical realizations of Q-bits are by two-level atoms or by photons with their polarization, to mention but a few examples. The concept of quantum information was mainly introduced in order to discuss and eventually construct quantum computers.

## 15.2 Phase and Decoherence

The general wave function corresponding to a Q-bit has the form

$$|\psi\rangle = \alpha|\uparrow\rangle + \beta|\downarrow\rangle\,. \tag{15.11}$$

As is well known from quantum theory, in order to extract numerical measurable values from (15.11) one has to form expectation values of the corresponding operators. Let us consider, as an example, the expectation value of the $x$-component of the spin. We denote the corresponding spin operator by $s_x$. Then in the bra- and ket-notation, which we use everywhere in this chapter, the expectation value is given by

$$\langle\psi|s_x|\psi\rangle = |\alpha|^2\langle\uparrow|s_x|\uparrow\rangle + |\beta|^2\langle\downarrow|s_x|\downarrow\rangle + \alpha^*\beta\langle\uparrow|s_x|\downarrow\rangle + \alpha\beta^*\langle\downarrow|s_x|\uparrow\rangle\,. \tag{15.12}$$

Because $\alpha$ and $\beta$ are complex quantities we can write them in the form

$$\alpha = r_1 e^{i\chi_1},\ \beta = r_2 e^{i\chi_2} \tag{15.13}$$

where $r_1$, $r_2$ are real amplitudes and $\chi_1$, $\chi_2$ phases. As can be seen from (15.12), what matters is the relative phase

$$\chi = \chi_1 - \chi_2\,. \tag{15.14}$$

Now assume that the spin under consideration is coupled to an environment which pushes the spin all the time so that one may speak of spin-diffusion. What happens to the coefficients which appear in (15.12) when we now make an average over these different spin orientations? Because this phase does not appear in the absolute values

$$|\alpha|^2,\ |\beta|^2, \tag{15.15}$$

where we denote the average by a bar, a phase average leaves them unchanged. On the other hand because during spin diffusion the relative phases may take all kinds of orientations the average vanishes so that

$$\overline{\alpha^*\beta} \to 0\,. \tag{15.16}$$

The terms $\alpha^*\beta$, $\alpha\beta^*$ are typical for a quantum mechanical superposition. However, when the spin is coupled to its environment these terms vanish and the quantum mechanical coherence gets lost. Thus we must speak of decoherence. In the following it shall be understood that the coupling of our quantum system to the environment is zero so that no decoherence effects appear.

## 15.3 Representation of Numbers

One of the main goals of the theory of quantum information is to construct a quantum computer. This is a device which processes numbers. In order to learn how this can be done I briefly remind the reader of some basic facts of mathematics. Our conventional number system is a decimal system in which each number can be represented as a polynomial of powers of 10, i.e. in the form

$$N = a_n 10^n + a_{n-1} 10^{n-1} + \ldots + a_0\,, \tag{15.17}$$

where it is understood that the coefficients obey

$$0 \leq a_j \leq 9,\; j = 0, 1, \ldots, n\,.$$

E.g. the number $N = 137$ is then represented in the form

$$137 = 1 \cdot 10^2 + 3 \cdot 10^1 + 7\,. \tag{15.18}$$

Instead of the basis 10 we may use any other basis. The most common and important basis is the number 2 in which case $N$ can be written as

$$N = b_m 2^m + b_{m-1} 2^{m-1} + \ldots + b_0\,. \tag{15.19}$$

Consider as an example

$$7 = 1 \cdot 2^2 + 1 \cdot 2 + 1\,, \tag{15.20}$$

which can be written in analogy to the decimal system in the form

$$111\,. \tag{15.21}$$

Let us take as another example the above number 137, then one can easily check that

$$137 = 1 \cdot 2^7 + 0 \cdot 2^6 + 0 \cdot 2^5 + 0 \cdot 2^4 + 1 \cdot 2^3 + 0 \cdot 2^2 + 0 \cdot 2 + 1 \tag{15.22}$$

holds so that in this binary system the number $N$ is written as

$$10001001\,. \tag{15.23}$$

## 15.4 Register

How can we store these numbers physically? We consider a set of non-interacting spins. Consider as a simple example three spins which are described by their wave functions

$$\phi_1\, \phi_2\, \phi_3\,. \tag{15.24}$$

Because these spins are non-interacting we may simply form their product in order to represent the total system. Consider the spin configuration given by

$$\phi_1 = |\uparrow\rangle,\ \phi_2 = |\downarrow\rangle,\ \phi_3 = |\uparrow\rangle \tag{15.25}$$

so that (15.24) acquires the form of the left hand side of

$$|\uparrow\rangle|\downarrow\rangle|\uparrow\rangle = |\uparrow\downarrow\uparrow\rangle\,. \tag{15.26}$$

Now theoretical physicists as well as mathematicians are lazy people, at least what the writing of formulas is concerned. Thus instead of writing the left hand side one introduces the right hand side of (15.26). Note that the right hand side is just a shorthand of the left hand side and that the sequence of arrows indicates the index of the corresponding spin, i.e. more precisely the left hand side of (15.26) reads

$$|\uparrow\rangle_1|\downarrow\rangle_2|\uparrow\rangle_3\,. \tag{15.27}$$

Now a simple but remarkable trick can be used time and again, namely we can represent any number in a variety of ways, be it in the decimal system, or in the binary system, or eventually by means of physical states represented by (15.26), e.g.

$$\begin{aligned}\text{(decimal) } 5 &\to 1 \cdot 2^2 + 0 \cdot 2^1 + 1 \cdot 1\\ &\to 101 \text{ (binary)}\\ &\to |\uparrow\downarrow\uparrow\rangle \text{ (spins)}\end{aligned} \tag{15.28}$$

Or we may use as a shorthand the bra- and ket-notation and use

$$|5\rangle \to |101\rangle \to |\uparrow\downarrow\uparrow\rangle\,. \tag{15.29}$$

In this way, any number can be represented by a sufficiently large number of spins.

## 15.5 Entanglement

One of the most surprising facts is that of entanglement which has been strongly attacked by Einstein jointly with Rosen and Podolsky but which is now well established experimentally. As an example consider the wave function

$$\phi_1(\uparrow)\phi_2(\downarrow) + \phi_1(\downarrow)\phi_2(\uparrow) . \tag{15.30}$$

It implies that when spin no. 1 is measured with "spin up" then necessarily spin no. 2 must be in the "spin down" state. When we use photons instead of spins where the polarization direction replaces the spin direction, then experiments show that two photons exhibit this entanglement even if they are kilometres apart from each other; first the two photons are produced in an entangled state in a crystal, then they propagate in different directions. When the polarization of one photon is measured, the corresponding polarization of the other photon is fixed.

# 16. Quantum Computation

The present situation of quantum computation reminds me of the early days of laser physics. At that time it was felt that the concept of the laser is a wonderful idea but hardly any application of the laser was seen at that time. It was said that the laser was a solution to problems which still have to be found. As we all know, this situation has dramatically changed in the meantime with the overwhelming variety of applications of the laser. At present it is said that the quantum computer is a fascinating concept but still there are only few applications where it can show its superiority to classical computers and even here the really convincing applications are still ahead of us. At present, one of the few outstanding possible applications of quantum computers is Shor's algorithm for the factoring of large numbers. We postpone the presentation of the motivation to tackle this problem to Section 16.4 below. In this chapter we proceed as follows. We first remind the reader of the basic elements of classical computers, i.e. gates. Then we discuss their quantum analogues. Finally we present Shor's approach to factor large numbers by means of the quantum computer.

## 16.1 Classical Gates

Gates, be them classical or quantum, process information. Let us consider first of all logical operations and let us illustrate the whole procedure by simple physical systems. Let us consider two wires A, B in sequence which can be switched to be active or non-active for an electric current. Clearly at the output an electric current arrives only if both wires are in their active state. This result can be represented by a truth table representing the logical operation AND.

**Table 16.1**

| A\B | 0 | 1 |
|---|---|---|
| 0 | 0 | 0 |
| 1 | 0 | 1 |

When we arrange the wires in parallel, we can readily convince ourselves that this device realizes the operations OR.

**Table 16.2**

| A\B | 0 | 1 |
|---|---|---|
| 0 | 0 | 1 |
| 1 | 1 | 1 |

An important operation is the exclusive OR (XOR) which is represented by

**Table 16.3:**

| A\B | 0 | 1 |
|---|---|---|
| 0 | 0 | 1 |
| 1 | 1 | 0 |

Here a non-vanishing output appears only if either $A$ or $B$ are on but not both. These truth tables reveal a very important aspect of conventional classical computers, namely in each case two input numbers are used to produce only one output number. This implies that these processes are not invertible mathematically or, physically speaking, that the processes are irreversible. The irreversibility implies heat production. Thus in the past a number of schemes referring to classical computers were proposed to minimize heat production which becomes quite a problem when the devices are miniaturized. However, the price to be paid for a reversible computer is that additional information must be stored.

It is well known mathematically that the logical operations such as AND and OR form a Boolean algebra, which physically can be realized by a network of correspondingly connected gates. With respect to quantum computers and classical computers as well we note that there is an equivalence between logic operations such as represented by AND and OR and algebraic manipulations such as addition and multiplication (or their combinations). E.g. the operation AND corresponds to a multiplication in the binary system as is easily checked by comparing

$$0 \cdot 0 = 0,\ 0 \cdot 1 = 0,\ 1 \cdot 0 = 0,\ 1 \cdot 1 = 1 \tag{16.1}$$

with the truth Table 16.1.

## 16.2 Quantum Gates

In the quantum domain the gates described in the preceding section are replaced by quantum gates which are reversible and can be mathematically described by unitary transformations. We already encountered the action of a quantum gate acting on an individual spin in Section 15.1. Writing instead of "spin down" and "spin up" the numbers 0 and 1, respectively, we may rewrite the action of the gate represented by (15.6) with $\vartheta = \pi/2$ in the form

$$U|0\rangle = |0\rangle + |1\rangle \tag{16.2}$$

$\left(\text{up to a normalization factor } \frac{1}{\sqrt{2}}\right)$. Consider the simplest example of a register composed of only two spins and consider gates transforming simultaneously the individual Q-bits (spins). We introduce the product of operators

$$U_{\text{tot}} = U_1 \times U_2 \tag{16.3}$$

with the actions

$$U_j|0\rangle_j = |0\rangle_j + |1\rangle_j \,. \tag{16.4}$$

(The sign $\times$ indicates the direct product.)

Then we may form

$$\begin{aligned} U_{\text{tot}}|00\rangle &= U_1|0\rangle_1 U_2|0\rangle_2 \\ &= (|0\rangle + |1\rangle)(|0\rangle + |1\rangle) \end{aligned} \tag{16.5}$$

or by multiplying the products out we obtain

$$\begin{aligned} &= |00\rangle + |01\rangle + |10\rangle + |11\rangle \\ &\quad\ \downarrow \qquad\ \downarrow \qquad\ \downarrow \qquad\ \downarrow \\ &= \ 0 \ + \ 1 \ + \ 2 \ + \ 3 \,. \end{aligned} \tag{16.6}$$

In our example the operation (16.5) with (16.4) has produced out of the state zero four states representing the numbers from 0 to 3 in binary representation. In proceeding from the upper to the lower row in (16.5) we dropped the indices referring to the individual spins because these indices can be read off, at least implicitly, by means of the sequence of factors (remember the laziness principle!). Quite generally, by flipping simultaneously all the $N$ spins of a register from the down state into the superposition (16.4), we obtain a superposition of states

$$U_{\text{tot}}|00\ldots0\rangle = \sum_{n=0}^{M} |n\rangle \,, \tag{16.7}$$

where $M = 2^N - 1$.

Quite generally we may state that a quantum logic gate is a device which performs a fixed unitary operation on Q-bits in a fixed period of time. A quantum network is a device consisting of quantum logic gates whose computational steps are synchronized in time (D. Deutsch, Proc. R. Soc. Lond. **A 425**, 73 (1989)).

Let us consider examples of important quantum gates. The Hadamard gate is represented by the matrix

$$H = \frac{1}{\sqrt{2}} \begin{pmatrix} 1 & 1 \\ 1 & -1 \end{pmatrix} \tag{16.8}$$

acting in the state space

$$|0\rangle = \begin{pmatrix} 0 \\ 1 \end{pmatrix}$$

$$|1\rangle = \begin{pmatrix} 1 \\ 0 \end{pmatrix}. \tag{16.9}$$

It acts on the vectors (16.9) according

$$H|v\rangle = \frac{1}{\sqrt{2}}((-1)^v|v\rangle + \langle 1 - v\rangle) \tag{16.10}$$

or more explicitly

$$H|0\rangle = \frac{1}{\sqrt{2}}(|0\rangle + |1\rangle)$$

$$H|1\rangle = \frac{1}{\sqrt{2}}((-1)|1\rangle + |0\rangle)\,. \tag{16.11}$$

In a pictorial representation it is given by Fig. 16.1.

In terms of spins, this gate flips a spin from its –z or z-direction into the horizontal plane. The phase shift gate is represented in the same space by

$$U_\phi = \frac{1}{\sqrt{2}} \begin{pmatrix} 1 & 0 \\ 0 & e^{i\phi} \end{pmatrix} \tag{16.12}$$

with the result

$$U_\phi|v\rangle = e^{i\phi}|v\rangle \tag{16.13}$$

and represented by Fig. 16.2.

The Hadamard and phase shift gates allow us to generate the most general state of a single Q-bit as shown in Fig. 16.3.

$|v\rangle$ —[H]— $(-1)^v|v\rangle + |1 - v\rangle$

**Fig. 16.1.** Representation of the Hadamard transform

$\phi$

$|v\rangle$ —●— $e^{iv\phi}|v\rangle$

**Fig. 16.2.** Representation of the phase shift operator

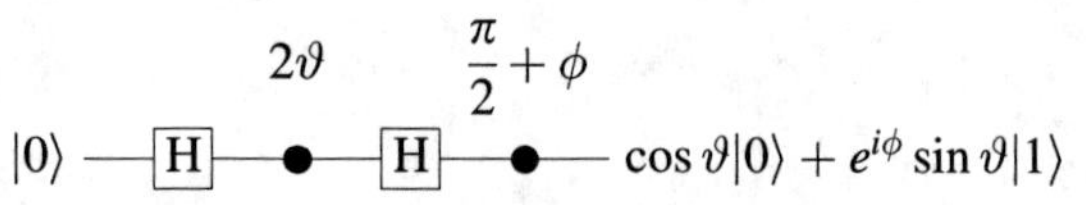


**Fig. 16.3.** A combination of Hadamard and phase shift operation

Note that the most general wave function can always be written in the form shown on the right hand side of Fig. 16.3 because the coefficients $\alpha$ and $\beta$ of the representation (15.11) are only determined up to a common phase factor.

The most important two-Q-bit-gate is the controlled-Not or XOR-gate which can be considered as the extension of the classical XOR-gate into the quantum domain. This gate flips the second Q-bit (target) if the first (control) Q-bit is $|1\rangle$ and does nothing if the control Q-bit is $|0\rangle$. If a representation of the basic vectors is chosen in the form

$$|00\rangle = \begin{pmatrix} 1 \\ 0 \\ 0 \\ 0 \end{pmatrix} \quad |01\rangle = \begin{pmatrix} 0 \\ 1 \\ 0 \\ 0 \end{pmatrix}$$

$$|10\rangle = \begin{pmatrix} 0 \\ 0 \\ 1 \\ 0 \end{pmatrix} \quad |11\rangle = \begin{pmatrix} 0 \\ 0 \\ 0 \\ 1 \end{pmatrix} \tag{16.14}$$

the matrix operation corresponding to XOR is given by

$$U_{\text{XOR}} = \begin{pmatrix} 1 & 0 & 0 & 0 \\ 0 & 1 & 0 & 0 \\ 0 & 0 & 0 & 1 \\ 0 & 0 & 1 & 0 \end{pmatrix} . \tag{16.15}$$

So far we have shown the most important steps which allow us to build a quantum computer. To repeat, such a computer consists of a register composed of individual Q-bits represented e.g. by spins, each in a specific initial state, e.g. "spin down", and using a network of devices, the logical gates, these spins are subjected to physical operations so that eventually a specific final state of the register is reached.

## 16.3 Calculation of the Period of a Sequence by a Quantum Computer

In this section we want to explain the core of Shor's approach where the concept of the quantum computer enters in a decisive fashion. This relies on the fact that the quantum computer allows the calculation of a very large number of initial inputs in parallel. The task to be solved is this: We start from positive integers $n = 0, 1, 2, \ldots$ and form functions $f(n) = x^n$ where $x$ is some integer number. Then by a simple algorithm one can change this function so that it becomes periodic, i.e. $f(n+L) = f(n)$. If $n$, $L$ are large numbers the determination of $L$ becomes an extremely time consuming problem for conventional computers which at present represents an insurmountable problem for numbers with, say, 250 digits.

We start from a register with $K$ spins so that there are $M = 2^K$ states. In other words, the register can represent numbers from zero till $2^K - 1$. We use a second register which is large enough so that the results of the evaluation of the function $f(n)$ can be stored. We start with a state in which both registers are in their zero state, i.e. "all spins down"

$$|0,0\rangle \equiv |0,0,\ldots;0,0,\ldots\rangle\,. \tag{16.16}$$

We now apply the spin flip operator (16.7) to the first register

$$U_{\text{tot}}|0,0\ldots;0,0,\ldots\rangle \tag{16.17}$$

so that we obtain a new state

$$|\Psi\rangle = \frac{1}{\sqrt{M}}\sum_{n=0}^{M-1}|n;0\rangle\,. \tag{16.18}$$

Now we let the computer do its first calculation which is achieved by a series of transformations which are altogether a unitary transformation $U_C$. This can be done to each individual state according to

$$U_C|n;0\rangle = |n;f(n)\rangle\,. \tag{16.19}$$

But now the crucial point of the quantum computer enters, namely we may apply $U_C$ also to the total wave function in (16.18). Because of the linearity of quantum mechanics we obtain

$$U_C|\Psi\rangle = \frac{1}{\sqrt{M}}\sum_{n=0}^{M-1}U_C|n;0\rangle \equiv |\Psi'\rangle \tag{16.20}$$

and using (16.19) the result

$$|\Psi'\rangle = \frac{1}{\sqrt{M}} \sum_{n=0}^{M-1} |n; f(n)\rangle \,. \tag{16.21}$$

Though the sum over $n$ is very large, sending the wave function (16.18) through the computer according to $U_C$, we obtain the results of all the numerous calculations simultaneously.

In order to find the period, we subject the state $|n\rangle$ to a quantum Fourier transformation which can be achieved by unitary transformations

$$|n\rangle = \frac{1}{\sqrt{M}} \sum_{k=0}^{M-1} e^{2\pi i n k/M} |k\rangle \,. \tag{16.22}$$

Inserting (16.22) into (16.21) we obtain

$$\frac{1}{M} \sum_{n=0}^{M-1} \sum_{k=0}^{M-1} e^{2\pi i n k/M} |k; f(n)\rangle \,. \tag{16.23}$$

It is quite crucial that during the whole process quantum coherence is preserved as is witnessed by the superpositions in (16.18), (16.20), (16.21), and (16.23). This requires that no measurements are made during this process and, of course, that the whole process is not disturbed by the environment. Now in the final step a measurement is made on the first register by measuring all the spins with respect to their z-components. What is the probability of finding a specific spin configuration, i.e. finding a specific number $k$? A clear-cut answer can be given if the function $f$ is periodic. As we shall show in detail, the sum over $n$ yields a constructive interference from the exponential functions only when $(k/M)$ is a multiple of the reciprocal period $\frac{1}{L}$.

Periodicity of $f(n)$ implies that for

$$n = 0, ..., L-1 \tag{16.24}$$

the functions $f(n)$ are all different from each other but that for larger $n$ the corresponding repetitions occur. Therefore we write

$$n = n' + L \cdot l, \; l = 0, 1, ..., \; n' = 0, ..., L-1 \tag{16.25}$$

and split (16.23) into the sums

$$\sum_{n=0}^{M-1} = \sum_{n'=0}^{L-1} \sum_{l=0}^{M'-1} + \text{rest} \tag{16.26}$$

where $M' = 1 \left[\frac{M}{L}\right]$ is the largest integer $\leq \frac{M}{L}$ and the rest is given by

$$\text{rest} = \sum_{n=LM'-1}^{M-1} . \tag{16.27}$$

We may assume that for sufficiently large $M$ the rest can be neglected. Inserting (16.26) into (16.23), splitting $n$ according to (16.25), and rearranging sums we obtain

$$\frac{1}{M}\sum_{k}\sum_{n'=0}^{L-1} e^{2\pi i n' k/M} |k; f(n')\rangle \cdot \sum_{l=0}^{M'-1} e^{2\pi i L l k/M} . \tag{16.28}$$

The last sum is a geometric series which can be evaluated to yield

$$\frac{1 - e^{2\pi i L M' k/M}}{1 - e^{2\pi i L k/M}} . \tag{16.29}$$

According to quantum mechanics, the probability of finding a specific configuration $k = k'$ is given by the absolute square of the coefficient of the state $|k'\rangle$. This coefficient is essentially determined by (16.29). It tells us that the maxima of its absolute square lie at values $k = k_d$

$$k_d = \frac{M}{L} d, \quad d = 1, 2 ... \tag{16.30}$$

and that the probability distribution has a comb structure. From this we can deduce the length (minimal) according to

$$L = \frac{M}{k_1} . \tag{16.31}$$

As is well known the peaks of the absolute value of (16.29) are the more pronounced the larger the ratio $M/L$. This allows precise estimates of the accuracy of this approach.

Note that the results of the measurements are of a probabilistic nature so that the measurement of the spins in the first register must be made sufficiently often (for precise estimates cf. references).

## 16.4 Coding, Decoding and Breaking Codes

Presently quantum information is playing an increasingly important role in coding and possibly decoding. In order to show the relevance of the quantum approach we briefly remind the reader of results in classical information. We start with some simple mathematical concepts.

### 16.4.1 A Little Mathematics

An integer $N \geq 2$ is said to be prime if it is divisible only by 1 and $N$. The greatest common divisor of two integers $N$ and $N'$ is the greatest positive integer $D$ which divides both $N$ and $N'$. We denote $D$ by

$$D = gcd(N,N')\,. \tag{16.32}$$

Two integers $N$, $N'$ are coprime if

$$gcd(N,N') = 1\,. \tag{16.33}$$

*The modulo-formalism*
Consider two positive integers $a$, $N$. We put

$$a \bmod N = r \tag{16.34}$$

where $r$ is a positive integer. It can be found by the formula

$$a = mN + r \tag{16.35}$$

where $m$ is a positive integer and

$$0 \leq r \leq N-1\,. \tag{16.36}$$

In other words, in order to find $r$ we divide $a$ by $N$ and retain only the rest. The modulo-formalism has a number of rules. One of their most important ones are

$$(a \cdot b) \bmod N = (a \bmod N)(b \bmod N) \bmod N \tag{16.37}$$

and

$$(a \pm b) \bmod N = (a \bmod N + b \bmod N) \bmod N\,.$$

### 16.4.2 RSA Coding and Decoding

In our time cryptography is based on computational complexity, i.e. the difficulty of breaking codes numerically. One of the most popular codes is that of RSA (Rivest, Shamir, Adleman). If someone wants to receive an encoded message, this person can publish a so-called public key which consists of two numbers $N$ and $e$. In it $N$ is a product of two large prime numbers, $p$ and $q$

$$N = p \cdot q\,. \tag{16.38}$$

$e$ must be coprime to both $p-1, q-1$. In order to send a message, in the first step it is encoded by numbers, e.g. a number is attributed to each letter of the alphabet. Then the message is broken into pieces each representing a number with

$$m < N \,. \tag{16.39}$$

To encode this piece the formula

$$c = m^e \bmod N \tag{16.40}$$

is used. Decoding can be done again by a simple formula, namely

$$m = c^d \bmod N \,. \tag{16.41}$$

Here the exponent $d$ is determined by

$$ed = 1 \bmod (p-1)(q-1) \,. \tag{16.42}$$

The solution of (16.42), i.e. the determination of $d$, can be done by conventional procedures and does not require too much effort. However, the difficulty consists in knowing the factors $p-1, q-1$, or, in other words, the factors $p$ and $q$ of $N$. According to present computational algorithms the decomposition of a known number $N$ into its prime factors according to (16.38) requires a very large time which even for moderately large numbers $N$ can last longer than the age of the universe. Thus one of the main objectives of quantum computers is to solve this problem. The first crucial steps, which involve quantum computation, were described in Section 16.3. Now we want to show how the determination of the length of a period can be used to solve (16.38).

### 16.4.3 Shor's Approach, Continued

As we have seen, Shor's formalism of the quantum computer allows us to determine the period of functions which are generated by some algorithm. Following the same author we want to show how we are now able to factor a number $N$. To this end we choose an integer number $x$ and assume that $N$, $x$ are coprime. We consider the function

$$f(n) = x^n \bmod N \tag{16.43}$$

for $n = 0, 1, 2, \ldots$. This series acquires the form

$$1, x, x^2, \ldots, x^{L-1};\ 1, x, x^2, \ldots, x^{L-1}; \ldots \tag{16.44}$$

where $L$ is the first number where

$$x^L = 1 \bmod N \tag{16.45}$$

holds. $L$ is the smallest length of the period. We assume that $L$ has been determined by the quantum computer. More technically speaking, we require $N^2 \le M = 2^K$ for a sufficient resolution of the Fourier transform step according to Shor (1994), Ekert and Josza (1996). Provided $L$ is even we may write (16.45) in the form

$$(x^{L/2}+1)(x^{L/2}-1) = 0 \bmod N\,. \tag{16.46}$$

If $L$ is not even we must choose another $x$. To draw further conclusions from (16.46), we write it in the form

$$\frac{a \cdot b}{N} = m \tag{16.47}$$

where according to the modulo-formalism $m$ must be an integer. Here

$$a = (x^{L/2}+1) \tag{16.48}$$

and

$$b = (x^{L/2}-1)\,. \tag{16.49}$$

Invoking the modulo-formalism we may assume

$$0 \le a < N, \quad 0 \le b < N\,. \tag{16.50}$$

Can $a$ and $b$ be equal to zero? The relation

$$b = 0 \bmod N \tag{16.51}$$

implies

$$x^{L/2} = 1 \bmod N \tag{16.52}$$

which says that the period is $L/2$ but not $L$ in contradiction to the assumption that $L$ is the shortest period. In the case $a = 0$ we have

$$x^{L/2} + 1 = 0 \bmod N \tag{16.53}$$

which implies that $a$ is divisible by $N$ and we cannot deduce any factors from $N$ by the formalism. Thus we must try the procedure with another $x$. Note that the whole procedure fails, of course, if $N$ is prime. But in this case we do not need any factoring at all. In the following we assume

$$x^{L/2} + 1 \ne 0 \bmod N\,. \tag{16.54}$$

Both $a$ and $b$ must have common factors with $N$, which is illustrated by the following example where $N$ is a product of two prime factors

$$N = pq\,. \tag{16.55}$$

Then (16.47) acquires the form

$$\frac{a \cdot b}{p \cdot q} = m\,, \tag{16.56}$$

which means that $a$ must be divisible by $p$ (or $q$) and $b$ must be divisible by $q$ (or $p$). Generally, in order to find factors of $N$ we have to calculate the greatest common divisors of $a$ and $b$ individually with $N$, i.e.

$$d_1 = gcd(a, N), \quad d_2 = gcd(b, N)\,. \tag{16.57}$$

Actually both divisors (16.57) must be factors of $N$ we have sought.

## 16.5 The Physics of Spin 1/2

In the context of the mathematics of quantum computation only the $z$ component of the spin, i.e. $s_z$, appears explicitly. When we wish to deal with the underlying physics, especially with respect to physical realizations of a quantum computer, we must take care of the fact that $s_z$ is just a component of a 3-dimensional vector

$$\boldsymbol{s} = (s_x, s_y, s_z)\,. \tag{16.58}$$

In a suitable quantum mechanical description the components are operators described by Pauli matrices

$$s_x = \frac{\hbar}{2}\begin{pmatrix} 0 & 1 \\ 1 & 0 \end{pmatrix}, \quad s_y = \frac{\hbar}{2}\begin{pmatrix} 0 & -i \\ i & 0 \end{pmatrix}, \quad s_z = \frac{\hbar}{2}\begin{pmatrix} 1 & 0 \\ 0 & -1 \end{pmatrix}, \tag{16.59}$$

where $\hbar$ is Planck's constant, divided by $2\pi$. Because these operators do not commute (actually they obey commutation relations of the quantum mechanical angular momentum) we can determine eigenfunctions just to one of these components. Usually one chooses the eigenfunctions to $s_z$. They are given by two-dimensional vectors

$$\phi_\uparrow \equiv |\uparrow\rangle = \begin{pmatrix} 1 \\ 0 \end{pmatrix} \tag{16.60}$$

and

$$\phi_\downarrow \equiv |\downarrow\rangle = \begin{pmatrix} 0 \\ 1 \end{pmatrix} \tag{16.61}$$

and thus fulfil the equations

$$s_z\phi_\uparrow = \frac{\hbar}{2}\phi_\uparrow, \quad s_z\phi_\downarrow = -\frac{\hbar}{2}\phi_\downarrow. \tag{16.62}$$

As we shall see below, we have to consider also superpositions of the form

$$\phi = \alpha\phi_\uparrow + \beta\phi_\downarrow = \begin{pmatrix}\alpha\\ \beta\end{pmatrix} \tag{16.63}$$

with the normalization condition

$$|\alpha|^2 + |\beta|^2 = 1. \tag{16.64}$$

$|\alpha|^2$ and $|\beta|^2$ are the probabilities for finding the +z and –z direction, respectively, when the z-component of the spin-state (16.63) is measured. In order to visualize the physical content of the superposition (16.63), we form expectation values which in the notation of vectors and matrices are defined by

$$\langle s_x\rangle = (\alpha^*, \beta^*)s_x\begin{pmatrix}\alpha\\ \beta\end{pmatrix}. \tag{16.65}$$

A simple evaluation using (16.63), (16.60), (16.61), and (16.59) yields

$$\langle s_x\rangle = \frac{\hbar}{2}(\alpha^*\beta + \alpha\beta^*), \tag{16.66}$$

$$\langle s_y\rangle = \frac{\hbar}{2}i(\alpha\beta^* - \alpha^*\beta), \tag{16.67}$$

$$\langle s_z\rangle = \frac{\hbar}{2}(|\alpha|^2 - |\beta|^2) \tag{16.68}$$

whose verification we leave as a little exercise to the reader. We note that all physically relevant results, e.g. those expressed by the expectation values, are independent of a common phase factor $e^{i\chi}$ of $\alpha$ and $\beta$

$$\phi \to e^{i\chi}\phi. \tag{16.69}$$

In order to elucidate the meaning of the results (16.66)–(16.68), we write the coefficients $\alpha, \beta$ in the form

$$\alpha = \cos\frac{\vartheta}{2}, \quad \beta = e^{i\chi}\sin\frac{\vartheta}{2}. \tag{16.70}$$

We then obtain

$$\langle s_x \rangle = \frac{\hbar}{2} \sin\frac{\vartheta}{2} \cos\frac{\vartheta}{2} \cdot 2\cos\chi \tag{16.71}$$

or using a trigonometric identity we find

$$\langle s_x \rangle = \frac{\hbar}{2} \sin\vartheta \cos\chi\,. \tag{16.72}$$

Similarly after a little calculation we obtain

$$\langle s_y \rangle = \frac{\hbar}{2} \sin\vartheta \sin\chi \tag{16.73}$$

and

$$\langle s_z \rangle = \frac{\hbar}{2} \cos\vartheta\,. \tag{16.74}$$

Quite clearly $\vartheta$ is an angle which the spin forms with the z direction whereas $\chi$ is an angle between the spin projection on the $x$-$y$ plane with the corresponding axes.

## 16.6 Quantum Theory of a Spin in Mutually Perpendicular Magnetic Fields, One Constant and One Time Dependent

A number of important experiments on spin have been carried out with the following arrangement: both a constant, spatially homogeneous magnetic field in the z direction and an oscillating field in the $x$-$y$ plane are applied. This leads to the interesting phenomenon of spin flipping which is fundamental for quantum computation.

We shall see that we can easily solve these problems using the spin formalism introduced in Sect. 16.5. We write the magnetic field expressed as a time-dependent and a time-independent part:

$$\boldsymbol{B} = \boldsymbol{B}_0 + \boldsymbol{B}^S(t)\,, \tag{16.75}$$

where the vectors of the magnetic fields are defined as

$$\boldsymbol{B}_0 = (0, 0, \boldsymbol{B}_z^0) \tag{16.76}$$

and

$$\boldsymbol{B}^s(t) = (\boldsymbol{B}_x^s(t), \boldsymbol{B}_y^s(t), 0)\,. \tag{16.77}$$

The spin wave functions must obey, as always in quantum mechanics, a Schrödinger equation in which a Hamilton operator acts on the wave function and thus determines its time evolution. This Hamilton operator is obtained by "translating" classical observables in the energy function into their corresponding operators. In the present case of spins, these are the Pauli matrices and the Schrödinger (or Pauli) equation reads

$$\frac{e}{m_0}(\boldsymbol{Bs})\phi = i\hbar\frac{d\phi}{dt} \tag{16.78}$$

($e$: charge of spin-particle, $m_0$ its mass).

We write the solution of the Schrödinger equation in the general form

$$\phi(t) = c_1(t)\phi_\uparrow + c_2(t)\phi_\downarrow \equiv \begin{pmatrix} c_1(t) \\ c_2(t) \end{pmatrix}. \tag{16.79}$$

To arrive at equations for the still unknown coefficients $c_1$ and $c_2$ we substitute (16.79) in (16.78). Observing the decomposition (16.75–16.77) and using the matrix form of $\hat{s}_x$, $\hat{s}_y$, and $\hat{s}_z$ we obtain the Schrödinger equation in the form

$$\mu_B \begin{pmatrix} B_z^0 & B_x^s - iB_y^s \\ B_x^s + iB_y^s & -B_z^0 \end{pmatrix} \cdot \begin{pmatrix} c_1 \\ c_2 \end{pmatrix} = i\hbar \begin{pmatrix} \dot{c}_1 \\ \dot{c}_2 \end{pmatrix} \tag{16.80}$$

where $\mu_B = e\hbar/(2m_0)$. If we multiply the matrix, we obtain more explicitly

$$\left(\frac{1}{2}\hbar\omega_0\right)c_1 + \mu_B(B_x^s - iB_y^s)c_2 = i\hbar\dot{c}_1\,, \tag{16.81}$$

$$\mu_B(B_x^s + B_y^s)c_1 - \frac{1}{2}\hbar\omega_0 c_2 = i\hbar\dot{c}_2\,. \tag{16.82}$$

Here we have introduced the frequency

$$\omega_0 = 2\mu_B B_z^0/\hbar \tag{16.83}$$

as an abbreviation. In order to simplify the following calculation, let us think of the transverse magnetic field as rotating with the frequency $\omega$. In other words, the magnetic field has the form

$$B_x^s = F\cos\omega t\,,$$
$$B_y^s = F\sin\omega t\,. \tag{16.84}$$

Since $B_x^s$ and $B_y^s$ appear in (16.81, 16.82) in a combined form, let us first consider these expressions. We can express them as an exponential function, due to elementary relationships between sines and cosines:

$$B_x^s \pm iB_y^s = F(\cos\omega t \pm i\sin\omega t) = F\exp(\pm i\omega t)\,. \tag{16.85}$$

Then (16.81, 16.82) simplify to

$$(\hbar\omega_0/2)c_1 + \mu_B F\exp(-i\omega t)c_2 = i\hbar\dot{c}_1\,, \tag{16.86}$$

$$\mu_B F\exp(i\omega t)c_1 - (\hbar\omega_0/2)c_2 = i\hbar\dot{c}_2\,. \tag{16.87}$$

We shall solve these two equations in two steps. In the first, we put the coefficients $c_j(t)$ into the form

$$c_1(t) = d_1(t)\exp(-i\omega_0 t/2);\quad c_2(t) = d_2(t)\exp(i\omega_0 t/2)\,. \tag{16.88}$$

If we substitute this in (16.86), (16.87), perform the differentiation and rearrange, we obtain

$$\mu_B F\exp[-i(\omega-\omega_0)t]d_2 = i\hbar\dot{d}_1\,, \tag{16.89}$$

$$\mu_B F\exp[i(\omega-\omega_0)t]d_1 = i\hbar\dot{d}_2\,. \tag{16.90}$$

These equations become very simple when we set the rotational frequency of the magnetic field $\omega$ equal to the spin frequency $\omega_0$:

$$\omega = \omega_0\,, \tag{16.91}$$

which corresponds to a typical experimental arrangement. We then obtain

$$\mu_B F d_2 = i\hbar\dot{d}_1\,, \tag{16.92}$$

$$\mu_B F d_1 = i\hbar\dot{d}_2\,. \tag{16.93}$$

With the abbreviation $\mu_B F/\hbar = \Omega$, the solution to (16.92), (16.93) reads

$$d_1 = a\sin(\Omega t + \Phi)\,, \tag{16.94}$$

$$d_2 = ia\cos(\Omega t + \Phi)\,, \tag{16.95}$$

where the amplitude $a$ and phase $\Phi$ are free to vary. The normalisation condition for the spin wavefunction requires that $a = 1$. If we substitute (16.94), (16.95) in (16.88) and this in (16.79), and do the same with (16.94), we obtain the desired spin wavefunction

$$\phi(t) = \sin(\Omega t)\exp(-i\omega_0 t/2)\phi_\uparrow + i\cos(\Omega t)\exp(i\omega_0 t/2)\phi_\downarrow\,. \tag{16.96}$$

The spin functions and the spin formalism naturally seem very unintuitive. In order to see the meaning of the above equations, let us remember that the immediate predictions of quantum mechanics can be read from the corresponding expectation values (Sect. 16.5). We shall first develop the expectation value of the spin operator in the z direction. A comparison of (16.63) with (16.96) shows that we can now express the $\alpha$ and $\beta$ of (16.63) in the form

$$\begin{aligned} \alpha &= \sin(\Omega t)\exp(-i\omega_0 t/2)\,, \\ \beta &= i\cos(\Omega t)\exp(i\omega_0 t/2)\,. \end{aligned} \tag{16.97}$$

These can be immediately substituted into the end results (16.66–16.68), however, to give

$$\begin{aligned} \langle \hat{s}_z \rangle &= (\hbar/2)\sin^2(\Omega t) - \cos^2(\Omega t) \\ &= -(\hbar/2)\cos(2\Omega t)\,. \end{aligned} \tag{16.98}$$

According to (16.98), the z component of the spin oscillates with the frequency $2\Omega$. If the spin is originally down at $t = 0$, it flips up, then down again, and so on.

For the other components,

$$\langle \hat{s}_x \rangle = -\frac{\hbar}{2}\sin(2\Omega t)\sin(\omega_0 t)\,, \tag{16.99}$$

$$\langle \hat{s}_y \rangle = -\frac{\hbar}{2}\sin(2\Omega t)\sin(\omega_0 t)\,. \tag{16.100}$$

These equations indicate that the spin motion in the $x$-$y$ plane is a superposition of two motions, a rapid rotational motion with the frequency $\omega_0$ and a modulation with the frequency $2\Omega$. The entire result (16.98–16.100) can be very easily interpreted if we think of the expectation value of the spin as a vector $\boldsymbol{s}$ with the components $\langle \hat{s}_x \rangle$, $\langle \hat{s}_y \rangle$, and $\langle \hat{s}_z \rangle$. Obviously the projection of the vector on the z axis is $(\hbar/2)\cos(2\Omega t)$, while the projection in the $x$-$y$ plane is $(\hbar/2)\sin(2\Omega t)$. As can be seen from the formulae, the spin gradually tips out of the –z direction toward the horizontal, and then further into the +z direction, while simultaneously precessing. The spin thus behaves exactly like a top under the influence of external forces.

We shall consider this process again, in more detail. At a time $t = 0$,

$$\langle \hat{s}_z \rangle = -\hbar/2\,. \tag{16.101}$$

We now ask when the spin, considered intuitively, is in the horizontal position, i.e. when

$$\langle \hat{s}_z \rangle = 0\,. \tag{16.102}$$

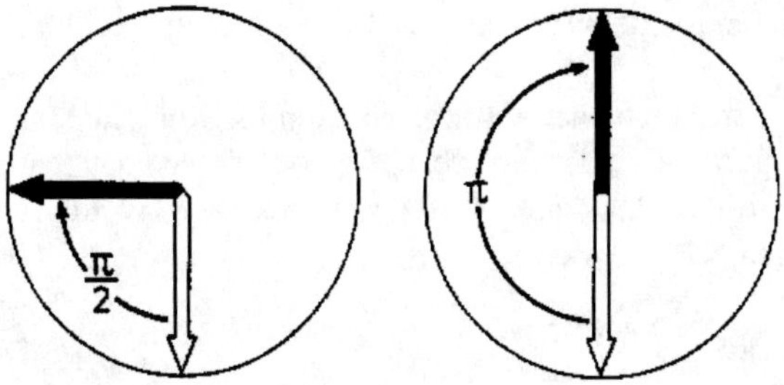


**Fig. 16.4.** Visualization of spin flips

This is clearly the case when the cosine function vanishes, that is when

$$2\Omega t = \pi/2 \tag{16.103}$$

holds, or when the time

$$t = \pi/(4\Omega) = \pi\hbar/(4\mu_B F) \tag{16.104}$$

has passed. If one allows the transverse magnetic field to act upon the spins for this time, they will be pointing in the horizontal position (Fig. 16.4). In other words, they have been rotated by an angle $\pi/2$. We therefore speak of a $\pi/2$ or of a 90° pulse.

Naturally, we may allow the magnetic field to act for a longer time, for example until the spins are pointing up, i.e.

$$\langle \hat{s}_z \rangle = \hbar/2\,. \tag{16.105}$$

This occurs when

$$\cos(2\Omega t) = -1 \tag{16.106}$$

is fulfilled, i.e. after the time

$$t = \pi\hbar(2\mu_B F)\,. \tag{16.107}$$

In this case, we speak of a $\pi$ or of a 180° pulse (Fig. 16.4).

With these considerations, we have sketched the most important traits of spin resonance. By applying a rotating magnetic field, we can cause the spin to flip from one direction to another. In practice, of course, one does not apply a magnetic field rotating with the spin frequency, but a linearly oscillating magnetic field. This can be pictured as a superposition of two fields rotating in opposite directions. Then one of the fields rotates with the spin, as before, while the other rotates with twice the frequency, as seen from the point of view of the rotating spin system.

The corresponding equations have practically the same form as those above, except for an additional, rapidly oscillating term, which comes from the "oppositely rotating" magnetic field. To a good approximation, this can be ignored; the result is the "rotating wave approximation".

In conclusion of this section we make contact with the notation used in quantum information and quantum computation. To this end we make use of the fact that wave functions are fixed only up to (constant) phase factors, which we now adjust appropriately. The result (16.96) can be expressed as

$$\phi = e^{-i\omega_0 t/2}\begin{pmatrix} \sin(\Omega t + \Phi) \\ i\cos(\Omega t + \Phi)e^{i\omega_0 t} \end{pmatrix}. \tag{16.108}$$

We write the wave function in such a form that at an initial time $t_0$ it coincides with the "spin up" or "spin down" wave functions, respectively, which we used in chapter 15 and the previous section. We want to count the time in such a way that the initial states are given by $t = 0$. Thus we make the replacement

$$t \to t_0 + t \tag{16.109}$$

in (16.108) which yields

$$\phi(t) = e^{-i\omega_0 t/2}e^{-i\omega_0 t_0/2}\begin{pmatrix} \sin(\Omega t + \Omega t_0 + \phi) \\ i\cos(\Omega t + \Omega t_0 + \phi)e^{i\omega_0 t + i\omega_0 t} \end{pmatrix}. \tag{16.110}$$

To meet the condition (note the new phase factor!)

$$\phi(0) \overset{!}{=} \begin{pmatrix} 0 \\ 1 \end{pmatrix} \tag{16.111}$$

we drop the constant phase factor in front of (16.110) and make the choice

$$\Omega t_0 + \Phi = 0, \quad e^{i\omega_0 t_0} = -i. \tag{16.112}$$

Thus our result with the initial condition (16.111) reads

$$\phi = e^{-\omega_0 t/2}\begin{pmatrix} \sin\Omega t \\ \cos\Omega t \cdot e^{i\omega_0 t} \end{pmatrix}. \tag{16.113}$$

In order to fulfil

$$\phi(0) = \begin{pmatrix} 1 \\ 0 \end{pmatrix} \tag{16.114}$$

we make the choice

$$\Omega t_0 + \Phi = \pi/2, \quad e^{i\omega_0 t_0} = -i \tag{16.115}$$

and obtain

$$\phi = e^{-i\omega_0 t/2} \begin{pmatrix} \cos \Omega t \\ -\sin \Omega t \cdot e^{i\omega_0 t} \end{pmatrix} . \tag{16.116}$$

From (16.113) and (16.116) we can construct a unitary transformation that when applied to (16.111) or (16.114) just yields (16.113) or (16.116), respectively. This unitary transformation reads

$$U = e^{-i\omega_0 t/2} \begin{pmatrix} \cos \Omega t & \sin \Omega t \\ -\sin \Omega t e^{i\omega_0 t} & \cos \Omega t e^{i\omega_0 t} \end{pmatrix} . \tag{16.117}$$

At this instant we can make contact with the unitary operator (15.6) which has been introduced in the context of the quantum computer above. When choosing a vanishingly small constant field we may put

$$\omega_0 = 0 . \tag{16.118}$$

Equivalently and more realistic, we may measure the spin in a corotating frame. Identifying $\Omega \cdot t$ with $\vartheta/2$ in (16.70) we realize that by these operations we obtain the flip of the spin from the –z direction into the horizontal plane or into the +z direction. In this way we may see the physical realization of the unitary transformation (15.6). Choosing somewhat different initial conditions with other phase factors one may convince oneself that the Hadamard transformation introduced in (16.8) is also just a spin flip operator from the –z direction into the horizontal plane.

## 16.7 Quantum Computation and Self-Organization

In conclusion of this chapter let us discuss whether quantum computation has something to do with self-organization as studied in this book. In my approach self-organization is understood as a process in which systems without specific interference from the outside are able to produce spatial or temporal patterns in physical systems or, more generally speaking, are compressing information in a specific way. This kind of selection process requires competition mechanisms which in turn require dissipative processes in one way or another. These dissipative processes are at least in general introduced by coupling a system to the environment or, if we are dealing with large systems, by internal thermo-dynamical processes. This is surely in contradiction to the basic assumptions underlying quantum computation which abhors any dissipative processes. The concepts of quantum information have revolutionized our understanding of information and especially show that we cannot ignore the physical basis of information, be it in the classical, be it in the quantum domain. We expect that the future of quantum computation will show a careful combination of purely quantum coherent processes with dissipative processes so that maximum speed, efficiency, and reliability can be obtained. Such a combination will surely be necessary when quantum computers comprise more and more components and thus become more and more sensitive to disturbances. Quite clearly, a fascinating development of this field is still ahead of us.

# 17. Concluding Remarks and Outlook

In this concluding chapter I wish to discuss what we have achieved in this book and to point out some areas where open questions remain.

In Chap. 1 we dealt with the nature of complex systems. Their most salient feature seems to be that they are practically inexhaustible with respect to our efforts to fully understand them. We must be content to study specific aspects of these systems which seem to us particularly suited for our purposes or interests. Accordingly, we have made an attempt in this book to cope with complex systems by choosing a rather general feature: we have focussed our attention on those situations in which complex systems change their macroscopic behavior qualitatively. More precisely, we have studied situations in which a so-called nonequilibrium phase transition occurs. This approach does not only allow us to deal with physical systems such as lasers and fluids but also with far more complex systems, such as biological systems. Undoubtedly a great variety of further applications can be found. As is witnessed by the laser example, our approach also enables us to treat oscillatory phenomena, but the extension to chaotic systems remains an open question. It is certainly a field for further research. The vehicle we have used in our approach is information. As we have seen, the word information may have quite different meanings so that we first had to discuss various definitions of the word. In particular we saw that the concept of *information* in the Shannon sense seems to be more appropriate in some cases than that of *entropy.* This is based on the fact that, in physics, at least entropy has a quite specific meaning and is defined for systems in thermal equilibrium. When we deal with physical systems far from thermal equilibrium or biological systems, we have to interpret the results anew and we must use new constraints in order to maximize the information or (statistical) entropy.

In particular we saw that, in a certain sense, a new type of information arises, a type of information that refers to the collective variables or order parameters. This suggests that we call the part of information that refers to the order parameters, and which mirrors the collective properties of the system, "synergetic information". At the same time the order parameters adopt a new meaning, namely that of "informators".

It is important to discuss what kind of information, in the usual sense of the word, can be revealed by our general approach based on the use of moments of variables. As we have seen, in this way we can determine the order parameters and their stationary distribution functions. Whether this kind of information is sufficient or not, depends on what we are aiming to achieve. Let us consider the example of a laser. The stationary distribution function is quite sufficient to characterize the steady-state intensity and intensity fluctuations of laser light. But this distribution function does not tell us anything about a very important property of laser light, namely its temporal coherence. To learn about this one has to deal in much greater detail with the dynamics. In the framework of our approach this kind of dynamics can be guessed from the time-dependent moments and from the transitional probabilities that can be derived from them by means of the maximum information entropy principle. Generally speaking, once averages over specific macroscopic data are known, the whole procedure allows us to guess microscopic events and processes.

It is tempting to conclude that in systems far from thermal equilibrium or even in nonphysical systems, (Shannon) information plays the same role as entropy in systems in thermal equilibrium or close to it, namely as the *cause of processes*. I am reluctant to follow this interpretation for the following reason: First of all we found that the results of our analysis depend on properly chosen constraints. Here we were able to define adequate constraints for a class of phenomena which can be characterized as nonequlibrium phase transitions. And we make guesses on the stationary distribution function. In that respect we may say that we can define a kind of potential which drives the system to specific stationary states. But this stationary potential does not give us a unique prescription for deriving for instance the Fokker-Planck equation which determines the dynamics, an aspect stressed by R. Landauer time and again. Thus it seems that the maximization of information (or entropy) is not a fundamental law which drives systems in a unique way. On the other hand there have been very remarkable studies by Graham and Tél and others on the construction of potentials for stochastic processes in situations where bifurcations occur. We have not attempted here to compare our approach with theirs, and this remains an interesting problem for further research.

These ideas are of basic importance for questions concerning biological evolution or development, or in other words, phylogenesis and ontogenesis, of animals. The question, of course, is whether evolution and development are governed by extremal principles, especially extremal principles connected with a single function, such as entropy or information. Our results might be a hint that such a function exists, but the price to be paid is that nothing or only little can be said about the dynamics. But whether such a function really exists, also remains an open question. On the other hand the fact that the choice of moments provided us with the adequate constraints for information in nonequilibrium phase transitions demonstrates the power of our approach. The behavior of the moments reflects the tendency of systems to become coherent. So what we have been studying here is the emergence of coherence and macroscopic order in complex systems via self-organization. But whether biology may be viewed totally from this point of view is another unanswered question. As we said before, complex systems seem to be inexhaustible.

Thus although a deep and fundamental question remains open, the method we have outlined in this book is certainly a powerful tool to study the coherence properties of complex systems close to nonequilibrium phase transitions from a unifying point of view.

# References

## Chapter 1

### Sections 1.5.1,2

L. D. Landau, E. M. Lifshitz: In *Course of Theoretical Physics*, Vol. 5: Statistical Physics (Pergamon, London-Paris 1952)
R. Becker: *Theory of Heat* (Springer, Berlin, Heidelberg 1967)
A. Münster: *Statistical Thermodynamics*, Vol. 1 (Springer, Berlin, Heidelberg 1969)
H. B. Callen: *Thermodynamics* (Wiley, New York 1960)
P. T. Landsberg: *Thermodynamics* (Wiley, New York 1961)
R. Kubo: *Thermodynamics* (North Holland, Amsterdam 1968)

### Section 1.5.3

H. Haken: *Synergetics, An Introduction*, Springer Ser. Synergetics, Vol. 1, 3rd ed. (Springer, Berlin, Heidelberg 1983)
H. Haken: *Advanced Synergetics*, Springer Ser. Synergetics, Vol. 20, 2nd corr. printing (Springer, Berlin, Heidelberg 1987)
H. Haken: *Synergetics: Introduction and Advanced Topics* (Springer, Berlin, Heidelberg 2004)

### Section 1.6.1

C.E. Shannon: A Mathematical Theory of Communication. Bell System Techn. J. **27**, 370–423, 623–656 (1948)
C.E. Shannon: Bell System Techn. J. **30**, 50 (1951)
C.E. Shannon, W. Weaver: *The Mathematical Theory of Communication* (Univ. of Illin. Press, Urbana 1949)
L. Boltzmann: *Vorlesungen über Gastheorie*, 2 vols. (Leipzig 1896, 1898)

### Section 1.6.2

*H. Haken*: In *Thermodynamics and Regulation of Biological Processes*, ed. by I. Lamprecht, A.I. Zotin (Walter de Gruyter & Co, Berlin, New York 1984)

### Section 1.6.3

H. Haken: paper presented at international conference on information, Tokyo, 1986
H. Haken: In *Ordnung aus dem Chaos. Prinzipien der Selbstorganisation und Evolution des Lebens*, Bernd-Olaf Küppers (Hrsg.) (Piper, München, Zürich 1987)

## Chapter 2

H. Haken: *Synergetics, An Introduction*, Springer Ser. Synergetics, Vol. 1, 3rd ed. (Springer, Berlin, Heidelberg 1983)
H. Haken: *Advanced Synergetics*, Springer Ser. Synergetics, Vol. 20, 2nd corr. printing (Springer, Berlin, Heidelberg 1987)
H. Haken: *Synergetics: Introduction and Advanced Topics* (Springer, Berlin, Heidelberg 2004)

For further texts and monographs on the Fokkler-Planck equation, see

H. Risken: *The Fokker-Planck Equation*, Springer, Berlin, Heidelberg (1989)
A.T. Bharucha-Reid: *Elements of the Theory of Markov Processes and their Applications*, Dover, Mineola, New York (1997)
C.W. Gardiner: *Handbook of Stochastic Methods*, 2nd ed. Springer, Berlin, Heidelberg, (1985)

## Chapter 3

### Section 3.2

S. Kulback: Ann. Math. Statist. **22**, 79 (1951)
S. Kulback: *Information Theory and Statistics* (Wiley, New York 1951)

### Section 3.3

E.T. Jaynes: Phys. Rev. **106**, 4, 620 (1957); Phys. Rev. **108**, 171 (1957)
E.T. Jaynes: In *Delaware Seminar on the Foundations of Physics* (Springer, Berlin, Heidelberg 1967)
E.T. Jaynes: Am. J. Phys. **33**, 391 (1965)
E.T. Jaynes: In *The Maximum Entropy Formalism*, ed. by R.D. Levine, M. Tribus (MIT Press, Cambridge, Mass. 1978)

## Chapter 5

An early application of Jaynes' principle to systems with "several temperatures" was given by R.S. Ingarden, his coworkers, and other authors in numerous papers. See for example:

R.S. Ingarden: Bull. Acad. Polon. Sci., Ser. Sci. Math. Astron. Phys. **11**, 541 (1963)
R.S. Ingarden: Fortschr. Phys. **12**, 567 (1964 b); **13**, 755 (1965 a)
R.S. Ingarden: Acta Phys. Polon. **27**, 179 (1965 b)
R.S. Ingarden: Ann. Inst. Henry Poincare **8**, 1 (1968 a)
R.S. Ingarden: Bull. Acad. Polon. Sci., Ser. Sci. Math. Astron. Phys. **16**, 57 (1968 b)
R.S. Ingarden: Acta Physica Polonica Vol. XXXVI, Fasc. **5** (11), 855 (1969)

In this chapter and Chap. 6 I essentially follow

H. Haken: Z. Phys. B-Condensed Matter **61**, 335 (1985)
H. Haken: Z. Phys. B-Condensed Matter **63**, 487 (1986)

### Section 5.2

The laser distribution function was derived by

H. Risken: Z. Phys. **186**, 85 (1965)
H. Haken: *Laser Theory*, Corr. printing (Springer, Berlin, Heidelberg 1984; originally published as *Handbuch der Physik*, Vol. 25/2c, ed. by S. Flügge, 1970)

### Section 5.3

H. Haken: *Laser Theory*, Corr. printing (Springer, Berlin, Heidelberg 1984; originally published as *Handbuch der Physik*, Vol. 25/2c, ed. by S. Flügge, 1970)

## Chapter 6

References as for Chap. 5

## Chapter 7

In this chapter (up to Sect. 7.8) I essentially follow

H. Haken: Z. Physik B-Condensed Matter **61**, 329 (1985)
H. Haken: Z. Physik, B-Condensed Matter **62**, 255 (1986)

### Section 7.9

J.A. Tuszynski: Z. Phys. B-Condensed Matter **65**, 375 (1987)
W. Witschel: Chem. Phys. **50**, 265 (1980)
W. Witschel, J. Bohmann: J. Phys. A **13**, 2735 (1980)

### Section 7.10

In this section I essentially follow

Yu.L. Klimontovich: Physica A **142**, 390 (1987)
Yu.L. Klimontovich: Z. Phys. B-Condensed Matter **65**, 125 (1987)

## Chapter 8

### Section 8.2

H. Haken, J. Weberruss: unpublished results

## Chapter 9

H. Haken: Z. Physik B-Condensed Matter **63**, 505 (1986)

## Chapter 10

### Section 10.1

H. Haken: In *Frontiers in Quantum Optics*, ed. by E.R. Pike, S. Sarkar (Adam Hilger, Bristol 1986)
H. Haken: *Laser Theory*, Corr. printing (Springer, Berlin, Heidelberg 1984; originally published as *Handbuch der Physik*, Vol. 25/2c, ed. by S. Flügge 1970)

### Section 10.2

H. Haken: In *Frontiers in Quantum Optics*, ed. by E.R. Pike, S. Sarkar (Adam Hilger, Bristol 1986)

## Chapter 11

For the experiments see:

J.A.S. Kelso: Bull. Psychon. Soc. **18**, 63 (1981)
J.A.S. Kelso: Am. J. Phys: Reg. Integr. Comp. **15**, R1000–1004 (1984)

A theory in terms of deterministic or stochastic equations was developed by

H. Haken, J.A.S. Kelso, H.H. Bunz: Biol. Cybern. **51**, 347 (1985)
G. Schöner, H. Haken, J.A.S. Kelso: Biol. Cybern. **53**, 247 (1986)

## Chapter 12

Pattern recognition by computers is presently an exploding field, see e.g.

H. Haken (ed.): *Pattern Formation by Dynamic Systems and Pattern Recognition*, Springer Ser. Syn., Vol. 5 (Springer, Berlin, Heidelberg 1979)
J.S. Denker (ed.): *Neural Networks for Computing*, AIP Conf. Proc. 151, New York 1986
M. Candill, S. Butler: IEEE First Int. Conf. on Neural Networks, Vols. I–IV, SOS Printing, San Diego 1987

### Section 12.1

P.A. Devijver, J. Kittler: *Pattern Recognition, A Statistical Approach* (Prentice Hall, Englewood Cliffs, N.J. 1982)

## Section 12.2

H. Haken: In *Computational Systems, Natural and Artificial*, ed. by H. Haken, Springer Ser. Syn., Vol. 38 (Springer, Berlin, Heidelberg 1987)

A comprehensive description is given at

H. Haken: Synergetic Computers and Cognition, Springer, Berlin, Heidelberg (1991)

## Section 12.3

For further work, see

H. Haken: Synergetic Computers and Cognition, Springer, Berlin, Heidelberg (1991)

## Section 12.4

For further work, see

H. Haken: Synergetic Computers and Cognition, Springer, Berlin, Heidelberg (1991)

For the special case spin glasses see D.H. Ackley, G.E. Hinton, T.J. Sejnowski: Cognitive Science **9**, 147 (1985)

## Section 12.5

See first edition of this book

## Section 12.6

H. Haken: unpublished manuscript

## Sections 12.7 and 12.8

For a somewhat different procedure, see

L. Borland, H. Haken: Z. Physik B – Condensed Matter **88**, 95 (1992)
L. Borland, H. Haken: Ann. Physik **1**, 452 (1992)
L. Borland: *Ein Verfahren zur Bestimmung der Dynamik stochastischer Prozesse.* Ph.D. Thesis, Institut für Theoretische Physik u. Synergetik, Universität Stuttgart (1993)
L. Borland, H. Haken: Rep. Math. Physics **33**, 35 (1993). This paper introduces negative powers of $q$ in the constraints

## Section 12.9

## Section 12.10

S. Siegert, R. Friedrich, J. Penike: Physics Lett. A **243**, 275 (1998)

S. Siegert: *Analyse raumzeitlicher Daten des menschlichen Elektroenzephalogramms.* Diplom Thesis, 3. Institut für Theoretische Physik, Universität Stuttgart (1998)

These authors start directly from the usual definition of the drift and diffusion coefficients. See also

R. Friedrich, J. Peinke: Physica D **102**, 147 (1997)

who derive a Fokker-Planck equation describing the statistical properties of a turbulent cascade

### Section 12.11

### Section 12.12

12.12.1
For an example of numerical analysis of data sets to validate the Markov assumption, see

R. Friedrich, J. Zeller, J. Peinke: Europhys. Lett. **41**, 153 (1998)

12.12.2
Further reading:

a) Time series analysis
H. Kantz, T. Schreiber: *Nonlinear Time Series Analysis.* Cambridge University Press, Cambridge, U.K. (1997)
H. Kantz, J. Kurths, G. Mayer-Kress, eds.: *Nonlinear techniques in physiological time series analysis.* Springer, Berlin, Heidelberg (1998)

b) Definitions of dimensions
H.G. Schuster: *Deterministic Chaos*, 2nd ed. VCH, Weinheim, (1998)
E. Ott: *Chaos in Dynamical Systems.* Cambridge University Press, Cambridge, U.K. (1993)

c) Embedding theory:
F. Takens: In *Dynamical Systems and Turbulence*, ed. by D.A. Rand, L.S. Young, Lecture Notes in Mathematics, Vol. 898, Springer, New York (1981)
R. Sauer, J. Yorke, M. Casdagli: J. Stat. Physics **65**, 579 (1991)

## Chapters 15 and 16

Quantum computers were first discussed by Benioff in the context of simulating classical Turing machines with quantum unitary evolution.

P. Benioff: J. Stat. Phys. **22**, 563 (1980)
P. Benioff: J. Math. Phys. **22**, 495 (1981)
P. Benioff: Phys. Rev. Lett. **48**, 1581 (1982)

Feynman considered the question of how well classical computers can simulate quantum systems and concluded that classical computers suffer from an exponential slowdown in trying to simulate quantum systems.

R.P. Feynman: Int. J. Theor. Phys. **21**, 467 (1982)
R.P. Feynman: Found. Phys. **16**, 507 (1986)

Deutsch suggested that quantum superposition might allow quantum evolution to perform many classical computations in parallel.
D. Deutsch: Int. J. Theor. Phys. **24**, 1 (1985)
D. Deutsch: Proc. Roy. Soc. London A **400**, 97 (1985)

Factorization of (large) numbers:
P.W. Shor: In *Proceedings of the 35th Annual Symposium on the Foundations of Computer Science*, ed. by S. Goldwasser, IEEE Computer Society Press, Los Alamitos, CA (1994)
A. Ekert, R. Jozsa: Rev. Mod. Phys. **68**, 733 (1996)

RSA Coding:
R. Rivest, A. Shamir, L. Adleman: Communications ACM **21**, 120 (1978)

Excellent revue articles are
A. Ekert: In *Fundamentals of Quantum Information*, ed. by D. Heiss, Springer, Berlin (2002) contains also detailed estimates on accuracy
S.L. Braunstein: In *Quantum Computing*, ed. by S.L. Braunstein, Wiley-VCH, Weinheim (1999)

an easy-to-read article
These articles contain many further references.

## Chapter 17

R. Graham, T. Tel: Phys. Rev. **35 A**, 1328 (1987)
R. Graham, T. Tel: Phys. Rev. Lett. **52**, 9 (1984)
R. Graham, T. Tel: J. Stat. Phys. **35**, 729 (1984); **37**, 709 (1984)
R. Graham, T. Tel: Phys. Rev. **A 31**, 1109 (1985)
R. Graham, D. Roekaerts, T. Tel: Phys. Rev. **A 31**, 3364 (1985)
M.I. Freidlin, A.D. Wentzell: *Random Perturbations of Dynamical Systems* (Springer, New York 1984)
R. Landauer: Relative Stability in the Dissipative State, Preprint 1978

# Subject Index


Absolute temperature 66, 202
Activator molecules 26, 28
Adiabatic approximation 30, 137
Adiabatic elimination technique 84
Algebraic complexity 4
Amount of information 53
Amplifying processes 24
Animal society 2
Associative action 165 ff
Associative memory 159, 163
Atomic dipole moment 110
Atomic polarization 37
Attractors 16 ff, 20 ff, 192 f
–, basin of 22
–, chaotic 16
–, relative importance of 18 f, 22
Axon 35

Belousov-Zhabotinsky reaction 9
Biased action 59
Binary system 54, 219 f
– –, a sequence of symbols 54
Bit 54
Boltzmann distribution function 5, 65
Boltzmann's constant 13, 65
Boolean algebra 223
Boundary condition 40
Brain 2, 28 f
Brownian motion of a particle 37, 181 f
Brownian particle 38

Capacity of a communication channel 16, 23
Cell 2, 28
–, differentiation of 28
Cell membrane 2
Chaos 52
– theory 189 f
Chaotic attractor 16
Chemical field 26
Chemical potential 67
Chemical reactions 2 ff, 24, 28, 37
– –, concentric waves 24
– –, continuous oscillations 24
– –, macroscopic patterns 8
– –, spatial spirals 24
Closed systems 16, 23
Code breaking 229 ff
Coding 229 ff
Coherent state 30
Complex systems 1 ff, 36 ff, 51
– – in biology 2
– – in chemistry 2
– –, concepts and methods 1 ff
– –, qualitative behavior of 1 ff
– –, quantitative methods 1, 36 ff
– – and synergetics 1
– –, unifying principles 5
Complexity 4 f, 45
–, measure of the algebraic degree of 5
–, reduction of 45
Compressibility of a crystal 2
Compression 5 ff, 25, 151
– of information 7, 25, 151
– of a string of data 5
Computer 3 f, 11, 15
– of the fifth generation 15
Conditional probability 49, 77, 119
– – for an enslaved mode amplitude 119
Constraints 13, 33, 56 ff, 70, 74
–, choice of 33 f
–, external 13
– and information 56
–, quantum mechanical 211
Contour line 22
Control 187 ff
Control parameter 13, 46, 83, 97, 108 ff, 111 ff, 114
Convection instability 138 f
Cooperation of cells 2
Coordination of muscles 3
Coprime 230
Correlation functions 35, 130 ff, 166
– – as constraints 130 ff, 166
Critical fluctuations 94 f, 115, 119, 147 f, 151
Crystal 2, 6
–, compressibility of 2
– as a conductor of electricity 2
–, order of 6
Cyclic adenosinemonophosphate (cAMP) 26

– –, spiral pattern of concentration of 26
Cytoplasm 2

Darwinism 4
Data 4 ff
–, initial 4
–, –, minimum number of 5
–, sequence of 4
–, string of 4
Decimal system 219
Decoding 229 ff
Decoherence 218 f
Deformation 199 ff
Dendrites 35
Density matrix 204, 211 f
– –, trace of 204
Detailed balance 41 f, 45, 50, 72, 102, 136
– –, first and second version of the principle of 42
Deterministic chaos 8
Diagonalization of a matrix 76
Diffusion coefficient 43, 79 f, 125, 173, 185 ff
Diffusion matrix 43
Diffusion of molecules 13
Diffusion term 40
Dirac's bra- and ket notation 216
Discrimination of vectors 160
Distribution function 5, 56, 69 ff
– – of a nonequilibrium system 72
– – of order parameters 69
– – of particles 56
DNA 5
Drift coefficient 79 f, 173, 185 ff
Drift term 40

Economy 2, 37
Efficiency of self-organizing systems 81 ff
Electric field strength 70
– – –, amplitude of 70
Elementary particle physics 6
Emotional expressions 199 ff
Energy 12, 65 f, 79
–, free 65, 68, 79
–, influx of 24
–, internal 65 f
– of state 66
Enslaved modes 35, 48, 74 ff
– –, amplitudes of 48
– –, determination of 74 ff
Entanglement 221
Entropy 5, 12, 65, 88, 95, 102 *see also* Information entropy
–, concept of 5
Events 53, 56
–, independent 53
–, scarcity of 56

Evolution equations 47, 83 ff
Expectation values, quantum mechanical 234
External constraints 13

Feature space 159
Features 28, 58, 153 ff
–, characteristic 28
– and information entropy 58
–, selection of 153 ff
Feynman's disentangling theorem 213
Fever curve 4
Filamentation 37
First principles 6
Fixed point 16 f
Fluctuating forces 16, 38, 46, 49, 106, 113, 147, 160
– –, Gaussian distributed 49
Fluctuating intensity of light from stars 4
Fluctuation times 70
Fluctuations 18 f, 33, 51, 70
Fluid dynamics 35
Fluids 8 f, 11, 28, 36 f
–, hexagons 11
–, oscillations of vortices 8
–, spatio-temporal patterns 8
–, specific spatial patterns 8 f
Fokker-Planck equation 40 ff, 48 f, 71 f, 79 f, 102, 113, 125 ff, 132 ff, 137, 146 ff, 160 ff
– – belonging to the short-time propagator 132
– –, determination of 79 f
– –, diffusion coefficient of 43, 79 f, 125
– –, diffusion term of 40, 125
– –, drift coefficient of 79 f
– –, drift term of 40
– –, exact stationary solution for systems in detailed balance 40 ff
– –, guessing of 79 f
– –, irreversible drift coefficients of 43
– –, Îto differential equation 40, 132, 166
– –, multimode 72
– –, multivariable 80
– –, path integral solution of 44 f, 129
– –, reversible drift coefficients of 43
– –, Stratonovich equation 41, 132
– –, time-dependent solutions of 44
Fourier transformation, quantum 228
Free energy 65, 68, 79
Fundamental laws 6, 13, 134
– – for elementary particles 134

Galileo's experiments 7
Gates, classical 222 f
–, quantum 223 ff
Gaussian approximation 115
Gaussian function 89 ff, 108, 113 ff
Gaussian white-noise force 80

Genetic code 27
Genetic information 28
Goedel's theorem 5
Greatest common divisor 230
Gun oscillator 37, 169

Hadamard gate 224 ff
Hamiltonian operator 236
Heat 67, 138
– conduction 13, 138
–, generalized 62, 67
Hebb's synapse 28
Helium-neon laser 114
Hierarchical levels of semantic information 28
Hierarchy 14, 27 f
– of information levels 27
– of instabilities 14
Hormones 27
Horse 4, 10
–, change of gaits 10
Human hand movements 10, 29, 35, 140 ff
– – –, fluctuations of the phase 147
– – –, in-phase mode 140 f
– – –, Kelso's experiments 10, 29, 140
– – –, oscillation frequency of the fingers 150
– – –, out-of-phase mode 140 f
– – –, transitions 10, 141 f
Hydra 8 f, 26
–, regeneration of 9

Information
–, amount of 29, 53
–, annihilation of 21
–, average 15
– carrier 26
–, compression of 25, 36, 151 f, 195 ff
–, concept of 14, 21, 23, 53, 58
–, conservation of 21
– and constraints 56, 74 ff
– content 21
– – of a symbol 58
– for continuous variables 63
–, creation of 25
– deficiency 19 f
–, definition of 81
–, of enslaved modes 81 ff, 214
–, exchange of 23
–, extremum of 57
– gain 57, 81 ff, 163 f
–, generation of 21
–, genetic 28
–, greatest 57
– to maintain an ordered state 29
–, maximization of 33
–, mean change of 58
–, meaningful 16, 56
–, meaningless 16, 56
–, measure for the amount of 53
– and measurement 85 ff
– and the method of Lagrangian multipliers 57
– and Morse alphabet 53
– of order parameters 81 ff, 214
–, production of 26
–, quantum theory of 203 ff, 216 ff
– receiver 26
– with respect of the interval of accuracy 63
– and the role of meaning and purpose 7
–, semantic 28
–, Shannon 15
– stored and created in the brain 28
–, synergetic 242
– per symbol 55
– transfer 27
–, transmission of 55
–, useful 56
–, useless 56
Information entropy 58 ff, 66 f, 95, 103, 113 ff *see also* Information
– – and constraints 58 f
– –, maximum of 59 ff
– – and mean values 61
– – and measurement 59
– – of systems below and above their critical point 115
– – and thermodynamics 65 ff
– – and unbiased estimates 59
Information processing 29, 58
– –, self-organization in 29
Inhibitor molecules 26
Initial time 16
Instabilities 13 f, 33
–, hierarchy of 14
Integrability condition 43
Intensity 53, 70 ff
– correlations 70, 73
–, first moment of 73
– fluctuations 53, 208
– of the light field 70
–, output 70, 73
–, second moment of 69, 73
– of a single mode laser 69 ff
Inversion 110 ff, 136 ff
–, critical 112
Irreversible processes 22

Jaynes's principle 33

Karhunen-Loeve expansion 158
Kullback information *see* Information gain

Lagrange equations of the first kind 61
Lagrange multipliers 57, 59, 61, 73 ff, 95, 115 ff, 130 ff, 137, 145, 155, 164, 194 ff

– –, explicit determination 169
– –, direct determination of 115ff
– –, method of 57, 59, 74f
Lagrange parameters *see* Lagrange multipliers
Landau equation 80
Landau functional 80
Landau theory of phase transitions 79f
Langevin equation 37ff, 50, 71, 79, 113, 125ff, 146ff, 160ff
– –, Îto equation 38f, 125ff
– –, Stratonovich equation 39
Laplace 7, 46
Laplace operator 46
Laser 2, 8, 13, 23ff, 27ff, 30, 33, 37, 46, 69ff, 95, 102f, 110ff, 135ff
– action 31
–, adiabatic approximation 30, 137
–, axial modes of 71
– condition 31
–, decay constant 30
–, electric field strength of 70, 135
– equations 110, 137ff
–, helium-neon 114
–, inversion of the atomic system 30f, 37, 110ff, 136f
–, loss rate of 70
–, multimode 69, 71f, 135f
–, optical transition frequency of the atoms 110
– a paradigm for the self-organization of coherent processes 8
–, polarization of the atomic system 136
– and the principle of detailed balance 72, 136
–, pump rate of 32, 110f
–, pump strength of 32
–, rate constant of 30
–, schematic drawing of 2
–, single mode 69ff, 110ff, 136ff
– theory 30, 70
– threshold 104
Lasing threshold 69
Limit cycle 17, 22, 52
Line integral 44
Linear stability analysis 46f
logical operation 222f
–, AND 222
–, OR 223
–, XOR 223, 226

Macro-observables 33
Macroscopic level 6, 16, 23, 33, 36
Macroscopic observations 69
Macroscopic pattern 13, 33, 73
Macroscopic phenomena 13
Macroscopic quantities 53, 69
Macroscopic state 13
Macroscopic structures 69
Macroscopic variables 13
Macroscopic world 36, 53
Macroscopically observed quantities 33
Markov property 189
Markovian process 45, 125ff, 165, 168
– –, conditional probability of 165
Master equation 45, 196
Maximum calibre principle 125
Maximum entropy principle 12
Maximum information entropy principle *see* Maximum information principle
Maximum information principle (MIP) 1, 33f, 53ff, 69ff, 72f, 74ff, 79, 87, 117, 125, 154f, 165
– – –, application of self-organizing systems 69ff
– – – and choice of constraints 33, 125
– – –, extension to systems far from thermal equilibrium 33f
– – – for nonequilibrium phase transitions 74
– – –, quantum mechanical formulation 205ff
Mean field approximation 99
Meaning 7, 23, 24
–, emergence of 23
–, self-creation of 23
Measure of ignorance 56
–, receipt of 53
–, reduncancy of 17
–, relative importance of 16, 18, 20, 22
Microscopic equations of motion 13
Microscopic level 6, 16, 33, 45, 118
Microscopic motion of molecules 38
Microscopic quantities 53
Microscopic theory 12, 69, 76f, 102, 113, 136
Microscopic world 36
Mode skeleton 51
Modes 47, 69ff, 76ff, 211ff
–, determination of 78
–, enslaved 49, 76ff, 107ff, 211ff
– of the laser 69ff
–, stable 47, 112, 119
–, unstable 47, 112, 119
Modulo-formalism 230
Moments 35, 69, 73ff, 117ff, 128f, 142ff, 162, 204
–, conditional 128f
– of fourth order 74, 117ff, 145, 162
– of observed variables 35
Morphogenesis 8ff, 37, 152
– of behavior 152
–, self-organization in 10
Morse alphabet 53f, 56
– – and messages 53
Motor program 29, 151

Muscles 2f
–, coordination of 3
Mutation 27

Nerve cells 2f, 35
– –, net of 3
– – and pattern recognition 2
Nervous system 28
Net production rate of signals 31
Networks 2, 164
Neuronal nets 37
Neurons 37, 151
–, firing rates of 37
Newtonian mechanics 7, 12f
– – and the predictability of the future 7
Newton's law 133f
Noise 16, 133
Nonequilibrium phase transitions 34, 49, 68, 73, 74ff, 81ff, 95, 125ff, 135ff, 155
– – – and symmetry 75
Nonequilibrium systems 69
Non-Markovian process 189ff
Normalization 20, 59, 77, 157, 197
– condition 77
– factor 40
– of a probability distribution 59, 157
– properties in case of no information deficiency 20

Occupation number 30, 110
Occupation probability 29, 31, 65
– – of a quantum state 65
Open systems 12f, 24, 33
operators, product of 224
Order parameters 13, 25ff, 28, 33, 35, 45, 47, 49, 51, 74ff, 79, 81ff, 95ff, 107ff, 115ff, 158
– –, calculation of the information of 88ff
– –, –, analytical results 95ff
– –, determination of 74ff
– –, distribution function of 81ff
– – as informators 26, 182
–, moments of 79
– –, space of 160
– – in quantum mechanics 211
Order parameter equation 49
Ordered state 29ff
– – of the laser 31
Orthogonality relation 117f

Parabolic cylinder functions 97
Particle numbers 53
Partition function 60, 66, 68, 110, 176
– –, quantum mechanical 208, 210
Path integral 44, 125ff
Pattern recognition 1, 2, 22, 28, 78, 153ff, 159ff
– –, an algorithm for 159
– –, as information compression 197ff
– –, parallel computer for 153
Patterns 22ff, 51f, 73ff, 78, 138, 159ff
–, determination of 74ff
–, formation of 51, 73
–, global 22
–, prototype 159f
–, spatial 78
–, time-dependent 52
Pauli matrices 233
Period of a sequence, calculation by quantum computers 227ff
Phase 218
Phase shift gate 225f
Phase transitions 24
Phenomenological equations 36
Phenotype 27
Pheromones 27
Photons 30f, 110
–, coherent 30f
–, flux of 31
–, incoherent 30
–, production rate of 30
–, – –, equation for 31
Physical systems 53
– –, closed 53
– –, open 53
Planck's constant 216
Plasma 37
Population dynamics 37
Population entropy 158
Possibilities 56
–, greatest number of 56
Possible events 53
Power input 86, 101
Pre-biotic evolution 37
Predictability 7f
– and classical mechanics 8
– and quantum mechanics 8
Prediction 187f
– of events 7
Principle of detailed balance 41f, 45, 50, 72
Probability 32f, 41f, 53, 63, 77, 81ff, 125ff, 154, 165ff
–, conditional 42, 77, 125ff, 165
– density 63
– distribution 32f, 45, 53, 154, 165
–, joint 41f, 81f
– theory 53
–, transition 42
Program 4f
–, minimum 5
–, minimum length of 4
Process modelling 187f
Psychology 3

Pulsar 4
Pulse 217
– $\frac{\pi}{2}$ 217
– $\pi$ 217
Purpose 7, 24

Quantum computation 222 ff
Quantum mechanics 8, 12, 29, 134
– – and the concept of predictability 8
– –, order parameters, enslaved modes and patterns 211
Quantum systems 203 ff
Quantum wells 203
Quasi periodic motion 52
q-bits 216 ff

Random variable 59, 70
Realizations 53, 56 *see also* Possible events
–, number of 56
Recognition of dominant structures in patterns 33, 78
Reducibility 6
Register 220
Relative frequency 15, 55 f, 58
– – of a letter 15
– – of the occurrence of a symbol 55
– – of a particle 56
Relaxation phenomena 68
Relaxation processes 13, 110, 125
–, incoherent 110 ff
Relaxation time 149
Relevance 24
Representation of numbers 219
Ring laser 135
Robotics 187 ff
RSA-coding 230 f

Schrödinger equation 195, 236
Second-order phase transition 94
Selection 27
Self-consistency conditions 124, 130
– – for the occurrence of a nonequilibrium phase transition 124
Self-creation of meaning 15
Self-organizing systems 1, 10, 36, 69, 81 ff
– –, efficiency of 81 ff
– – and spatial (temporal) structures 69
Semiconductors 195
Sensitivity to initial conditions 8
Shannon information 15 ff
Shor's approach 227 ff, 231 ff
Slaving principle 13, 25, 27, 33, 45, 48 f, 77, 81 ff, 113
– –, application to information 82 ff
Slime mold 10, 26
Smoothing schemes, additive noise 174 ff
– –, multiplicative noise 174 ff
– – trajectory 192 ff
Sociology 37
Soft mode instability 119
Soft single-mode instabilities 84 f
Solid state physics 37
Soma 35
Spin 216, 220, 233 ff
Spin-glass model 164
Spontaneous emission 30, 106
Stable point 18
State vector 38, 46, 161
– –, multidimensional 38
Stationary process 70
Statistical mechanics 5, 7, 33, 36 *see also* Thermodynamics
– –, derivation from microscopic laws 7
Steam engine 2
S-theorem of Klimontovich 102 ff
Stimulated emission 30, 106
Stirling's formula 55
Stochastic forces 162
Stochastic process 38 f
Stochastic variables 71
Strange attractor 17
Symmetry 75, 197 ff
Symmetry breaking 18, 78, 95, 115
Synaptic strength 163
Synergetic computers 37, 161 ff
– –, basic construction principle of 161 ff
Synergetic information 26, 208
Synergetics 1, 13 f, 33 ff, 36, 38
–, macroscopic approach to 1
–, microscopic approach to 1
–, the second foundation of 33
Systems 1 ff, 11 ff, 16, 22 ff, 27, 29, 41, 95, 102, 152, 169 ff, 182 *see also* Complex systems
–, biological 29, 152, 208
–, closed 16, 95
–, degeneracy of 27
–, efficiency of 27
–, error-correcting 22
– far from thermal equilibrium 11, 13 f, 24, 41, 95, 102
–, learning process of 22
–, man-made 4
– in nature 4
–, nonequilibrium 35
–, open 12 f, 24, 33, 95
–, quantum 169
–, reliability of 27
– in thermal equilibrium 11, 41

Taylor series 172
Thermal equilibrium 11 ff, 41, 69
Thermodynamic entropy 88

Thermodynamic potential 68
– –, generalized 43
Thermodynamics 1, 7, 11 f, 13, 53, 65 ff, 95, 110, 125
–, first law of 12
–, generalization of the concepts of 110
–, irreversible 13, 68, 73, 125
–, macroscopic phenomenological theory 12
– second law of 12, 23
– and statistical mechanics 65
Time-ordered operators 209 ff
Torus 17
Truth table 222
Turing machine 4 f
– –, schematic of 5

Unbiased estimates 59
Unit matrix 217
Unitary matrix 217
Universal computer 4
Universal laws 6

Variables 59, 66, 68, 70
–, dependent 66, 68
–, independent 66, 68
–, random 59, 70
Variance 62
Variational principle 63

Weber function 97 ff
Weight function 70

# Springer Series in Synergetics

**Synergetics** An Introduction 3rd Edition
By H. Haken

**Synergetics** A Workshop
Editor: H. Haken

**Synergetics** Far from Equilibrium
Editors: A. Pacault, C. Vidal

**Structural Stability in Physics**
Editors: W. Güttinger, H. Eikemeier

**Pattern Formation by Dynamic Systems and Pattern Recognition**
Editor: H. Haken

**Dynamics of Synergetic Systems**
Editor: H. Haken

**Problems of Biological Physics**
By L. A. Blumenfeld

**Stochastic Nonlinear Systems**
in Physics, Chemistry, and Biology
Editors: L. Arnold, R. Lefever

**Numerical Methods in the Study of Critical Phenomena**
Editors: J. Della Dora, J. Demongeot, B. Lacolle

**The Kinetic Theory of Electromagnetic Processes** By Yu. L. Klimontovich

**Chaos and Order in Nature**
Editor: H. Haken

**Nonlinear Phenomena in Chemical Dynamics** Editors: C. Vidal, A. Pacault

**Handbook of Stochastic Methods**
for Physics, Chemistry, and the Natural Sciences 2nd Edition
By C. W. Gardiner

**Concepts and Models of a Quantitative Sociology** The Dynamics of Interacting Populations By W. Weidlich, G. Haag

**Noise-Induced Transitions** Theory and Applications in Physics, Chemistry, and Biology By W. Horsthemke, R. Lefever

**Physics of Bioenergetic Processes**
By L. A. Blumenfeld

**Evolution of Order and Chaos**
in Physics, Chemistry, and Biology
Editor: H. Haken

**The Fokker-Planck Equation**
2nd Edition By H. Risken

**Chemical Oscillations, Waves, and Turbulence** By Y. Kuramoto

**Advanced Synergetics**
2nd Edition By H. Haken

**Stochastic Phenomena and Chaotic Behaviour in Complex Systems**
Editor: P. Schuster

**Synergetics – From Microscopic to Macroscopic Order** Editor: E. Frehland

**Synergetics of the Brain**
Editors: E. Başar, H. Flohr, H. Haken, A. J. Mandell

**Chaos and Statistical Methods**
Editor: Y. Kuramoto

**Dynamics of Hierarchical Systems**
An Evolutionary Approach
By J. S. Nicolis

**Self-Organization and Management of Social Systems** Editors: H. Ulrich, G. J. B. Probst

**Non-Equilibrium Dynamics in Chemical Systems**
Editors: C. Vidal, A. Pacault

**Self-Organization** Autowaves and Structures Far from Equilibrium
Editor: V. I. Krinsky

**Temporal Order** Editors: L. Rensing, N. I. Jaeger

**Dynamical Problems in Soliton Systems**
Editor: S. Takeno

**Complex Systems – Operational Approaches** in Neurobiology, Physics, and Computers Editor: H. Haken

**Dimensions and Entropies in Chaotic Systems** Quantification of Complex Behavior 2nd Corr. Printing
Editor: G. Mayer-Kress

**Selforganization by Nonlinear Irreversible Processes**
Editors: W. Ebeling, H. Ulbricht

**Instabilities and Chaos in Quantum Optics**
Editors: F. T. Arecchi, R. G. Harrison

**Nonequilibrium Phase Transitions in Semiconductors** Self-Organization Induced by Generation and Recombination Processes By E. Schöll

**Temporal Disorder in Human Oscillatory Systems**
Editors: L. Rensing, U. an der Heiden, M. C. Mackey

**The Physics of Structure Formation** Theory and Simulation Editors: W. Guttinger, G. Dangelmayr

**Computational Systems – Natural and Artificial** Editor: H. Haken

**From Chemical to Biological Organization** Editors: M. Markus, S. C. Müller, G. Nicolis

**Information and Self-Organization** A Macroscopic Approach to Complex Systems 2nd Edition By H. Haken

**Propagation in Systems Far from Equilibrium** Editors: J. E. Wesfreid, H. R. Brand, P. Manneville, G. Albinet, N. Boccara

**Neural and Synergetic Computers** Editor: H. Haken

**Cooperative Dynamics in Complex Physical Systems** Editor: H. Takayama

**Optimal Structures in Heterogeneous Reaction Systems** Editor: P. J. Plath

**Synergetics of Cognition** Editors: H. Haken, M. Stadler

**Theories of Immune Networks** Editors: H. Atlan, I. R. Cohen

**Relative Information** Theories and Applications By G. Jumarie

**Dissipative Structures in Transport Processes and Combustion** Editor: D. Meinköhn

**Neuronal Cooperativity** Editor: J. Krüger

**Synergetic Computers and Cognition** A Top-Down Approach to Neural Nets By H. Haken

**Foundations of Synergetics I** Distributed Active Systems 2nd Edition By A. S. Mikhailov

**Foundations of Synergetics II** Complex Patterns 2nd Edition By A. S. Mikhailov, A. Yu. Loskutov

**Synergetic Economics** By W.-B. Zhang

**Quantum Signatures of Chaos** 2nd Edition By F. Haake

**Rhythms in Physiological Systems** Editors: H. Haken, H. P. Koepchen

**Quantum Noise** 2nd Edition By C. W. Gardiner, P. Zoller

**Nonlinear Nonequilibrium Thermodynamics I** Linear and Nonlinear Fluctuation-Dissipation Theorems By R. Stratonovich

**Self-organization and Clinical Psychology** Empirical Approaches to Synergetics in Psychology Editors: W. Tschacher, G. Schiepek, E. J. Brunner

**Nonlinear Nonequilibrium Thermodynamics II** Advanced Theory By R. Stratonovich

**Limits of Predictability** Editor: Yu. A. Kravtsov

**On Self-Organization** An Interdisciplinary Search for a Unifying Principle Editors: R. K. Mishra, D. Maaß, E. Zwierlein

**Interdisciplinary Approaches to Nonlinear Complex Systems** Editors: H. Haken, A. Mikhailov

**Inside Versus Outside** Endo- and Exo-Concepts of Observation and Knowledge in Physics, Philosophy and Cognitive Science Editors: H. Atmanspacher, G. J. Dalenoort

**Ambiguity in Mind and Nature Multistable Cognitive Phenomena** Editors: P. Kruse, M. Stadler

**Modelling the Dynamics of Biological Systems** Editors: E. Mosekilde, O. G. Mouritsen

**Self-Organization in Optical Systems and Applications in Information Technology** 2nd Edition Editors: M.A. Vorontsov, W. B. Miller

**Principles of Brain Functioning A Synergetic Approach to Brain Activity, Behavior and Cognition** By H. Haken

**Synergetics of Measurement, Prediction and Control** By I. Grabec, W. Sachse

**Predictability of Complex Dynamical Systems** By Yu. A. Kravtsov, J. B. Kadtke

**Interfacial Wave Theory of Pattern Formation** Selection of Dentritic Growth and Viscous Fingerings in Hele–Shaw Flow By Jian-Jun Xu

**Asymptotic Approaches in Nonlinear Dynamics** New Trends and Applications By J. Awrejcewicz, I. V. Andrianov, L. I. Manevitch

**Brain Function and Oscillations**
Volume I: Brain Oscillations.
Principles and Approaches
Volume II: Integrative Brain Function.
Neurophysiology and Cognitive Processes
By E. Başar

**Asymptotic Methods for the Fokker–Planck Equation and the Exit Problem in Applications**
By J. Grasman, O. A. van Herwaarden

**Analysis of Neurophysiological Brain Functioning** Editor: Ch. Uhl

**Phase Resetting in Medicine and Biology**
Stochastic Modelling and Data Analysis
By P. A. Tass

**Self-Organization and the City** By J. Portugali

**Critical Phenomena in Natural Sciences**
Chaos, Fractals, Selforganization and Disorder: Concepts and Tools By D. Sornette

**Spatial Hysteresis and Optical Patterns**
By N. N. Rosanov

**Nonlinear Dynamics of Chaotic and Stochastic Systems**
Tutorial and Modern Developments
By V. S. Anishchenko, V. V. Astakhov, A. B. Neiman, T. E. Vadivasova, L. Schimansky-Geier

**Synergetic Phenomena in Active Lattices**
Patterns, Waves, Solitons, Chaos
By V. I. Nekorkin, M. G. Velarde

**Brain Dynamics**
Synchronization and Activity Patterns in Pulse-Coupled Neural Nets with Delays and Noise By H. Haken

**From Cells to Societies**
Models of Complex Coherent Action
By A. S. Mikhailov, V. Calenbuhr

**Brownian Agents and Active Particles**
Collective Dynamics in the Natural and Social Sciences By F. Schweitzer

**Nonlinear Dynamics of the Lithosphere and Earthquake Prediction**
By V. I. Keilis-Borok, A. A. Soloviev (Eds.)

**Nonlinear Fokker-Planck Equations**
Fundamentals and Applications
By T. D. Frank

**Patters and Interfaces in Dissipative Dynamics**
By L. M. Pismen